NEW DEVELOPMENTS IN MEDICINAL CHEMISTRY

NEW DEVELOPMENTS IN MEDICINAL CHEMISTRY

MARTA P. ORTEGA
AND
IRENE C. GIL
EDITORS

Nova Science Publishers, Inc.
New York

For permission to use material from this book please contact us:
Telephone 631-231-7269; Fax 631-231-8175
Web Site: http://www.novapublishers.com

LIBRARY OF CONGRESS CATALOGING-IN-PUBLICATION DATA

New developments in medicinal chemistry / [edited by] Marta P. Ortega and Irene C. Gil.
p. ; cm.
Includes bibliographical references and index.
ISBN 978-1-60456-810-3 (hardcover)
1. Pharmaceutical chemistry. I. Ortega, Marta P. II. Gil, Irene C.
[DNLM: 1. Chemistry, Pharmaceutical. 2. Drug Design. QV 744 N5325 2009]
RS403.N48 2009
615'.19--dc22

2009006386

Published by Nova Science Publishers, Inc. ✦ New York

CONTENTS

PREFACE

Medicinal or pharmaceutical chemistry is a scientific discipline at the intersection of chemistry and pharmacology involved with designing, synthesizing and developing pharmaceutical drugs. Medicinal chemistry involves the identification, synthesis and development of new chemical entities suitable for therapeutic use. It also includes the study of existing drugs, their biological properties, and their quantitative structure-activity relationships (QSAR). Pharmaceutical chemistry is focused on quality aspects of medicines and aims to assure fitness for the purpose of medicinal products. Medicinal chemistry is a highly interdisciplinary science combining organic chemistry with biochemistry, computational chemistry, pharmacology, pharmacognosy, molecular biology, statistics, and physical chemistry. This new book presents leading research from around the world in this frontal field.

Chapter 1 - Folate biosynthesis remains a key target for antimicrobial therapy. Folate is an essential vitamin (vitamin B9) that is required for many one-carbon transfer reactions and is a critical precursor for the biosynthesis of purines, pyrimidines, and amino acids. Unlike higher eukaryotes that scavenge preformed folates, prokaryotic and lower eukaryotic microorganisms are dependent on several enzymes for the *de novo* biosynthesis of folate. One of these enzymes, dihydropteroate synthase (DHPS), is the target of the first chemically-synthesized antimicrobial agents, the sulfadrugs, which date back to the 1940s. Others are essential enzymes that remain to be explored as drug targets. Resistance to the sulfadrugs rapidly emerges due to the ability of the microbe to alter its susceptibility to the drug by various means. Recently a number of new structures of the enzymes in the pathway has become available. We review the recent literature relating to these targets (the enzymes: 7,8-dihydroneopterin aldolase (DHNA), 6-hydroxymethyl-7,8-dihydropterin pyrophosphokinase (HPPK), dihydropteroate synthase (DHPS), dihydrofolate synthase (DHFS)), their mode of action and how current drugs may modulate this on a structural level. Furthermore, these data advance our understanding of the emergence of drug resistance and may aid efforts and play a major role in the design of new, more effective compounds as antimicrobial agents. To this end we also review the recent literature in the development of inhibitors of these enzymes. Future progress in this key area has the potential to benefit the war against devastating organisms such as drug-resistant *Staphylococcus aureus* and *Plasmodium falciparum.*

Chapter 2 - Polyphenols constitute the most abundant group of antioxidants of normal human food (tea, red wine, grapes, olive oil, chocolate, broccoli, cherries, pomegranates, peanuts, berries and other fruits or vegetables including Ginkgo *biloba*) that protect against oxidative stress and their associated pathologies such as inflammation, cancer and coronary heart diseases. [1] The presence of phenol functions in their structures confers stability against oxidation. Also the biological properties are related to the phenol groups and their disposition in the structure. These compounds that occurs naturally in various food and beverages of plant origin, were also named Vitamin P (more than 1500 compounds) and their main beneficial biological effects are: [2, 3] a) The diminution of reactive oxygen species related with the inflammation process, with the immune system by the recruitment of leucocytes and the blood homeostasis. b) Inhibition of growth of several tumors. Thus, following the American Cancer Society dietary guidelines [4] of "five or more pieces of fresh fruit and vegetables per day" to help prevent cancer and anti-inflammatory diseases. Also fruit and vegetable juices or tea might provide substances that help prevent cancer. c) Reduction of inflammatory effects such as coronary diseases related to the oxidation of LDL (light density lipoprotein). d) Treatment of skin aging in humans (Figure 1).

Chapter 3 - Although there have been large improvements in cancer treatment over the last twenty years, the lack of cancer chemotherapeutic drugs is still a major cause of death in this century. Medicinal plants play an important role in the treatment of cancer by offering unique active drugs or their templates for clinical uses, as exemplified by paclitaxel (Taxol®), vinca alkaloids (vincristine, vinblastine), and flavopiridol. The strategies for developing anticancer agents from medicinal plants have changed in the last decade for a number of reasons, including advances in technology, changes in the plant selection mode and biological activity testing. This review reflects and discusses the newest methods applied for searching anticancer agents from medicinal plants including plant selection and extraction, the active principle isolation, structure elucidation and biological testing. This review reflects and discusses the newest methods applied for searching anticancer agents from medicinal plants including plant selection and extraction, the active principle isolation, structure elucidation and biological testing.

Chapter 4 - Quantum dots have increasingly been incorporated into a wide variety of biological assays as improved fluorescent probes. Their photophysical properties permit the investigation of cellular processes and biological phenomena with unprecedented spatial resolution and temporal longevity. Consequently, quantum dots are poised to facilitate advances in future drug development applications. Multiplexed detection in whole cell assay format may ultimately provide added insight into the extremely complex biochemical mechanisms involved in drug receptor interactions. This article provides a detailed discussion of biological applications which have incorporated quantum dot detection, with a particular emphasis on their possible integration into drug discovery and medicinal chemistry applications.

Chapter 5 - Transition metals play a fundamental role in different living systems. In particular, in aerobic organisms, the presence of copper is essential for the function of many enzymes related to cellular respiration, iron homeostasis, neurotransmitter production, peptide biogenesis, connective tissue biosynthesis, and antioxidant defense. Copper compounds are reported to act as antioxidant, anti-inflammatory, antimicrobial, antiparasitic, anticonvulsant and antitumoral agents. In addition, in vertebrates, copper deficiency causes skeletal alterations. Vanadium, another transition metal, is present in trace amounts in higher animals. Even though its essentiality has not yet been clearly established, experimental results both in vivo and in vitro suggest that vanadium compounds may participate in important biological functions acting as insulinmimetic, osteogenic and antitumoral compounds. Once absorbed, vanadium and copper are distributed among tissues and stored mainly in bone. In this chapter, the behavior of these metal derivatives on bone-related cells in cultures is discussed in detail. Two cellular lines, MC3T3E1 derived from mouse calvaria, and UMR106 from rat osteosarcoma, have been used as a model for normal and tumoral bone processes. To expand the studies on antineoplastic metal drug activity, experiments with copper and vanadium compounds have been undertaken on two tumoral cell lines of human colon adenocarcinoma (Caco-2 and TC7). Different copper complexes with pharmacologically active ligands such as the antihypertensive drug losartan and a derivative of the antiparasitic santonin, santonic acid, were synthesized and tested in vitro in the mentioned cell lines. Both complexes improve the antitumoral effect of free copper ions. This behavior agreed with the morphological cellular alterations. On the other hand, the biological effects of vanadyl(IV) complexes with the flavonoids quercetin and hesperidin were discussed and compared with a vanadium(IV) derivative of the structural related ligand, the disaccharide trehalose. In the tumoral cell lines these compounds were deleterious. The effects on cellular differentiation (specific alkaline phosphatase activity and collagen type I production) were also described for the osteosarcoma cells. Moreover, for the complexes with quercetin and trehalose, the effect on the activation of ERK (extracellular regulated kinase) cascade was investigated using specific antibodies in order to identify one of the possible mechanisms of action. Altogether, these promising results of a first stage in medicinal chemistry on metal-based drugs merit further investigation in animal models.

Chapter 6 - Malaria remains on of the most prevalent and deadly of the insect-borne communicable diseases. Since the parasite *Plasmodium* developed resistance to classical drugs of quinine type, artemisinin-based peroxides have become the most important antimalarial drugs. Various artemisinin derivatives have been developed to improve the antimalarial performance, while insights into its mechanism of action reveal that the pharmacophore unit in artemisinin is a peroxide group bound to the 1,2,4-trioxane skeleton. However, it is not economical to synthesize artemisinin directly and currently, the global demand for artemisinin cannot be met by the low yield obtained from the cultivation of herb *Artemisia annua*. Therefore, current research focuses on the development of easily accessible, cheap and effective antimalarial drugs based on synthetic peroxides. The main synthetic analogues of artemisinin belong to a 1,2,4-trioxane type of organic peroxides, that together with structurally related 1,2,4,5-tetraoxanes and 1,2,4-trioxolanes constitutes the core of antimalarial synthetic peroxides, while various endoperoxides, cyclic and acyclic perketals are also active antimalarial agents. Another interesting concept in antimalarial chemotherapy are chimeric compounds or hybrids; these are composed of two different antimalarial agents joined in one molecule. The following chapter discusses synthetic antimalarial peroxides with an emphasis on the development of synthetic compounds from 2000 onwards.

Chapter 7 - NF-κB is an inducible transcription factor found in virtually all types of vertebrate cells, as well as in some invertebrate cells. While normal activation of NF-κB is required for cell survival and immunity, its deregulated expression is characteristic of cancer, inflammation, and numerous other diseases. Hence, NF-κB has recently become one of the major targets in drug discovery.

Several marine organisms use NF-κB (or analogues thereof), NF-κB inducers, or NF-κB inhibitors as chemical defence mechanisms, for parasitic invasion, for symbiosis, or for larval development. In particular, a wide range of marine natural products have been reported to possess NF-κB inhibitory properties, and some of these marine metabolites are currently in clinical trials as anticancer or anti-inflammatory drugs.

In the present review, we discuss the role of NF-κB inhibitors in marine chemical ecology, as well as in biomedicine. We also describe synthetic modifications that have been made to a range of highly promising marine NF-κB inhibitors, including the macrolide bryostatin 1 isolated from the bryozoan *Bugula neritina*, the lactone-γ-lactam salinosporamide A isolated from the actinomycete *Salinispora tropica*, the alkaloid hymenialdisine isolated from various sponges, the sesquiterpenoid hydroquinone avarol isolated from the sponge *Dysidea avara*, and the sesterterpene lactone cacospongonolide B isolated from the sponge *Fasciospongia cavernosa*, to increase their bioactivity and bioavailability, to decrease their level of toxicity or to lower the risk of other detrimental side-effects, and to increase the sustainability of their pharmaceutical production by facilitating their chemical synthesis.

Chapter 8 - Many efforts have been made in the past few years to build up, into the interlayer region of layered solids, supramolecular assemblies with special functionalities in the field of photochemistry, electrochemistry, molecular recognition, chiral recognition and catalysis.[1-5] Furthermore, the interlayer region of layered solids is starting to be used as a privileged reaction vessel to perform chemical reactions between the guest themselves (polymerisation reactions) or between the guests and the host (topotactic and grafting reactions).[6] In addition, the interlayer region of a layered solid may be considered a micro-container where guest species are stored, protected from oxidation or photolysis, and withdrawn for use by a chemical signal, i. e., by a deintercalation process.[4] Other interesting reactions performed with layered solids are "exfoliation reactions", that consist of separating the sheets of a layered compound into individual lamellae. This goal is reached with the aid of specific intercalation or de-intercalation reactions and leads to colloidal dispersion of lamellae. These dispersions can be used to obtain materials with a very high specific surface area useful in catalysis or films and thin layers with applications ranging from optical coating to microelectronics.[7]

In: New Developments in Medicinal Chemistry
Editors: Marta P. Ortega and Irene C. Gil
ISBN 978-1-60456-810-3

Chapter 1

FOLATE BIOSYNTHESIS – REAPPRAISAL OF OLD AND NOVEL TARGETS IN THE SEARCH FOR NEW ANTIMICROBIALS

James Swarbrick[1*], Peter Iliades[2♦], Jamie S. Simpson[1] and Ian Macreadie[3•]

1. Medicinal Chemistry and Drug Action,
Monash Institute of Pharmaceutical Sciences,
Monash University (Parkville Campus), Parkville, Victoria, Australia
2. The Walter and Eliza Hall Institute of Medical Research,
Parkville, Victoria, Australia
3. CSIRO Molecular and Health Technologies,
Parkville, Victoria, Australia

ABSTRACT

Folate biosynthesis remains a key target for antimicrobial therapy. Folate is an essential vitamin (vitamin B9) that is required for many one-carbon transfer reactions and is a critical precursor for the biosynthesis of purines, pyrimidines, and amino acids. Unlike higher eukaryotes that scavenge preformed folates, prokaryotic and lower eukaryotic microorganisms are dependent on several enzymes for the *de novo* biosynthesis of folate. One of these enzymes, dihydropteroate synthase (DHPS), is the target of the first chemically-synthesized antimicrobial agents, the sulfadrugs, which date back to the 1940s. Others are essential enzymes that remain to be explored as drug targets. Resistance to the sulfadrugs rapidly emerges due to the ability of the microbe to alter its susceptibility to the drug by various means. Recently a number of new structures of the enzymes in the pathway has become available. We review the recent literature relating to these targets (the enzymes: 7,8-dihydroneopterin aldolase (DHNA), 6-

* 381 Royal Parade, Parkville, Victoria 3052, Australia.
♦ 1G Royal Parade, Parkville, Victoria 3050 Australia.
• 343 Royal Parade, Parkville, Victoria 3052 Australia.

hydroxymethyl-7,8-dihydropterin pyrophosphokinase (HPPK), dihydropteroate synthase (DHPS), dihydrofolate synthase (DHFS)), their mode of action and how current drugs may modulate this on a structural level. Furthermore, these data advance our understanding of the emergence of drug resistance and may aid efforts and play a major role in the design of new, more effective compounds as antimicrobial agents. To this end we also review the recent literature in the development of inhibitors of these enzymes. Future progress in this key area has the potential to benefit the war against devastating organisms such as drug-resistant *Staphylococcus aureus* and *Plasmodium falciparum*.

ABBREVIATIONS

AMPCPP, AB-METHYLENEADENOSINE TRIPHOSPHATE;
DHNA, 7,8-DIHYDRONEOPTERIN ALDOLASE;
DHPS, DIHYDROPTEROATE SYNTHASE;
DHFR, DIHYDROFOLATE REDUCTASE;
DHFS, DIHYDROFOLATE SYNTHASE;
DHPP, 6-HYDROXYMETHYL-7,8-DIHYDROPTERIN-PYROPHOSPHATE;
FAS, FOLIC ACID SYNTHESIS;
FPGS, FOLYPOLYGLUTAMATE SYNTHASE;
GTP-CH, GTP CYCLOHYDROLASE;
HP, 6-HYDROXYMETHYL-7,8-DIHYDROPTERIN;
HPPK, 6-HYDROXYMETHYL-7,8-DIHYDROPTERIN PYROPHOSPHOKINASE;
PABA, *P*-AMINOBENZOIC ACID;
PCP, *PNEUMOCYSTIS CARINII* PNEUMONIA;
SMX, SULFAMETHOXAZOLE;
TMP, TRIMETHOPRIM;
TS, THYMIDYLATE SYNTHASE.

INTRODUCTION

Effective antimicrobials are essential for the maintenance of our 21st century lifestyle. Without antimicrobials, death rates from simple and common infectious diseases would be high, epidemics would be rampant and advances in surgery and immunosuppressive therapies would amount to nil. Good targets for antimicrobials are essential enzymes (for the microorganism) that are not present in the host organism. Furthermore, it is preferable that the enzyme has a track record as a target for drugs and its properties are well documented. The folate biosynthetic pathway fits the criteria of being an ideal target for antimicrobial therapy and is the focus of this review.

Our review commences with a historical perspective of the clinical application of the sulfadrugs that target folate synthesis to the inevitable rapid spread of antibiotic resistance in several relevant pathogenic organisms. Understanding of resistance is explained at the molecular and structural level of the folate biosynthesis pathway. An insight into the catalytic mechanism and function is provided by virtue of a multitude of structures for all the relevant enzymes often in many catalytic states. Therefore, we describe current strategies that tap into this structural data resource and review structural biology combined with medicinal chemistry, assay technologies, modern structure-based approaches and methods for inhibitor design. The folate biosynthesis pathway is shown to have considerable untapped potential to be exploited for the rational design of new antibiotics that can slow the onset of resistance and combat current resistance isolates that are threatening epidemics across the world.

HISTORICAL ASPECTS LEADING TO RESISTANCE

Drug pressure on the folate biosynthetic pathway to date has only occurred on DHPS so it is not surprising that mutations in DHPS, DHPS gene duplication and the confounding effects of DHPS metabolites, pABA or environmentally acquired folates, are the only clinically-verified resistance modalities reported. From the very beginning, sulfonamides were used to treat diverse infections and consequently, resistance has been observed in all circumstances.

Sulfanilamide (the active component of Prontosil – an *azo* sulfonamide dye) was developed in 1932. Domagk demonstrated that it was capable of protecting mice from lethal streptococcal infections, rabbits from staphlylococcal infections, and in one case, cured an infant from a life-threatening infection. This work led to Domagk being awarded a Nobel prize in 1939. Prontosil was successfully trialled for the treatment of malaria in 1937 (Hill and Goodwin 1937) and thousands of sulfonamides / sulfones compounds, collectively called sulfadrugs, have been synthesised and tested since.

The use of sulfonamide and sulfones as antifolates predates the demonstration of their mode of action. As early as 1940 however, pABA was found to antagonize the bacteriostatic action of sulfonamides (Woods 1940). Enzymes involved in bacterial folate biosynthesis were identified in the 1960s as were the first resistance mechanisms. It was determined that increased pABA and folic acid synthesis could lead to sulfonamide resistance in Staphylococci (White and Woods 1965a, b).

Sulfonamides were found to act on the folate biosynthetic pathway as competitive (with pABA) inhibitors of DHPS. Furthermore, the sulfonamides tested were more inhibitory in cell-free enzymatic systems than as inhibitors of cell growth indicating limited cell permeability (Brown 1962). One year later, *E. coli* mutants were selected that had mutant enzymes that changed the cell's permeability to sulfonamides (Pato and Brown 1963).

In the 1950s it became evident that the combination of sulfonamides and the 2,4 diaminopyrimidine class of compounds (later shown to be DHFR inhibitors) were more effective than either drug alone to treat malaria patients infected with *P. falciparum* (Greenberg and Richeson 1950; Hurly 1959). The use of sulfonamides as a monotherapy to treat malaria infections was discontinued due to the low efficacy and high toxicity (Michel 1968; Rieckmann *et al.* 1968; Sheehy *et al.* 1967). In various studies in the 1960s, combination therapy proved more efficacious than the traditional anti-malarial (chloroquine) which they replaced owing to emerging parasite resistance (Chin *et al.* 1966; Chulay *et al.* 1984; Degowin and Powell 1964; Harinasuta *et al.* 1967; McGregor *et al.* 1963). However, antifolate resistance to SP (Sulfadoxine / Pyrimethamine) emerged almost immediately following its introduction in 1967 in Thailand (Peters 1987).

The dogma until recently was that sulfonamides exerted all their effects by competing with pABA to deplete the intracellular folate-cofactor pool, thus starving the cell. The sulfa-pterin analog diffused out of the cell and was no consequence to the growth of bacteria exposed to it (Roland *et al.* 1979). However, later work by Patel and colleagues showed that in *Saccharomyces cerevisiae*, the sulfa-containing folate analogs were growth inhibitory (Patel *et al.* 2003b).

Antifolate Drug Resistance Mechanisms

Numerous antifolate resistance mechanisms have been reported. Numerous reports have described multiple amino acid mutations, amino acid insertions/duplications in DHPS that confer drug resistance to sulfonamides from many organisms and some organism-specific examples will be reviewed. In some cases, these mutant DHPS forms have been detected in organisms where resistance has been acquired by horizontal transfer of resistance plasmids (Fermer *et al.* 1995). Furthermore, DHPS resistance elements have remained long-lived even when sulfonamide pressure has been withdrawn owing to drug pressure exerted by other drugs (Enne *et al.* 2001). Gene amplification was demonstrated as a potential sulfonamide drug resistance mechanism in model systems (Iliades *et al.* 2003; Nichols and Guay 1989) and recently shown to be a clinically valid resistance mechanism by following chromosomal amplification of genetic elements encoding DHPS in *S. agalactieae* (Brochet *et al.* 2007). The competing effect of increased levels of the DHPS substrate pABA was shown to confound the effects of sulfonamides (Woods 1940) both clinically and in model systems (Castelli *et al.* 2001) as was the acquisition of environmental folates (Carter *et al.* 2005; van Hensbroek *et al.* 1995). Drug resistance mutations have long been known to confer a fitness compromise. It has been shown in both model systems (Bouma and Lenski 1988; Iliades *et al.* 2004a; Schrag *et al.* 1997) and clinically (Fermer *et al.* 1997) that compensatory mutations or adaptation mechanisms are capable of facilitating enzymatic improvements or altered regulation of metabolic pathways to counter the fitness compromise.

Diseases Targeted by Antifolates

Sulfonamides have been used to treat a large number of fungal, bacterial and parasitic infections including; *Pneumocystis* pneumonia, malaria, *Pneumococcal* pneumonia, urinary tract infections, candidiasis, tuberculosis, leprosy, meningitis, toxoplasmosis and many others. As with most antibiotics which function as anti-metabolites, resistance evolved rapidly making these compounds of limited utility and consequently reducing man's armoury of effective antibiotic countermeasures.

Fungal Pathogens

Pneumocystis Jiroveci (Formerly Carinii)

Pneumocystis jiroveci is a major opportunistic pathogen that results in *Pneumocystis* pneumonia (PCP) of AIDS patients and immunocompromised individuals. It accounts for 40% of all AIDS-defining conditions and is the major cause of mortality of children with

AIDS in Africa (Ansari *et al.* 2003; Armstrong *et al.* 2000). Clinically, PCP has been treated with antifolates including combination therapy with sulfamethoxazole (SMX) and trimethoprim (TMP) as the preferred first-line treatment (Kovacs *et al.* 2001). In *P. jiroveci*, however, there is some evidence to suggest that TMP is ineffective and that such treatment is actually sulfamethoxazole monotherapy (Meshnick 1999; Walzer *et al.* 1992).

The widespread use of sulfamethoxazole-trimethoprim (SMX-TMP for prophylaxis against *P. jiroveci* pneumonitis in HIV-infected patients) has been implicated as the cause of the increase in SMX-TMP –resistant bacteria (Martin *et al.* 1999). Several studies have demonstrated point mutations in the *P. jiroveci* DHPS gene and have found an association between the use of sulfonamide or sulfone drugs for *P. carinii* prophylaxis and DHPS mutations (Kazanjian *et al.* 1998; Mei *et al.* 1998) but not in the DHFR gene (Ma *et al.* 1999).

The failure of prophylaxis and treatment of PCP patients has been associated with mutations in DHPS (similar to those that confer sulfa resistance in other organisms by a large number of epidemiological studies) (Ansari *et al.* 2003; Armstrong *et al.* 2000; Demanche *et al.* 2002; Helweg-Larsen *et al.* 1999; Kazanjian *et al.* 1998; Kovacs *et al.* 2001; Lane *et al.* 1997; Ma *et al.* 2002; Mei *et al.* 1998; Navin *et al.* 2001; Takahashi *et al.* 2000; Visconti *et al.* 2001). However more direct evidence demonstrating that such mutations confer resistance in *P. jiroveci* only recently emerged (Iliades *et al.* 2005a, b) using models systems.

In fungi, mutations involved in resistance to sulfonamides are shown in Table 1. It was demonstrated that (i) mutants having the single amino acid substitution T557A were more sensitive than the WT, (ii) mutants having two amino acid substitutions, T557A and P559S, had increased sulfadrug resistance relative to the wild type (WT), and (iii) there was cooperativity between individual mutations that led to the increased sulfadrug resistance of the double mutants.

Importantly, in the model studies of *P. jiroveci* DHPS mutations, it was observed that mutants having two amino acid substitutions were initially compromised for growth due to an increased requirement for pABA. Prototrophs that could grow in the absence of pABA could be isolated via continual passage on low pABA medium. These double mutants were found to be capable of improved growth vigor and consequently increased sulfadrug resistance indicating an adaptive response to the initial growth compromise presumably conferred by the DHPS mutations. This suggested that the DHPS mutations resulted in decreased enzyme activity thus reducing folate synthesis. The observed adaptation to low pABA medium implicated pABA up-regulation with sulfamethoxazole resistance mutations. Thus increased pABA synthesis probably reflects an adaptive response that compensates for the reduced pABA binding affinity by the double amino acid substitutions at the catalytic site of DHPS. Such an observation was similar to observations made by Schrag and Perrot who noted an adaptive response that resulted in secondary mutations to compensate for the fitness compromise conferred by the primary mutations that led to drug resistance (Schrag *et al.* 1997).

Table 1. Changes in P. jiroveci and S. cerevisiae DHPS that are associated with sulfonamide resistance.

Allele name	Amino acid position in DHNA-HPPK-DHPS	
	#557 in S. cerevisiae; #517 in P. jiroveci	#559 in S. cerevisiae; #519 in P. jiroveci
Wild Type	T	P
VRS	V	S
ARS	A	S
TRS	T	S
ARP	A	P

Numbering is from the first methionine of the primary translation product

Conclusive evidence that the DHPS mutations resulted in sulfadrug resistance emerged following the cloning of the trifunctional *P. jiroveci* DHNA-HPPK-DHPS (Iliades *et al.* 2004b) genes and their heterologous complementation in a DHPS-disrupted *E. coli* host strain (Iliades *et al.* 2004a). Site-directed mutagenesis of the *P. jiroveci* trifunctional DHNA-HPPK-DHPS genes (Iliades *et al.* 2005b) to reverse engineer the mutations suspected by epidemiological data to cause drug resistance provided an assay method that permitted a direct assessment of sulfadrug resistance. Thus the mutations observed clinically (T517A and P519S) in the PjFAS genes were found to result in a threefold increase in sulfamethoxazole resistance relative to the wild type clone and furthermore conferred significant cross resistance to a range of sulfadrugs.

Parasitic (Protozoan) Pathogens

Plasmodium Falciparum

Malaria is one of the primary causes of infant mortality in the developing world and is caused largely by the protozoan parasite *Plasmodium falciparum.* It accounts for 80% of malaria infections and 90% of deaths estimated to be between 1-3 million per year of the 500 million infections worldwide (Snow *et al.* 2005). The largest mortality and morbidity burden is suffered in sub-Saharan Africa and in particular by infants. To date, malaria has been targeted therapeutically and preventatively with numerous antimalarials including chloroquine (quinines), artemisinin, atovaquone, doxycline and antifolates. All have led to resistance and multiple drug resistance is common in many parts of the world where malaria is endemic.

P. falciparum drug resistance has spread very rapidly after the introduction of antimalarial therapeutic and prophylaxis measures in Asia, South America and Africa. Mutations that cause antifolate resistance have been found in DNA sequences of DHPS associated with resistance to sulfadoxine (Brooks *et al.* 1994; Reeder *et al.* 1996; Triglia *et al.* 1997; Wang *et al.* 1995; Wang *et al.* 1997). In DHPS the acquisition of multiple point mutations (Table 2) has been shown to increase sulfonamide resistance (Berglez *et al.* 2004; Triglia *et al.* 1997). The relative contribution of the mutations has been shown to be predictive of treatment failure in particular combinations (Dorsey *et al.* 2004). For example,

E540 was a strong predictor of treatment failure as was DHFR mutant R59 in combination with DHPS E540.

Table 2. Changes in P. falciparum DHPS that are associated with sulfodoxine (SDX) resistance. Numbering is from the first methionine of the primary translation product

Isolate	SDX response	Amino acid position				
		436	437	540	581	613
Wild-type	Sensitive	S	A	K	A	A
3D7	Resistant	G	G			
TN-1	Resistant	G	G	E		
K1	Resistant	G	G		G	
W2	Resistant	F	G			S
V1/S	Resistant	F	G			T

Alignments of *P. falciparum* DHPS with other DHPS enzymes indicate that changes that result in sulfadrug resistance lie in regions that are close to the site of catalysis. It would appear that changes involved in DHPS associated with sulfadrug resistance lead to reduced affinity of the sulfadrug for DHPS.

Resistance to antifolates by eukaryotes can be further achieved via acquisition of environmental (exogenous) folates via active folate-specific transporters (Wang *et al.* 2007; Zhang *et al.* 1992). It has been demonstrated that antimalarial therapy using antifolates can be significantly compromised by the concomitant dietary supplementation of folic acid (Carter *et al.* 2005; van Hensbroek *et al.* 1995). In addition, increased expression levels of DHPS increase or decrease sulfadrug resistance depending on whether pABA is high or low, respectively (Iliades *et al.* 2003; Iliades *et al.* 2004a). pABA overexpression as a potential mode alleviating the inhibitory action of sulfonamides has been demonstrated *in vivo* using a *S. cerevisiae* model (Castelli *et al.* 2001). Drug resistant mutants to DHFR inhibitors (Pyr) were found to be selected when the *in vivo* concentration fell below the effective therapeutic concentration (Nzila *et al.* 2000; Watkins and Mosobo 1993).

Bacterial Pathogens

Escherichia Coli

Multi-drug resistance has eventuated as a particular problem for the treatment of urinary tract infections. 25-80% of such infections were characterised as being multi-drug resistant by 1975 (Wise and Abou-Donia 1975). These resistant isolates showed sulfonamide resistance with Minimum Inhibitory Concentration (MICs) 100-1000 fold higher than sensitive strains. The resistance can be chromosomally encoded and include a single F28L mutation (Dallas *et al.* 1992) or can be carried on horizontally transmissible R plasmids SulA, B, C, D (DHPS, DHFS, GTP-CH and HPPK respectively]. These resistance plasmids have persisted undiminished despite the prolonged withdrawal of selection pressure following formal introduction of prescribing restrictions on co-trimoxazole, (Bean *et al.* 2005; Grape *et al.* 2003).

Neisseria Meningitidis

Neisseria meningitidis is a gram-negative diplococcal bacterium that infects and inflames the protective membranes covering the central nervous system resulting in the rapid progression from fever, headache and neck stiffness to coma and death in 10% of patients or higher if left untreated.

The massive use of sulfonamides for both prevention and treatment of meningococcal disease (Feldman 1986; Peltola 1983) led to the isolation of resistant strains of *Neisseria meningitidis* as early as 1937 (Feldman 1966), only a couple of years after the discovery of this class of drugs. Resistance has been shown to be mediated by altered forms of the chromosomal *dhps* gene (Fermer *et al.* 1995; Raadstroem *et al.* 1992). As a result the World Health Organisation now states "Although oral *cotrimoxazole* (trimethoprim sulfamethoxazole) is inexpensive and has good CSF penetration, sulfa resistant strains have become common and sulfadrugs are not recommended unless sulfa sensitivity testing has been done. In unfavourable conditions, the drug of choice is oily chloramphenicol" (World Health Organization and Control 2003).

Two DHPS mutations F31L and G194C were shown by MIC determinations to effect resistance (Fermer *et al.* 1997) as well as K*m* and the K*i*. A third mutation (P84S) did not have any obvious effect on the resistance phenotype, however analysis of its enzyme kinetics showed altered K*m* thus acting as a compensatory mutation for the first two mutations which mediate sulfonamide resistance.

Streptococcus Pneumoniae

Streptococcus pneumoniae, or pneumococcus, is a Gram-positive, alpha-haemolytic diplococcus bacterium and is a significant human pathogen recognized as a major cause of pneumonia worldwide. *S. pneumoniae* is the most common cause of acute respiratory infections in adults and children which infects an estimated 2.6 million children under five years of age annually. Pneumococcus causes over 1 million of these deaths, most of which occur in developing countries, where pneumococcus is probably the most important pathogen of early infancy (World Health Organization 2003). Antifolates are no longer used, due to resistance conferred by R plasmids encoding (i) plasmid borne SulA where a 2 amino acid duplication in DHPS confers drug resistance (Akiba *et al.* 1960; Skold 1976; Swedberg and Skoeld 1980; Watkins and Mosobo 1993) and (ii) plasmid borne TMP resistance (Fleming *et al.* 1972; Pattishall *et al.* 1977; Then 1993).

Summary of Sulfadrug Resistance

Numerous drug resistance mechanisms have emerged in all cases where sulfonamides have been used, and as is the case for all antibiotics this is not a new phenomenon. Resistance has emerged due to overprescribing, ineffective use of synergistic compounds, poor patient compliance with drug regime and the use of the drugs in the animal industry. Antifolates however have proved to be successful (until the emergence of resistance) in treating a broad range of infection thus validating the pathway as a source of important target. The folate biosynthetic pathway has multiple unique enzymes to prokaryotes and lower eukaryotes that recognize pterin-like molecules and as a result there could be potential for cross reactive compounds within the pathway. Such cross-reactivity could potentiate the effectiveness of antifolate compounds and possibly act in a powerful and synergistic fashion similar to the synergism seen between sulfadrugs and DHFR inhibitors.

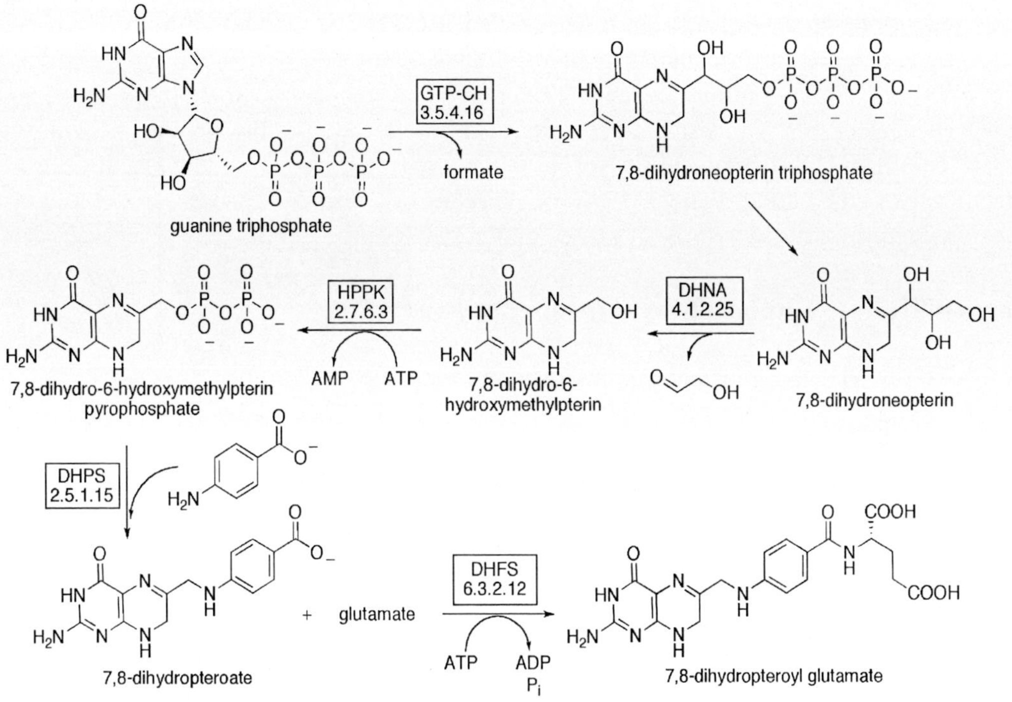

Figure 1. The folate biosynthetic pathway in plants and microbes.

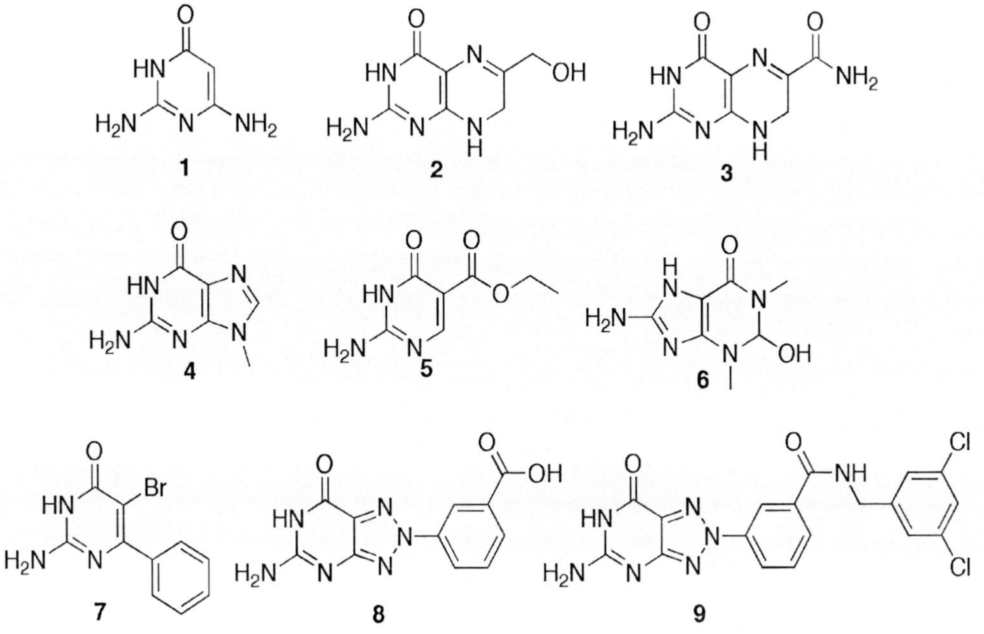

Figure 2. Structures of inhibitors of GTP-CH and DHNA.

Folate Biosynthesis Pathway

Folate is an essential metabolite required for many one-carbon transfer reactions and is a critical precursor for the synthesis of purines, pyrimidines and amino acids. Prokaryotic and lower eukaryotic micro-organisms are dependent on several enzymes for de novo folate biosynthesis (Figure 1). In contrast, higher eukaryotes (including mammals) do not possess these enzymes and are entirely dependent on folate derived from their diet and active folate transport mechanisms. Consequently enzymes of the *de novo* pathway may provide ideal targets for therapeutic intervention for treatment of infections and serious diseases and infections caused by prokaryotic and lower eukaryotic pathogens.

Targets from the Folate Biosynthesis Pathway

The *de novo* folate biosynthesis pathway comprises six committed enzymes GTPCH, a NUDIX pyrophosphatase, DHNA, HPPK, DHPS and DHFS that collectively converts GTP into dihydrofolate with the incorporation of pABA (Figure 1). The inactive dihydrofolate from this pathway is subsequently reduced into the active form tetrahydrofolate via DHFR. This step is not part of the *de novo* pathway *per se*, but rather is part of the subsequent folate utilization step. DHFR is common to mammalia, protozoa and bacteria. Nevertheless, DHFR active sites are both highly conserved within each of these three classes and also subtly different between mammalia, protozoa and bacteria. Consequently, DHFR is a successful target for the competitive inhibitors methotrexate, trimethoprim and pyrimethamine, which are in use for cancer therapy, bacterial and protozoal infections respectively. As described, the existing antifolate drugs tend to be synergic cocktails of a DHPS inhibitor, (*e.g.* sulfadioxine, sulfalene or dapsone) with the DHFR inhibitor, pyrimethamine.

GTP Cyclohydrolase

The first enzyme, GTP-CH, catalyses the conversion of GTP into 7,8-dihydroneopterin triphosphate using two molecules of GTP. Although the enzyme is found in a wide range of prokaryotic and eukaryotic species, dihydroneopterin triphosphate is either processed into tetrahydrofolate or, in the case of animals lacking the biosynthesis pathway, is converted into tetrahydrobiopterin. However the pterin binding site is very well conserved across all species and thus probably not ideally suited to antimicrobial development.

The compounds reported to be inhibitors of GTP-CH are close structural mimics of the reaction substrate or product. For example, although the monocyclic 2,4-diamino-6-hydroxypyrimidine (Figure 2, **1**), was thought to be an inhibitor of GTP cyclohydrolase, Xie *et al.* (1998) recently demonstrated that this compound actually acts through an indirect mechanism involving binding to GFRP (GTP-CH feedback regulatory protein), rather than direct inhibition.

Removal of Pyrophosphate ~ the Mystery Solved

The removal of the triphosphate in the second step seems to occur via two phosphatases, one removing pyrophosphate and the other a phosphate. The identification of a specific enzyme committed to the pyrophosphatase function was until recently a mystery.

Nevertheless, an *E. coli* NUDIX hydrolase was confirmed to convert DHNTP into DHNMP with the loss of pyrophosphate *in vivo* and the structure has been solved by X-ray analysis (Gabelli *et al.* 2007). In a knockout strain, the level of folate dropped considerably, and was restored with the NUDIX plasmid complementation, showing specificity of this enzyme. Despite this observation, there was no reduction in growth rate compared to the wild type, suggesting that conversion into DHNMP can take place via other non-specific enzymes or in a non-enzymatic manner. Accordingly, the NUDIX enzyme may not represent an attractive drug target. No enzyme to date that removes the final phosphate has been identified and the reaction most likely is mediated by non-specific phosphatases.

The four remaining enzymes of the folate pathway, DHNA, HPPK, DHPS and DHFS represent either current targets (DHPS) or potential sources for drug intervention. Therefore considerable attention has been given to these enzymes from sequence through to structure to both understand and interfere with the catalytic function, and to rationalize a molecular basis behind antibiotic resistance.

Formation of HP ~ DHNA

DHNA catalyses both the epimerization of 7,8-dihydroneopterin (DHNP) to 7,8-dihydromonapterin (DHMP) and the conversion of DHNP or DHMP to 6-hydroxymethyl-7,8-dihydropterin (HP) with the generation of glycoaldehyde (Figure 1). The DHNA structure from three organisms has been solved by X-ray crystallography (*S. aureus* (Hennig *et al.* 1998), *S. pneumonia* (Garcon *et al.* 2006) and *M. tuberculosis* (Goulding *et al.* 2005)). Much information has been derived using the *S. aureus* enzyme from structures of the holoenzyme (Goulding *et al.* 2005) and in complex with the product HP (Blaszczyk *et al.* 2007b), and substrate analogs, neopterin (NP) and monopterin (MP) (Blaszczyk *et al.* 2007b), and product (Goulding *et al.* 2005).

The epimerization mechanism proceeds through an active site lysine that functions as a general base and a bound water as a proton donor. Four structural snapshots along the catalytic trajectory from the X-ray analysis of *S. aureus* DHNA allow a detailed structural understanding of this mechanism.

Generally speaking, DHNA holoproteins are octameric in the absence or presence of substrate which sits between the protomers (Figure 3a). For example, the oligomeric state is supported by sedimentation equilibrium studies of *E. coli* and *Haemophilus influenza* enzymes (Haußmann *et al.* 1998). Interestingly, the structure of DHNA from *M. tuberculosis* is tetrameric in the "apo" enzyme and is octameric with product. Four helices occupy the substrate position in the inactive tetrameric enzyme. Hence, a substrate-induced conformational change has to take place which results in the displacement of these helices and the formation of the active octamer. This substrate-induced oligomerisation was supported using sedimentation equilibrium measurements (Goulding *et al.* 2005). Kinetic analysis of substrate binding gave a Hill coefficient of 2 indicating positive cooperativity and thus a substrate driven model for the formation of the octamer was proposed. The model involved an initial binding to a single low affinity tetrameric site followed by subsequent binding to the higher affinity octomeric sites. In this way DHNA from *M. tuberculosis* is allosterically regulated through oligomerisation. This unique characteristic of the *M. tuberculosis* enzyme was explained based on structural differences within the tetramer-tetramer interface and the pterin binding pocket as compared to the other DHNA structures.

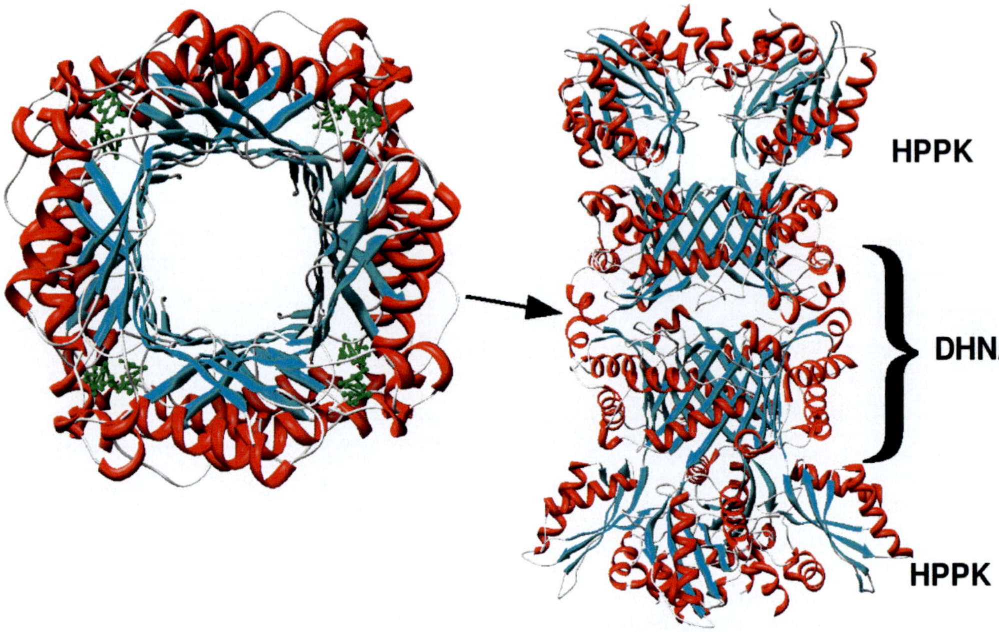

Figure 3. Left: Top view looking down through the pore of a ribbon representation of the octomeric structure of DHNA in complex with the product HP in green. Right: Orthogonal view showing the assembly of the bifunctional DHNA-HPPK enzyme from *S. pneumoniae*.

It has been shown in mammals that 7,8-dihydroneopterin acts as a powerful antioxidant by inhibiting low density lipoprotein oxidation, peroxyl radical formation and super-oxide generation. Hence the authors propose that, under stressful conditions, the inactive tetramer switches the role of the substrate to an antioxidant thereby protecting *M. tuberculosis* from the damaging effects of hydrogen peroxide from the host. When oxidative stress diminishes and the levels of substrate rises further, the octameric species forms and folate biosynthesis proceeds in the usual manner.

DHNA Inhibitors

Early reported inhibitors of dihydroneopterin aldolase were dihydropterins closely related to substrate and product structures, including the alcohol **2** and the amide 3 (Figure 2), reported by the Merck group (Zimmerman *et al.* 1977). These compounds had IC_{50} values with the enzyme of 1 and 5 μg/mL, respectively.

More recently, DHNA from *S. aureus* has been the focus of a very elegant high throughput crystallographic screening study for lead compounds (crystalLEAD) (Sanders *et al.* 2004). A library of 10,000 compounds was screened to detect initial hits, and these included methylguanine (4), pyrimidines 5 and 6 and aminopurine 7 (Figure 2). A subsequent round of screening used a focused library based on the observed three-point hydrogen-bonding motif common to the initial hits. Interestingly, none of these inhibitors gave structures from soaking experiments but co-crystallisation was successful in all cases. This is consistent with the observation of sigmoidal inhibition curves and the notion of a conformational shift taking place during ligand binding in a cooperative manner.

This smaller targeted library led to the identification of triazole 8, which displayed an IC_{50} of 1.5μM. A focused library based on carboxylic acid 8 was synthesized to evolve fragments to potentially fit into an extended nearby hydrophobic binding groove using simple amide linkage chemistry. The groove displayed a range of sizes in the different lead complexes, consequently both large and small hydrophobic substituents were added and gave a number of more potent inhibitors, the most potent of which (dichlorobenzyl substituted **9,** Figure 2) showed an IC_{50} of 69 nM. Although the compounds showed good *in vitro* inhibition, none showed *in vivo* activity when tested against several pathogens and were therefore not suitable as an antimicrobial. Furthermore, none was active against a hypersusceptible permeability mutant and Acr- efflux pump mutants of both *E. coli* and *H. influenza.* Similarly, they were also not active in the presence of permeability-enhancing agents and none inhibited bacterial uptake of radiolabelled pABA. This suggests that permeability is not the issue. It was proposed that inhibitors were inactivated by either intracellular enzymes or by non-specific binding to intracellular membrane proteins both of which could cause this lack of activity *in vivo.* An alternative possibility put forward is that the intracellular concentration of the DHNA substrate can be high relative to the K_m of the enzyme in which case the inhibitors would have to be significantly more potent to be effective.

This study is the most elegant and detailed structure-based effort published to date in the design of antimicrobials from the enzymes of the *de novo* folate biosynthesis pathway. It showcases the utility of the CrystaLEAD method to screen and deliver highly potent inhibitors. It reveals the shortcomings of ligand soaking over co-crystallisation to overlook potential lead compounds when ligand-induced conformational shifts occur. Conformational motion such as this may be very pertinent given the significant range of loop motions encountered in all the structures discussed herein. Finally, despite compound **9** being a good *in vitro* inhibitor, this work shows that elements other than structural design will have to be overcome before a suitable inhibitor can be a good *in vivo* lead compound.

Pyrophosphoryl Transfer ~ HPPK

HPPK catalyzes the transfer of pyrophosphate from a bound ATP to 6-hydroxymethyl-7,8-dihydropterin (HP), Figure 1. Over twenty HPPK structures have been deposited from five different species to date *.e.g. E. coli* (Xiao *et al.* 1999), *H. influenza* (Hennig *et al.* 1999), *S. pneumoniae* (Garcon *et al.* 2006), and *S. cerevisiae* (Lawrence *et al.* 2005), *Y. pestis* (Blaszczyk *et al.* 2007a)). Those from *S. cerevisiae* and *S. pneumoniae* have been solved as bifunctional enzymes with either DHPS or DHNA respectively. Generally speaking, there is little biophysical or structural evidence to suggest that the biological functional unit of HPPK is anything other than a monomer.

High sequence identity (33-61%) combined with active site structural similarity (Figure 4) for the determined structures, suggestes that broad spectrum anti-infective agents can be developed. The *E. coli* enzyme is both small (153 residues) and has a high thermal stability (Blaszczyk *et al.* 2000). Accordingly, a host of both NMR and X-ray structures has been deposited, largely from the work out of the H. Yan group, providing a convenient system to study the molecular basis behind pyrophosphoryl transfer and establishing a platform for structure-based drug discovery strategies.

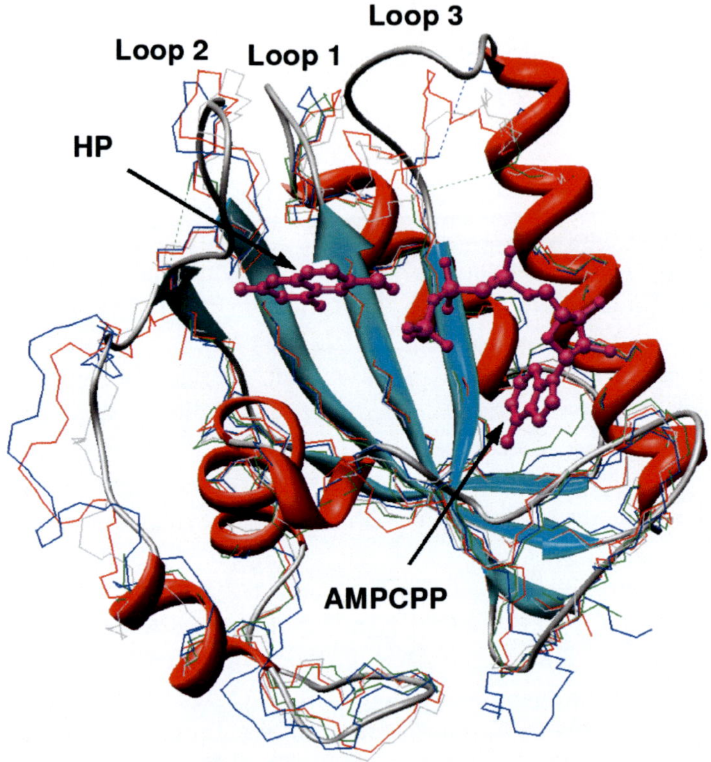

Figure 4. Superposition of E. coli (grey), H. influenza (red), S. pneumoniae (green), S. cerevisiae (blue) and Y. pestis HPPK structures. A ribbon representation of the fold is drawn along the Y. pestis structure. The bound HP and the ATP analogs are displayed as ball and stick.

The structure of the active site has been well characterized from a range of substrate, substrate analogue and product complexes (Blaszczyk *et al.* 2000; 2003; 2004a,b; Li *et al.* 2005; Shi *et al.* 2001; Stammers *et al.* 1999; Xiao *et al.* 1999; 2001). The structure comprises a three-layered αβα fold which creates a valley some 26Å long, 10Å wide and 10Å deep. Much understanding has evolved from X-ray structures in complex with HP and the non-hydrolysable ATP analogue MgAMPCPP. The pterin and ATP are fixed within their sub-sites either end of the valley by a multitude of hydrogen bonds and they interacts with a total of 26 residues, 13 of which are conserved. From the structures, an in-line single displacement mechanism with associative character in the transition state has been proposed.

The studies on the *E. coli* system show a range of loop motions that surround the active site. Interestingly, highly conserved side chains on loop 3 (L3) appear to point away from the active site in the apo structure but come into play during a series of large loop motions during the cycle. Thus loop dynamics (L1, L2 and L3, Figure 4) are shown to be an integral part of the catalytic mechanism. Loop L3 for example has to open up twice, moving up to 20Å and 28Å from the apo position during the cycle and repositions critical arginine side chains in the process. All three critical loops seal the active site for chemical transformation to occur. Within the active site only the adenosine binding site is rigid which is responsible for the observed high specificity. For example the binding of MgGTP is 260-fold less than MgATP (Blaszczyk *et al.* 2000). As mentioned, loop 3 is critical for product release, but is not necessary to bind ATP. Its role has been investigated elegantly from X-ray and biochemical methods using a loop deletion (Blaszczyk *et al.* 2004a). From thermodynamic and transient kinetic data, the full reaction trajectory has been established (Blaszczyk *et al.* 2004b), in which HPPK goes through six distinct steps and X-ray structures are available for each intermediate.

NMR spectroscopy is an invaluable tool to investigate loop dynamics in enzymes. In an NMR study the three loops of the binary complex of HPPK-AMPCPP were shown to assume multiple conformations (Li *et al.* 2006) based on the observation of multiple resonances for the loop residues, although multiple resonances are often observed when a proline is in the loop, as in loop 3, due to the *cis* - *trans* interconversion in slow exchange on the NMR time scale. In the ternary complex HPPK-PPK-AMPCPP-DMHP, only one set of NMR signals was observed suggesting that binding of the second substrate shifts the multiple conformations of the binary complex to a single "fixed" conformation during active site closure.

Although an induced fit model is implied based on X-ray 'snapshots' of almost all stages in the cycle, the NMR data favour an equilibrium conformational model, typical of enzymes (Swarbrick *et al.* 2005). The combined NMR and X-ray approach is optimal to characterize both motions and precise structural details in small enzymes.

HPPKs are generally monomeric in nature, however, a recent X-ray structure from *Y. pestis* (the causative agent of bubonic, pneumonic and septicaemic plagues) (Blaszczyk *et al.* 2007a) shows that the enzyme crystallises as a head to head dimer. In this, the ligands are in close contact with each other and a C2 symmetry operation relates the two practically identical monomers. Although other biophysical methods were not sought to report on this rather unique oligomeric state, a reduced surface area (2851Å^2) compared to typical (3900Å^2) for dimers of this size was noted and this casts doubt on the catalytic relevance of the dimer. Studies from the same group (either solution NMR or biophysical methods) are hopefully in progress to resolve this rather interesting anomaly.

HPPK Inhibitors

Bermingham *et al.* (2000) reported that the pterin substrate did not bind in the absence of ATP or a stable analogue (AMPCPP, 10) (Figure 5), and showed that the fluorescent derivative 11 bound similarly to ATP, shedding further light on the catalytic cycle. Both analogs (10 and 11) were competitive inhibitors with respect to ATP. Shi *et al.* (2001) reported the interesting bifunctional molecules (12 and 13), which, although not transition state mimics, consist of both ATP and the pterin substrates linked through phosphates. IC_{50} values of 1.27 μM and 0.44 μM were found for 12 and 13, respectively. Compound 13 was 116 and 76 times more potent than MgADP and 6-hydroxymethylpterin respectively. Similarly, ester 12 was 12 and 7 times more so and $MgHP_2A$ (12a) had no detectable affinity. Hence, linker length is very important (7Å); both tetra phosphate 13 and triphosphate 12 have approximately the correct length linkers, albeit with fewer favourable hydrogen bonds in the latter, while the corresponding bis-phosphate linked compound (12a) is too short to connect the pterin and ATP sub-sites. Crystallographic studies shows inhibitor 13 occupies both the ATP and pterin binding pockets, and causes conformational changes.

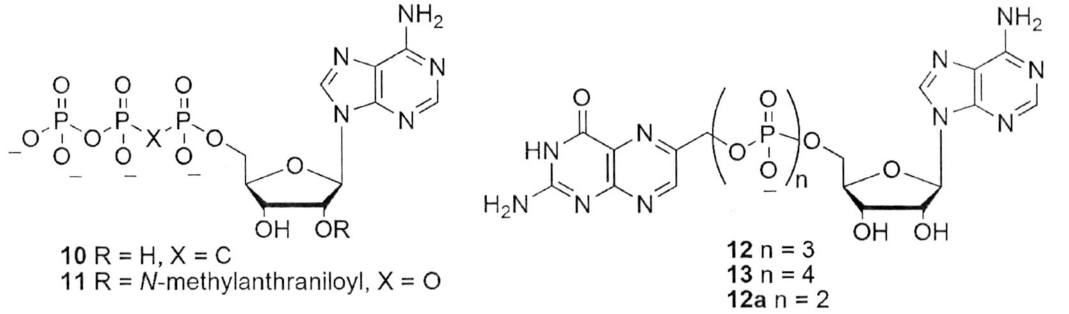

Figure 5. Inhibitors of HPPK.

Thanks to a range of structural, mechanistic and dynamic data, the stage is set for HPPK to shine as a new player to assist in the design of novel inhibitors of the *de novo* folate pathway.

Old Target with New Potential ~ DHPS

DHPS catalyses the conversion of 6-hydroxymethyl-7,8-dihydropterin-pyrophosphate (DHPP) and pABA to 7,8-dihydropteroate (Figure 1). It is the target for the well known antimicrobial and antifungal 'sulfadrug' class of compounds that act as competitive inhibitors and dead end substrate analogs of pABA. The structures of DHPS from five different organisms (*E. coli* (Achari *et al.* 1997), *S. aureus* apo form (Hampele *et al.* 1997), *M. tuberculosis* (Baca *et al.* 2000), *S. cerevisiae* (Lawrence *et al.* 2005) *and B. anthracis* (Babaoglu *et al.* 2004)) have been solved using X-ray crystallography. Structures include holoenzymes and binary complexes with pterin substrate, product analog and 6-methylamino-5-nitroisocytosine (MANIC 14) (Figure 6), a pterin-like inhibitor. A binary pABA complex or pABA analog has not yielded crystals, which is consistent with kinetic data (Vinnicombe and Derrick 1999) that showed the pABA site assembly can only take place subsequent, and in addition to, the binding of the pterin moiety. Functional units are homo-dimers that display a classic $(\beta/\alpha)_8$ TIM barrel fold which is well conserved over all structures (Figure 6). Each monomer is approximately 30-35kDa and a structure superposition reveals a highly conserved pterin binding pocket located deep towards the centre of the barrel. In contrast, the pABA site is situated in the variable loop region at the C-terminal pole and is seemingly more exposed and mobile (Babaoglu *et al.* 2004). In total 15 structures containing substrate and/or product analogs have been solved to date.

Recently, five structures of DHPS from *B. anthracis* have been determined (Babaoglu *et al.* 2004) in a variety of states along the catalytic cycle, including substrate, a bound pterin-like inhibitor, MANIC and a product analogue (pteroic acid). The work greatly advances the understanding of the catalytic mechanism and describes the architecture of the pterin and the pABA site, nestled within the loops.

Loop motions are critical for function in DHPS. During the cycle, a conserved arginine sidechain on loop 2, that was bound in the pterin site, is withdrawn from the barrel core to allow the ligand to bind (see Figure 7). As part of this process loop 1 moves into the active site and conserved aspartates and asparagines interact with the phosphate binding site and assemble the pABA channel. The authors further allude to details of the transition state geometry, a desirable quality for drug discovery, in which a potential for shuffling of the α phosphate into tight binding β pocket was rationalized. This shuffling would both impart strain on the bond to be cleaved and also optimize the S_N2 geometry, facilitating the reaction.

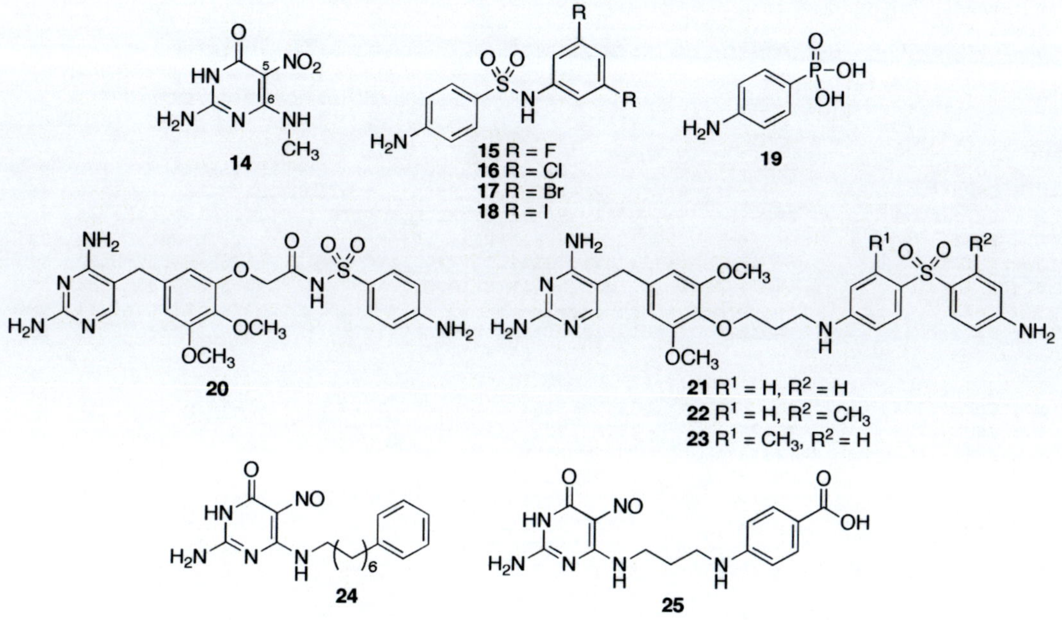

Figure 6. Structures of DHPS inhibitors

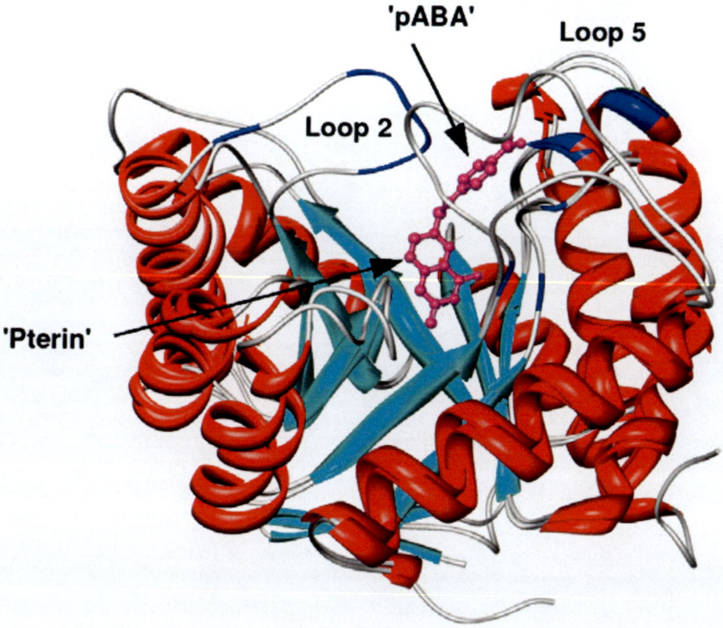

Figure 7. Superposition of the B. anthracis DHPS structure showing the bound product analog pteroic acid (magenta) and the E. coli structure bound to pterin monophosphate (not shown) structure. Loop mutations that confer sulfonamide resistance (Bacca et al., 2000) are coloured blue on the E. coli ribbon.

```
N.g DHFS-FPGS       ------------------------------------------------------------

E.c DHFS-FPGS       ------------------------------------------------------------
M.t DHFS-FPGS       ---------------------------------------------------------MNS    3
L.c FPGS            ------------------------------------------------------------
Human FPGS          ------------------------------------------------------------
S.c FPGS            MHKGKKNYPNLITSFRMNLKKIILNHDRFSHPERWKTNALLRFTFVYIKFLFDLMIIKNP   60
S.c DHFS            ------------------------------------------------------------
Pf DHFS-FPGS        -------------------------MEKNQNDKSNKNDIIHMNDKSGNYDKNNINNFID   34


N.g DHFS-FPGS       --------------MKTLQDWLSHLETAHSGG------LIDMGLERVSEVKKRMNLTPQC   40
E.c DHFS-FPGS       -----MIIKRTPQAASPLASWLSYLENLHS-K------TIDLGLERVSLVAARLGVLKPA   48
M.t DHFS-FPGS       TNSGPPDSGSATGVVPTPDEIASLLQVEHLLDQRWPETRIDPSLTRISALMDLLGSPQRS   63
L.c FPGS            -----------MNYTETVAYIHSFPRLAKTGDHRR-----------ILTLLHALGNPQQQ   38
Human FPGS          -----MEYQDAVRMLNTLQTNAGYLEQVKRQRGDPQ--TQLEAMELYLARSGLQVEDLDR   53
S.c FPGS            LRMVGKTYRDAVTALNSLQSNYANIMAIR-QTGDRKNTMTLLEMHEWSRRIGYSASDFNK  119
S.c DHFS            ------------------------------------MAIELGLSRITKLLEHLGNPQNS   23
P.f DHFS-FPGS       KNDEHDMSDILHKINNEEKKYEEIKSYSECLELLYKTHALKLGLDNPKKLNESFGHPCDK   94

                           ---P---                       -----Ω----
N.g DHFS-FPGS       P-VVVVAGTNGKGSVCAYLTQIYKQAG---FKTGTLTSPHLLHYNERIAINAEPVSDDTI   96
E.c DHFS-FPGS       PFVFTVAGTNGKGTTCRTLESILMAAG---YKVGVYSSPHLVRYTERVRVQGQELPESAH  105
M.t DHFS-FPGS       YPSIHIAGTNGKTSVARMVDALVTALH---RRTGRTTSPHLQSPVERISIDGKPISPAQY  120
L.c FPGS            GRYIHVTGTNGKGSAANAIAHVLEASG---LTVGLYTSPFIMRFNERIMIDHEPIPDAAL   95
Human FPGS          LNIIHVTGTKGKGSTCAFTECILRSYG---LKTGFFSSPHLVQVRERIRINGQPISPELF  110
S.c FPGS            LNIVHITGTKGKGSTAAFTSSILGQYKEQLPRIGLYTSPHLKSVRERIRINGEPISEEKF  179
S.c DHFS            LRVLHIAGTNGKGSVCTYLSSVLQQKS---YQIGKFTTPHLVHVTDSITINNKPIPLERY   80
P.f DHFS-FPGS       YKTIHIAGTNGKGSVCYKIYTCLKIKK---FKVGLFSSPHIFSLRERIIVNDEPISEKEL  151

N.g DHFS-FPGS       IASFERIEAARGEISLT---------------YFEFNALAAVDIFIREQVDVMILEVGLGG  142
E.c DHFS-FPGS       TASFAEIESARGDISLT---------------YFEYGTLSALWLFKQAQLDVVILEVGLGG  151
M.t DHFS-FPGS       VATYREIEPLVALIDQQ-SQASAGKGGPAMSKFEVLTAMAFAAFADAPVDVAVVEVGMGG  179
L.c FPGS            VNAVAFVRAALERLQQQ-------QADFNVTEFEFITALGYWYFRQRQVDVAVIEVGIGG  148
Human FPGS          TKYFWRLYHRLEETKD--GSCVSMPP-----YFRFLTLMAFHVFLQEKVDLAVVEVGIGG  163
S.c FPGS            AKYFFEVWDRLDSTTSSLDKFPHMIPGSKPGYFKFLTLLSFHTFIQEDCKSCVYEVGVGG  239
S.c DHFS            QNIRLQLEALNKSHSLK------------CTEFELLTCTAFKYFYDVQCQWCVIEVGLGG  128
P.f DHFS-FPGS       IHLVNEVLNKAKKLYIN------------PSFFEIITLVAFLHFLNKKVDYAIIETGIGG  199


N.g DHFS-FPGS       RLDAVNGFD-G--DCAVVTSVDLDHQAFLGDTVEQVGFEKAGVFRSG------KPAICGQ  193
E.c DHFS-FPGS       RLDATNIVD-A--DVAVVTSIALDHTDWLGPDRESIGREKAGIFRSE------KPAIVGE  202
M.t DHFS-FPGS       RWDATNVIN-A--PVAVITPISIDHVDYLGADIAGIAGEKAGIITRAPDGSPDTVAVIGR  236
L.c FPGS            DTDSTNVIT-P--VVSVLTEVALDHQKLLGHTITAIAKHKAGIIKRG------IPVVTGN  199
Human FPGS          AYDCTNIIRKP--VVCGVSSLGIDHTSLLGDTVEKIAWQKGGIFKQG------VPAFTVL  215
S.c FPGS            ELDSTNIIEKP--IVCGVTLLGIDHTFMLGDTIEEIAWNKGGIFKSG------APAFTVE  291
S.c DHFS            RLDATNVIPGANKACCGITKISLDHESFLGNTLSEISKEKAGIITEG------VPFTVID  182
P.f DHFS-FPGS       RLDATNILTKP--EVIVITSIGYDHLNILGDNLPIICNEKIGIFKKD------------A  245


N.g DHFS-FPGS       NPAPASLVAHAEAIGAKLLMVQRDFEFHAMEN------IQWNYRFRPQHSDGPARNRNAL  247
E.c DHFS-FPGS       PEMPSTIADVAQEKGALLQRRGVEWNYSVTD---------HDWAFSDAH--GTLEN---L  248
M.t DHFS-FPGS       QVPKVMEVLLAESVRADASVAREDSEFAVLRR-----QIAVGGQVLQLQGLGGVYSDIYL  291
L.c FPGS            LVPDAAAVVAAKVATTGSQWLRFDRDFSVPKA-----KLHGWGQRFTYEDQDGRISDLEV  254
Human FPGS          Q-PEGPLAVLRDRAQQISCPLYLCPMLEALEEGGPPLTLGLEGEHQRSNAALALQLAHCW  274
S.c FPGS            KQPPQGLTILKERAEERKTTLTEVPPFKQLEN----VKLGIAGEFQKSNASLAVMLASEI  347
S.c DHFS            GTNEASVINVVKERCKALG-----SELSVTDS--------QLNGNMIDTNSWG-CFDLAKL  229
P.f DHFS-FPGS       NVVIGPSVAIYKNVFDKAKELNCTIHTVVPEP--------RGERYNEENSRIALRTLEIL  297

                                                             ----linker---
N.g DHFS-FPGS       PFPALRGAYQLSNAACALTVLECLDDRLPVDIGAIKRGLLLVENPG-------RFQVLPG  300
E.c DHFS-FPGS       PLPLVP----QPNAATALAALRASG--LEVSENAIRDGIASAILPG-------RFQIVSE  295
M.t DHFS-FPGS       PLHGEHQAHNAVLALASVEAFFGAGAQRQLDGDAVRAGFAAVTSPG-------RLERMRS  344
L.c FPGS            PLVGDYQQRNMAIAIQTAKVYAKQ-TEWPLTPQNIRQGLAASHWPA-------RLEKISD  306
Human FPGS          LQRQDRHGAGEPKASRPGLLWQLPLAPVFQPTSHMRLGLRNTEWPG-------RTQVLRR  327
S.c FPGS            LHTSN------------ILEEKIKCSSNASIPEKFIIGLQNTKWEG-------RCQVLEK  388
S.c DHFS            PLNGEYQIFNLRVAMGMLDYLQMN-ELIDITKNEVSTRLAKVDWPGRLYRMDYRFDKVSN  288
P.f DHFS-FPGS       NIS-------IDYFLKSIIPIKPPLRIQYLATEQIQH-IKKKFSPD-------NLEHNVQ  342
```

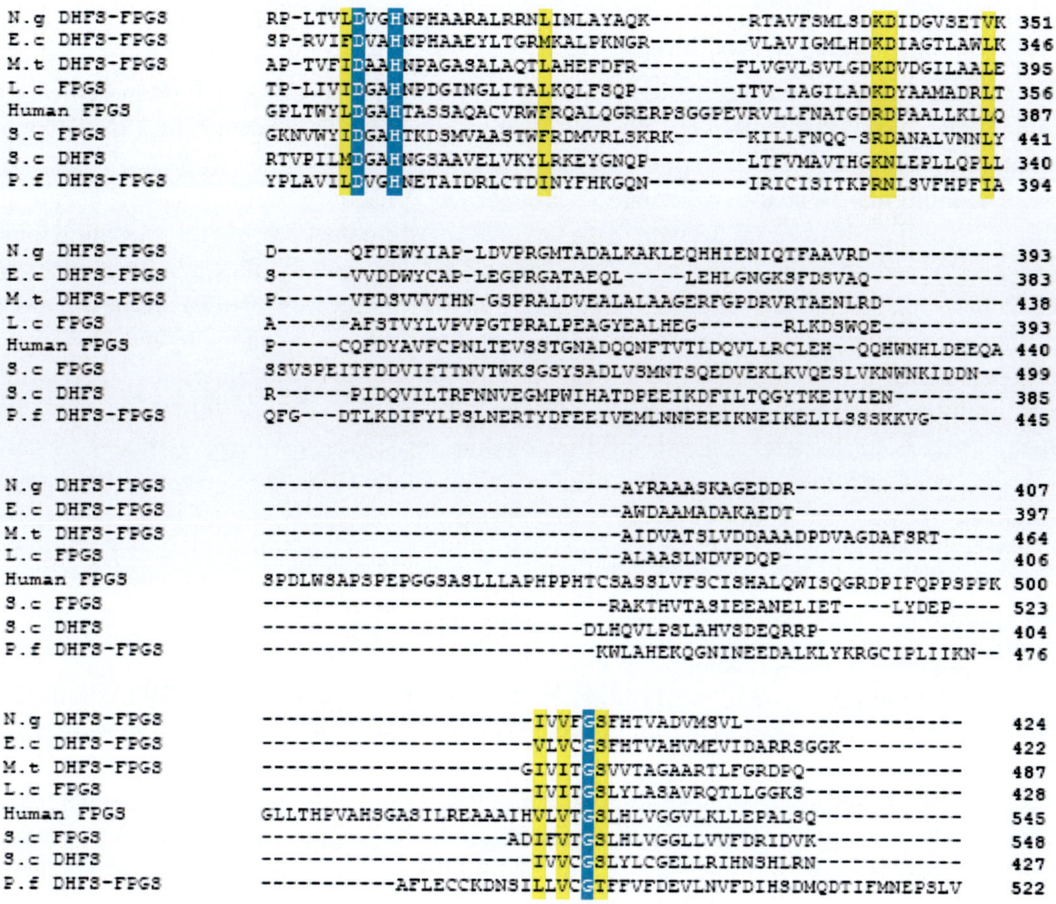

Figure 8. Multiple sequence alignment of S. cerevisiae DHFS against other DHFS and FPGS enzymes. The S. cerevisiae DHFS protein sequence is aligned with S. cerevisiae FPGS, P. falciparum DHFS-FPGS, L. casei FPGS, E. coli DHFS-FPGS, N. gonorrhoeae DHFS-FPGS, Mycobacterium tuberculosis DHFS-FPGS and human FPGS protein sequences using the Biology WorkBench program. The identical residues are highlighted in blue and conservation of strong groups is shown in yellow.

DHPS and Sulfonamide Resistance

Owing to the flexible and exposed nature of the pABA binding loop, DHPS can withstand a range of mutations without severely compromising the structural integrity or catalytic viability of the protein. In this manner, mutations in loops 2 and 5 confer sulfonamide resistance (Figure 7). The *Bacillus anthracis* structure in complex with pteroic acid identifies the much awaited pABA site, nestled within the flexible loops. It thus reveals a plausible rationale behind the rapid onset for the emergence of resistant isolates to the sulfadrugs.

An interesting proposal (Babaoglu *et al.*, 2004; Baca *et al.*, 2000) is that the highly conserved pterin site, in contrast, is particularly inflexible, due to the buried nature and is therefore less likely to tolerate mutations without compromising structural and functional integrity. This work therefore suggests that DHPS may well be a good drug target with a new class of drugs interacting with the highly conserved pterin binding site. This would offer substantial advantages over the sulfadrugs that target the more open, variable pABA binding.

To this end, they further report the binary complex with the pterin like inhibitor MANIC (**14**) (Figure 7) to illustrate the potential for pterin-like inhibitors. MANIC (**14**) displays all the critical electrostatic interactions of the substrate but the additional *N*-methyl group fits snugly into a small hydrophobic pocket providing further binding affinity compared to the native ligand.

The molecular basis for sulfonamide resistance in DHPS of *S. pneumoniae* has also been further investigated using molecular dynamics (MD) (Giordanetto *et al.* 2005). Simulations were performed both for wild type and mutant strains conferring sulfonamide resistance in complex with sulfonamide and DHPPP. Interestingly, the two structures display a very different structural response during the MD simulations; the sulfanilamide ligand rearranged to a favourable position for catalysis to occur in the wild type enzyme but was expelled in the mutant during the course of the MD run. The study illustrates the combined role of MD with X-ray analysis to investigate the mechanism of drug resistance that involves rather flexible binding loops and loop motions, pertinent to sulfonamide resistance. Loop motions are a recurring theme within the catalytic mechanisms in the *de novo* folate enzymes, an understanding of which is conceptually interesting and may further benefit in the design of effective inhibitors.

DHPS Inhibitors

As discussed, the sulfadrugs have long been known as clinically useful DHPS inhibitors. Although recent years have still seen efforts to improve on these agents, there have been no new clinical agents developed in recent decades. A selection of sulfadrugs is presented in Table 3, with reported values for their binding to DHPS from various organisms. Interestingly, the nanomolar affinities reported for dapsone and sulfamethoxazole against mycobacterial strains of DHPS are far more potent that those reported in other microorganisms (usually micromolar) (Nopponpunth *et al.* 1999). QSAR studies have been performed on a large number of sulfone and sulfonamide derivatives, showing that the electronic nature of substituents has a significant effect on binding affinity and efficacy (Coats *et al.* 1985; De Benedetti *et al.* 1989a; De Benedetti *et al.* 1989b; De Benedetti *et al.* 1987; Ho *et al.* 1975; Hopfinger *et al.* 1987; Koehler *et al.* 1988; Lopez de Compadre *et al.* 1987, 1988; Wiese *et al.* 1987). Sulfones and sulfonamides are still being investigated as inhibitors of the folate pathway For example, Chio *et al.*,(Chio *et al.* 1996) reported the di-halo sulfanilamides (14 - 17) had lower IC50 values than sulfamethoxazole, particularly for *T. gondii* and *P. carinii.* Eagon (Eagon and McManus 1989), reported the related, interesting phosphanilic acid (23) as an inhibitor, however there have been no more reports of related phosphonates as inhibitors.

The use of combination therapies has spurred investigations into linked molecules (Dhople 1999; Hyde *et al.* 1983; Wiese *et al.* 1996), incorporating both sulfonamide and pyrimidine subunits to act as inhibitors of both DHPS and DHFR. An early report from Hyde *et al.*(Hyde *et al.* 1983), showed that hybrids such as 21 were potent inhibitors of both DHPS and DHFR, however showed little antibacterial activity, possibly due to problems in cell penetration. More recently, Dhople (Dhople 1999) and Wiese *et al.*(Wiese *et al.* 1996), reported very similar compounds (*e.g.* 22 - 24) which were also potent inhibitors of both DHPS and DHFR, however these molecules showed greater antibacterial activity against *M. lufu* than the corresponding pyrimidine or sulfonamide alone, suggesting cell penetration and possibly a synergistic activity within the cell. In yeast studies sulfa-containing DHPS analogs have been shown to exhibit inhibition of growth and of DHFR (Patel *et al.*, 2003b), while they previously did not inhibit bacteria (Roland *et al*, 1979).

The only significant report to date of inhibitors of DHPS that are not close pterin analogs, sulfones or sulfonamides are those of Lever *et al.* (1985; 1986a,b), who report a range of monocyclic pteridine analogs (isocytosine derivatives), including (MANIC) already mentioned. MANIC (14) is the simplest of these compounds, however Lever *et al.* (1985), investigated a wide range of substituents, predominantly at the 6 position, although a smaller number of substituents at the 5 position were investigated. In general few of these derivatives were more potent than MANIC (14) itself, including derivatives extended from the 6-position (*e.g.* 24) or with pABA linked onto the nitroso derivative of MANIC (*e.g.* isocytosine 25). Unfortunately, these inhibitors were not found to inhibit the growth of whole bacterial cells, which may result from them not penetrating the cell membrane.

Addition of Glutamates - The Enzymes FPGS, DHFS and DHFS-FPGS

The final step in the folate synthesis pathway is the addition of glutamate to dihydropteroate, a reaction catalysed by the enzyme dihydrofolate synthase [E.C. number 6.3.2.12]. Given that DHFS activity is essential for bacteria and is absent in humans, DHFS has the potential to be a target for selective inhibitors. However, targeting DHPS is complicated by the fact that it shares a high degree of similarity with the enzyme, folylpolyglutamate synthase (FPGS) which is found in all organisms. Therefore, in the design of inhibitors for DHFS one needs to consider their effect on FPGS. For example, it would be futile to have an antimicrobial targeting DHFS, only to have it also target FPGS and host folate utilisation.

In some organisms such as *E. coli* and *Neisseria gonorrhoeae* addition of L-glutamate to DHF (*i.e.* DHFS activity) and the subsequent addition of further polyglyutamates (*i.e.* FPGS activity) are combined and catalysed by a single bifunctional enzyme, DHFS-FPGS, for which there is a recent structure. Although the sequence similarity is high between FPGS and DHFS, there is a growing body of structural evidence to suggest that it may well be feasible to design selective inhibitors against DHFS at least. FPGS may represent a potential target in its own right; compounds targeting FPGS may have utility in preventing cell proliferation and cancers (Clarke *et al.* 1987; DeMartino *et al.* 2006; Leil *et al.* 2007; Rosowsky *et al.* 1992; Wettergren *et al.* 2005).

FPGS adds multiple glutamates onto folate to make polyglutamated folates which lack the ability to cross membranes. The number of glutamates can often range from 5-10. Their addition produces a more stable folate that remains within cells. Given the similarity of the substrates of FPGS and DHFS it is perhaps not all that surprising that these enzymes share a high degree of sequence similarity as shown in Figure 8.

Of considerable interest is the specificity of these enzymes for particular reactions. For example, in humans there is only one enzyme of this class, hFPGS, and it is a specific FPGS since humans don't synthesise their own folate. Likewise the bacterium, *L. casei*, obtains its folate from milk and only requires an FPGS. However, the protozoal malaria parasite, *Plasmodium falciparum* is like the bacterium *Escherichia coli*, in having a single polypeptide enzyme that is bifunctional; the enzyme in these species has both FPGS and DHFS activities.

Table 3. Selected sulfadrugs and binding to DHPS

Compound	*E. coli*	*N. meningitidis*	*P. chabaudi*	*T. gondii*	*P. carinii*	*P. falciparum*	*M. tuberculosis*	*M. leprae*	C. albicans
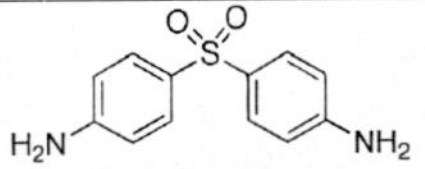 Diaminodiphenylsulfone (Dapsone)	IC_{50} 20 µM (McCullough and Maren 1973) I_{50} 34 µM (Otzen *et al.* 2004)	K_I 0.7 µM (Ho *et al.* 1975)		IC_{50} 0.25 µM (Allegra *et al.* 1990) K_I 0.81 µM (Allegra *et al.* 1990)	K_I 9 µM (Merali *et al.* 1990)	K_I 6 µM (Zhang and Meshnick 1991)	K_I 13 nM (Nopponpunth *et al.* 1999)	K_I 11 nM (Nopponpunth *et al.* 1999)	I_{50} 1.78 µM (Otzen et al. 2004)
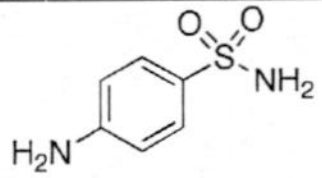 Sulfanilinamide	K_I 67 µM (Poe 1976)	K_I 6 µM (Poe 1976)	K_I 130 µM (Walter and Koenigk 1974)	IC_{50} 24 µM (Allegra *et al.* 1990) K_I 68 µM (Allegra *et al.* 1990)					
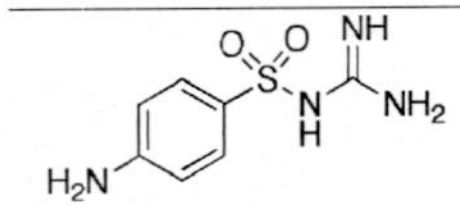 Sulfaguanidine	K_I 17 µM (Poe 1976)		K_I 69 µM (Walter and Koenigk 1974)						
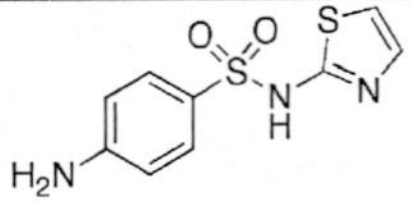 Sulfathiazole	K_I 2.3 µM (Poe 1976)		K_I 14 µM (Walter and Koenigk 1974)	IC_{50} 1.7 µM (Allegra *et al.* 1990) K_I 1.9 µM (Allegra *et al.* 1990)					
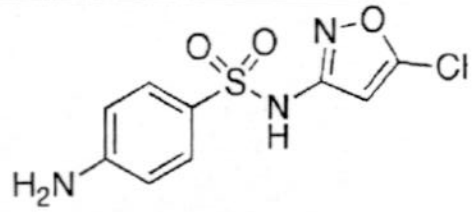 Sulfamethoxazole				IC_{50} 2.7 µM (Allegra *et al.* 1990) K_I 21 µM (Allegra *et al.* 1990)	K_I 59 µM (Merali *et al.* 1990)	K_I 36 µM (Zhang and Meshnick 1991)	K_I 28 nM (Nopponpunth *et al.* 1999)	K_I 30 nM (Nopponpunth *et al.* 1999)	
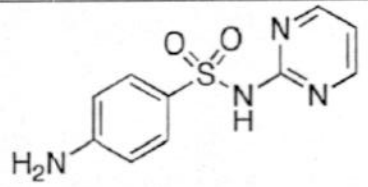 **Sulfadiazine**	IC_{50} 11 µM (McCullough and Maren 1973)	K_I 2 µM (Ho *et al.* 1975)		IC_{50} 11 µM (Allegra *et al.* 1990) K_I 19 µM (Allegra *et al.* 1990)		K_I 89 µM (Zhang and Meshnick 1991)			
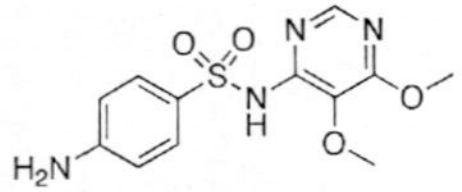 Sulfadoxine				IC_{50} 110 µM (Allegra *et al.* 1990)	K_I 140 µM(Merali *et al.* 1990)	K_I 39 µM(Zhang and Meshnick 1991)			

Yeast on the other hand is different again. It requires both FPGS and DHFS activities but encodes them on two different genes. DHFS is encoded by *FOL3* and is essential for viability on normal rich media. FPGS is encoded by *MET7* which is essential for methionine biosynthesis and respiratory function. It is also interesting that there is yet another similar protein in *S. cerevisiae*, described as a putative FPGS. Deletion of the gene, YKL132c, encoding this protein does not affect cell growth like the *MET7* or *FOL3* gene products; the main phenotype is reduced mating.

DHFS and FPGS Structures

How does an FPGS, a DHFS or an FPGS-DHFS have such narrow or broad specificities? Ways to answer these questions may come from detailed structure function studies, mutagenesis and from knowledge of the structures of enzymes from more species. The structures of two FPGSs are now known. The first FPGS structure is from the bacterium *L. casei* (Sun *et al.* 1998; Sun *et al.* 2001), an enzyme without cysteines, a property that may have aided its crytallisation. Although large amounts (tens of milligrams per litre of culture) of biologically active yeast DHFS have been produced (Patel *et al.* 2003a) none has yet crystallised, due to its aggregation during crystallisation (O. Patel, P. Pilling, R. Fernley, I. Macreadie, unpublished). This may be due to oxidation but it was not prevented by performing the crystallisation under anaerobic conditions or with a reducing agent. Mutagenesis of two "possibly exposed" cysteines as well as carboxylation of the cysteines did not lead to successful production of crystals (O. Patel, P. Pilling, R. Fernley, I. Macreadie, unpublished).

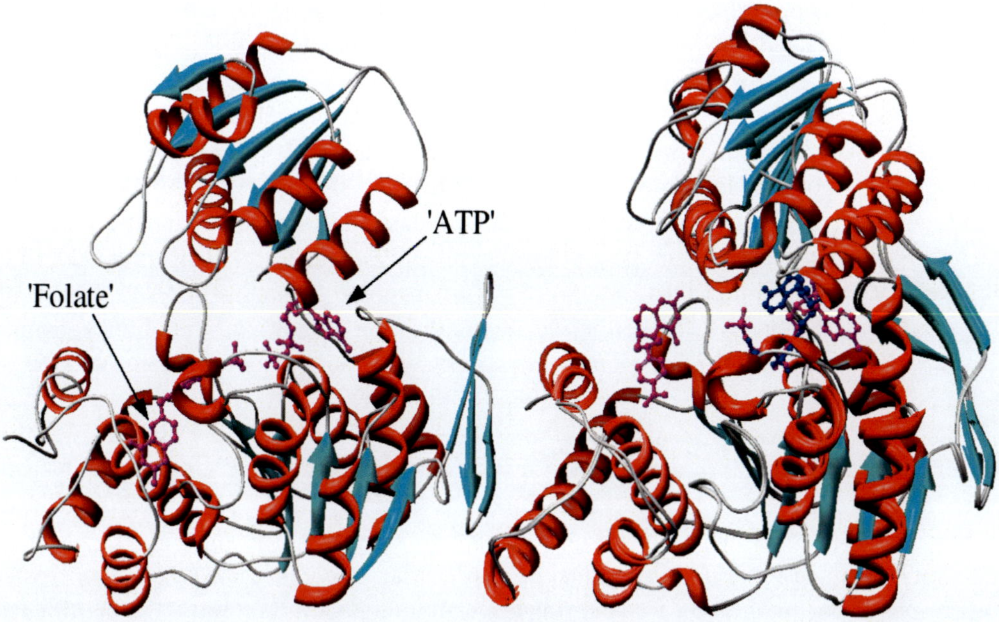

Figure 9. Ribbon representation of E. coli DHFS-FPGS (left) in complex with ADP and DHP and a superposition of L. casei FPGS (right) in the 'inactive', AMPPCP bound form and the 'active' ternary complex with AMPPCP and mHP (magenta). The superposition over the N terminal domain illustrates the rotation in the C terminal domain required for activation inactive (blue ligand) to active (magenta).

Much understanding of the mode of action of FPGS results from the *L casei* X-ray structure which has been solved in the apo, the ATP bound forms (Sun *et al.* 1998), as the binary with AMPPCP and as the ternary complex with AMPPCP and the preferred monoglutamate folate substrate 5,10-methylene-tetrahydrofolate (mTHF) (Sun *et al.* 2001). The structure comprises two domains in which ATP is bound between the two in a channel (Figure 9b). The protein is modular comprising the ATP-binding N terminal domain including the commonly found P-loop for phosphate binding, and a novel omega loop that is involved in the folate binding site. The binding of the first substrate (ATP) is not sufficient to create an active enzyme. A rotation of the C-terminal domain (Figure 9b) is needed upon binding of folate which then activates the enzyme to allow the binding of the third substrate, L-glutamate. During this activation process, the adenosine inserts into the 'specificity' pocket and undergoes a change in conformation from a C2' endo to a C3' endo and the base changes from the unusual *syn* to an *anti*. Interestingly, the ATP appeared to be further phosphorylated to adenosine tetraphosphate in which the δ phosphate may be mimicking a tetrahedral carbon intermediate. The putative glutamate-binding site has been further supported by site-directed mutagenesis and kinetic studies (Sheng *et al.* 2000). Finally it was noted, based on sequence conservation in the active site regions, that the *L. casei* FPGS structure provides an excellent model for the human enzyme (Sun *et al.* 2001).

The structure of the bifunctional bacteria DHFS-FPGS from *E. coli* has been solved from X-ray data (Figure 9a) (Mathieu *et al.* 2005). The structure displayed good similarity to the FPGS structure from *L casei* in the ATP binding site. Notably, the aryl moiety of the pterin binding site was substantially different between the two and consequently the *E. coli* structure provides the structural means to look for selective inhibitors of the pterin site.

The pterin binding site during DHFS activity is revealed in the ternary complex of ADP and the phosphorylated intermediate dihydropteroate bound to the *E. coli* DHFS-FPGS. Substantial rearrangements at the active site, or opening of a second pterin site is required in this model to accommodate further glutamate residues that are added to the dihydrofolate substrate. In this manner a structural basis for DHFS and FPGS specificity is thus put forward. A superposition of the *E. coli* DHFS-FPGS and the *L casei* structure revealed a probable second pterin site some 10Å distal to the pterin site in *E. coli* DHFS-FPGS that could accommodate a diglutamate molecule for FPGS activity. Unfortunately, FolC complexes with FPGS bound substrates were not amenable to crystallisation and this remains the current working model. Nevertheless, crystallisation conditions for FPGS from *M tuberculosis* have recently been reported (Young *et al.* 2006) in the presence of ADP and dihydrofolate.

From a drug discovery perspective, the pteridine in *E. coli* DHFS-FPGS showed high complementarity with the protein active site with all the polar heteroatoms of the pteridine ring involved in either hydrogen bond contacts to the protein or to a bound water. Consequently, the scope for the design of potent inhibitors that interact with the pterin site may be rather limited in terms of chemical diversity (Mathieu *et al.* 2005). However isosteric modifications in positions 5 and 8 of the ring are well tolerated. Importantly, the pterin pocket in *L. casei* (the human model) is very similar. However the aryl moiety binding site between the two are substantially different, indicating that selective inhibitors can in principal be designed.

DHFS Inhibitors

Roland *et al.* (Roland *et al.* 1979) showed that the pterin-sulfonamides that are formed as a result of sulfonamides replacing pABA in the reaction catalysed by DHPS bind to and inhibit DHFS. However, they suggest that the concentrations required for effective inhibition are unlikely to be reached in the intracellular environment. There are essentially no specific inhibitors of DHFS reported in the literature, and as such this may present a useful future target.

Multifunctional Enzymes

Several enzymes of the folate pathway exist as bi or tri-functional complexes, more so in eukaryotes and parasites. Several combinations are currently known (Figure 10). For example, in fungi, including *Candida albicans*, *Saccharomyces cerevisiae*, and *P. jiroveci*, the FAS genes are part of a single open reading frame that encodes a trifunctional, multidomain enzyme that includes DHNA, HPPK and DHPS (Iliades *et al.* 2004b; Volpe *et al.* 1993). We haven't included the DHFS/FPGS bifunctional ability observed in bacteria: for example, in *E. coli* DHFS can be thought of as bifunctional 'single enzyme'. The evolutionary drive for multifunctionality has led to the discovery of substrate channelling in many multifunctional enzymes and accordingly efforts to identify channelling mechanisms in the pathway have been undertaken.

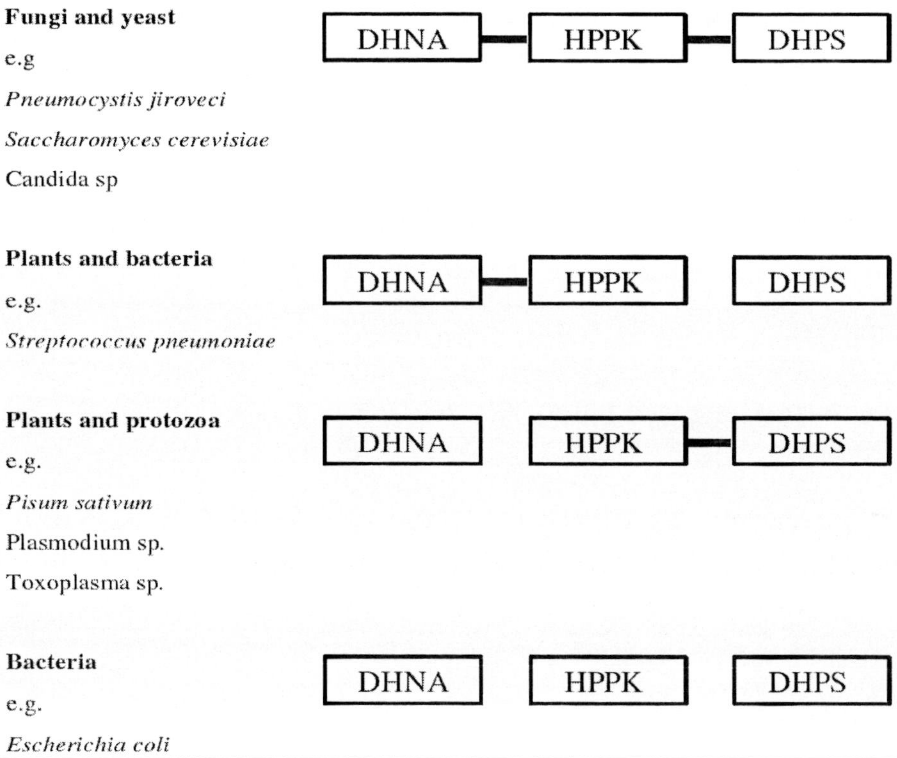

Figure. 10. Some examples of multifunctional enzymes within the folate biosynthesis pathway.

Substrate channelling is a when the reaction products of one enzyme are actively or passively transferred to another enzyme without escape into the bulk solvent. It occurs in many multi-functional enzyme systems (Then 1993) and has potentially several advantages over free diffusion of reaction products within the bulk solvent. These include unfavourable chemical equilibrium and excluding reaction products from competing enzymes. There is a substantial body of structural and biochemical evidence to illustrate the process of channelling substrates through tunnels or electrostatic channels between active sites. For example in the folate utilization steps, the TS-DHFR structure from *L. major* in complex with methotrexate and NADPH bound at the DHRF site and 5-fluoro-2prime-deoxyuridylate and 10-propargyl-5,8-dideazafolate bound at TS, shows a strong electrostatic positive potential that forms a surface channel that connects the two sites over the ~40 Å distance.

Two multi-functional enzyme structures from the *de novo* folate pathway have been structurally characterized: the bifunctional HPPK-DHPS enzyme from *S. cerevisiae (Lawrence et al. 2005)* and the bifunctional DHNA-HPPK from *S. pneumoniae* (Goulding *et al.* 2005). In the former, it was noted that the interfaces do not display the typical characteristics (buried surface area, surface complementarity, hydrophobic content and inter-species conservation) associated with typical protein-protein interactions. The physiological relevance of the domain-domain association was left unresolved and no structural evidence was put forward to be consistent with channelling of substrate from one enzyme to the other in this case.

In the octameric DHNA-HPPK structure (Figure 3b) (Goulding *et al.* 2005) the HPPK monomers do not make substantial contacts with DHNA and the stabilizing interaction that orient the HPPK domains are located on the interface within each HPPK tetramer. The active sites face out from each other and no evidence to support either a structural-based or charge-mediated tunnelling pathway of substrate between the two was made. Further evidence against tunnelling or a requirement for coupling was that the activity of DHNA dropped off with time whereas activity of HPPK was maintained, in accord with the belief that HPPK enzymes generally function in the monomeric state; DHNA is clearly multimeric.

We await other structural evidence from the trifunctional enzymes (Figure 10) to establish whether or not substrate channelling occurs in the enzymes of the *de novo* folate pathway.

Yeast and Bacterial Cell-based Methods for Analysis of Folate Synthesis Inhibitors

The screening of some organisms for sensitivity to inhibitors of the folate biosynthetic pathway can be difficult or dangerous. For example, with the malaria parasite, *Plasmodium falciparum*, the culture must take place in blood which contains endogenous folates. This can compromise the assay. Some organisms, such as the pathogen *P. jiroveci*, can't be cultured while others such as *Bacillus anthracis* are too dangerous to be cultured In these cases the genes encoding folate biosynthetic enzymes can be cloned and expressed in laboratory strains of *Escherichia coli* or *Saccharomyces cerevisiae* in which the endogenous genes have been deleted. The screening of the compounds can the be undertaken under simpler and safer conditions.

Yeast Screens

The major EUROSCARF yeast deletion collection lacks strains deleted for folate synthesis (except in heterozygous diploid knockouts, which retain one copy) since such deletion mutants are regarded as non-viable on rich medium. However, they are only non-viable on regular rich media and can be grown if an appropriate folate (e.g. folinic acid, folic acid) or folate end product (e.g. thymidine monophosphate) is added. This is an important observation in the context of screening for potential antifolates since it implies that the inhibition of an antifolate drug can be rescued by the addition of an appropriate folate or folate end product. In addition, some folate precursors may provide rescue of folate synthesis mutants. For example, yeast deleted for DHPS can be rescued by the addition dihydropteroate or dihydrofolate, the product formed by the action of DHFS on dihydropteroate and glutamate. Similarly the deletion of *FOL3*, the gene encoding DHFS, can be rescued by dihydrofolate, but not by dihydropteroate. Therefore, it is theoretically possible to distinguish between an inhibitor of DHPS and an inhibitor of DHFS by the pattern of rescue. In addition it is possible to rescue by competing with inhibitors. For example, sulfadrugs are classic examples of competitive inhibitors; and their inhibitory effect can be overcome by the addition of *p*-amino benzoic acid (Castelli *et al.* 2001).

Wild-type yeast strains can also be used for the analysis of folate synthesis inhibitors, however, the high level of endogenous folate synthesis in yeast can make such studies more difficult. Our experience is that wild-type yeast are less sensitive to many existing antimicrobials. With this in mind we would boldly suggest that an antifolate discovered using a yeast model could be even better for a clinical use!

Disruptions of the genes involved in folate synthesis, *FOL1* and *FOL3*, were performed in a *tup1* (thymidine monophosphate uptake) mutant strain. The strain employed, TH1, is able to take up thymidine monophosphate, also initially used for the recovery of the mutant deleted for the *DFR1* gene, which encodes dihydrofolate reductase (Huang *et al.* 1992). Phenotypic analysis of *fol1* and *fol3* mutants confirmed that in rich media (YEPD) there was a requirement for thymidine monophosphate, although folinic acid or folic acid can be used (Bayly *et al.* 2001). In minimal synthetic media it was established that in addition to the growth requirements inherited from the parental strain, *fol1* and *fol3* mutants had additional requirements for methionine, histidine, adenine and thymidine monophosphate (Bayly *et al.* 2001). This is consistent with the loss of folate synthesis or utilisation functions. The addition of dihydrofolic acid to minimal media was sufficient to restore growth to *fol1* and *fol3* mutants in the absence of any other requirement.

The use of yeast folate synthesis mutants has provided new insights into the mechanism of action of sulfadrugs in that the condensation product of sulfadrug with the pteroate has been shown to be itself inhibitory (Patel *et al.* 2003b). This clearly shows that sulfa drugs not only act by competition with pABA, but they have a multiple effect (Patel *et al.* 2004b). One target of the sulfa dihydropteroate has been shown to be the folate utilising enzyme, DHFR (Patel *et al.* 2004a). Increased production of DHFR led to an increase in resistance to sulfa DHP. However, it is possible that there are additional targets associated with folate utilisation (Patel *et al.* 2004b).

The construction of yeast strains with mutant DHPS alleles can provide useful information on antifolate resistance. For example, DHPS mutations found in sulfadrug-resistant isolates of the non-culturable fungus *Pneumocystis jiroveci*, were shown to confer sulfadrug-resistance to *S. cerevisiae* when genetically engineered into the yeast DHPS (Iliades *et al.* 2004a; Meneau *et al.* 2004). Total yeast gene replacement is another option - if a foreign DHPS provides complementation further insights into the mechanisms of drug resistance can be made. For example, loss of DHFS in yeast can be complemented by the *Pneumocystis jiroveci* DHFS gene (Hauser and Macreadie, 2006).

E. Coli Screens

E. coli strains with disruptions in genes involved in folate synthesis are also available. *folP* mutants have no DHPS, and do not grow on rich media (yeast extract + tryptone) unless thymidine is added. The requirement for thymidine can be complemented by a number of foreign DHPS genes including that from *Mycobacterium leprae* (Nopponpunth *et al.* 1999; Williams *et al.* 2000), *Mycobacterium tuberculosis* (Nopponpunth *et al.* 1999), *P. jiroveci* (Iliades *et al.* 2004b), and *Plasmodium falciparum* (Berglez *et al.* 2004). Of particular interest has been complementation by genes involved in sulfadrug resistance. For example, DHPS genes from *P. jiroveci* and *Plasmodium falciparum* that exhibit various degrees of resistance to sulfamethoxazole and sulfadoxine, have shown levels of sulfamethoxazole and sulfadoxine resistance comparable to the resistance seen in *P. falciparum* and *P. jiroveci* in an *E. coli folP* strain transformed and complemented with these mutant genes (Berglez *et al.* 2004; Iliades *et al.* 2005a).

In *E. coli, FOLC* encodes the bifunctional DHFS-FPGS (Bognar *et al.* 1985) and the *folC* mutant can be complemented by the *Plasmodium falciparum* bifunctional DHFS-FPGS (Salcedo *et al.* 2001). This mutant strain would therefore appear to have little utility to study a monofunctional DHFS. The remaining enzymes, HPPK and DHNA, are encoded individually in *E. coli*. They are potentially useful for a range of studies, but have not been exploited to any significant degree as drug targets.

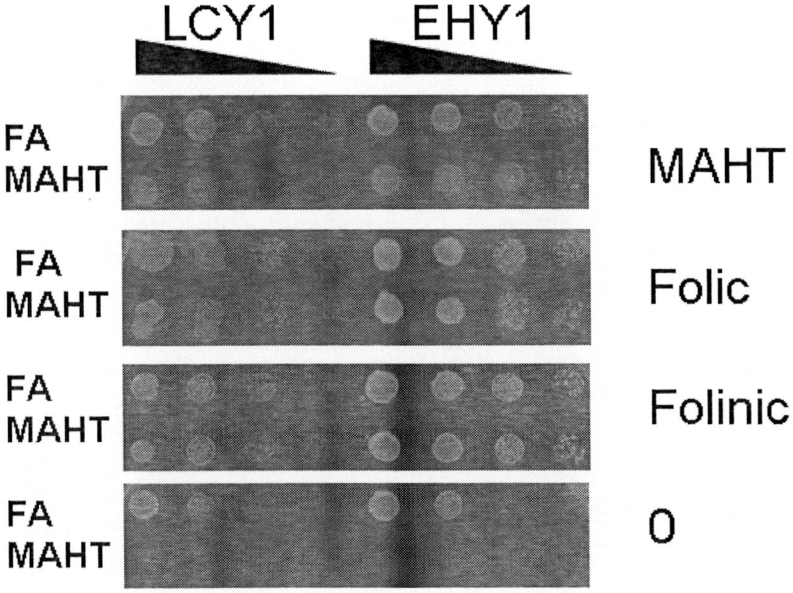

Figure 11. Cells lacking endogenous folate synthesis (LCY1 lacks DHFS and EHY1 lacks DHPS) were pre-grown on minimal media supplemented with folinic acid (denoted FA) or the folate end products, methionine, adenine, histidine and thymidine monophosphate (denoted MAHT) before inoculation onto solidified minimal medium with MAHT, folic acid, FA or no additive. Cells were serially diluted.

Practicalities in screens

From our previous screens on sulfadrugs and other folate synthesis inhibitors we have encountered a number of factors that require consideration lest they upset screens.

- Ensure media and culture vessels have minimal contaminating folate or pABA. Synthetic minimal media that is free of pABA and folic acid can be purchased or made.
- Inoculums should be small because excess cells can provide competing materials such as pABA or folate. It is good to trial a screen with a range of 10-fold serially-diluted cells; then the concentration of cells where no inhibition occurs can be clearly seen. See Castelli *et al.* (2001).
- Minimise folate levels by pre-growing cells on folate-deficient media. If folate is present cells may grow a significant number of generations before the lack of folate depletes growth. Thus the FA-grown cells may appear to grow almost as well in the absence of folate as in the presence of folate. See Fig, 11.
- It is worth trying a number of strains and species since there can be large variations in uptake and efflux. Permeabiity mutants are often used in yeast screens but we have not tried to use the usual uptake and efflux mutants. Probably the optimal yeast mutant for screening is one that is dependent on exogenous folate or a folate precursor for two reasons. First there is greater opportunity to competitively inhibit because folate levels will be low, and second, uptake of folate will be occurring so uptake of folate analogs might be expected also.

CONCLUSIONS

Antimicrobial resistance is major threat to our society that has been likened to having an impact on a par with global warming (Tapsall and Merlino, 2007). The World Health Organisation has summed up the situation as "Bad drugs, no drugs" while the Infectious Diseases Society of America has noted that "As antibiotic discovery stagnates a public health crisis advances". This review has re-examined the folate synthesis where we consider there are new hopes for major new drug discovery initiatives.

Our oldest chemically-synthesised antibiotics, the sulfadrugs, have taught us much about the rise of antibiotic resistance strains. More recently we have made considerable efforts to understand the catalytic and resistance mechanisms down to the molecular level using modern technologies. Such technologies and structural information pave the way for state-of-the-art screening, chemistry and novel *in vivo* screening methods that employ surrogate organisms, towards the design of selective high affinity inhibitors.

This review has focussed on the antimicrobial targets and our approaches to them. We have particularly addressed a structure-based understanding of the catalytic mechanism and function of the folate biosynthesis enzymes derived from information acquired over the last decade. We believe that this exquisite data resource combined with novel structure-based approaches will deliver inhibitors that will withstand the test of time, replacing the old antifolate drugs that are still widely used.

REFERENCES

Achari A, Somers DO, Champness JN, Bryant PK, Rosemond J, Stammers DK (1997) Crystal structure of the anti-bacterial sulfonamide drug target dihydropteroate synthase. *Nature Structural Biol.* 4:490-497

Akiba T, Koyama K, Ishiki Y, Kimura S, Fukushima T (1960) On the mechanism of the development of multiple-drug-resistant clones of Shigella. *Jpn. J. Microbiol.* 4:219-227

Allegra CJ, Boarman D, Kovacs JA, Morrison P, Beaver J, Chabner BA, Masur H (1990) Interaction of sulfonamide and sulfone compounds with *Toxoplasma gondii* dihydropteroate synthase. *J. Clin. Invest* 85:371-379

Ansari NA, Kombe AH, Kenyon TA, Mazhani L, Binkin N, Tappero JW, Gebrekristos T, Nyirenda S, Lucas SB (2003) Pathology and causes of death in a series of human immunodeficiency virus-positive and -negative pediatric referral hospital admissions in Botswana. *Pediatr. Infect. Dis. J.* 22:43-47

Armstrong W, Meshnick S, Kazanjian P (2000) *Pneumocystis carinii* mutations associated with sulfa and sulfone prophylaxis failures in immunocompromised patients. *Microbes Infect* 2:61-67

Babaoglu K, Qi J, Lee RE, White SW (2004) Crystal structure of 7,8-dihydropteroate synthase from *Bacillus anthracis* Mechanism and novel inhibitor design. *Structure* (Cambridge, MA, United States) 12:1705-1717

Baca AM, Sirawaraporn R, Turley S, Sirawaraporn W, Hol WGJ (2000) Crystal structure of *Mycobacterium tuberculosis* 6-hydroxymethyl-7,8-dihydropteroate synthase in complex with pterin mMonophosphate: New insight into the enzymatic mechanism and sulfa-drug action. *J. Mol. Biol.* 302:1193-1212

Bayly AM, Berglez JM, Patel O, Castelli LA, Hankins EG, Coloe P, Hopkins Sibley C, Macreadie IG (2001) Folic acid utilisation related to sulfa drug resistance in *Saccharomyces cerevisiae. FEMS Microbiol. Letts.* 204:387-390

Bean DC, Livermore DM, Papa I, Hall LM (2005) Resistance among *Escherichia coli* to sulphonamides and other antimicrobials now little used in man. *J. Antimicrob. Chemother.* 56:962-964

Berglez J, Iliades P, Sirawaraporn W, Coloe P, Macreadie I (2004) Analysis in *Escherichia coli* of *Plasmodium falciparum* dihydropteroate synthase (DHPS) alleles implicated in resistance to sulfadoxine. *Int. J. Parasitol.* 34:95-100

Bermingham A, Bottomley JR, Primrose WU, Derrick JP (2000) Equilibrium and kinetic studies of substrate binding to 6-hydroxymethyl-7,8-dihydropterin pyrophosphokinase from *Escherichia coli. J. Biol. Chem.* 275:17962-17967

Blaszczyk J, Li Y, Cherry S, Alexandratos J, Wu Y, Shaw G, Tropea JE, Waugh DS, Yan H, Ji X (2007a) Structure and activity of *Yersinia pestis* 6-hydroxymethyl-7,8-dihydropterin pyrophosphokinase as a novel target for the development of antiplague therapeutics. *Acta Crystallogr. D Biol. Crystallogr.* 63:1169-1177

Blaszczyk J, Li Y, Gan J, Yan H, Ji X (2007b) Structural basis for the aldolase and epimerase activities of *Staphylococcus aureus* dihydroneopterin aldolase. *J. Mol. Biol.* 368:161-9

Blaszczyk J, Li Y, Shi G, Yan H, Ji X (2003) Dynamic roles of arginine residues 82 and 92 of *Escherichia coli* 6-hydroxymethyl-7,8-dihydropterin pyrophosphokinase: crystallographic studies. *Biochem.* 42:1573-1580

Blaszczyk J, Li Y, Wu Y, Shi G, Ji X, Yan H (2004a) Essential roles of a dynamic loop in the catalysis of 6-hydroxymethyl-7,8-dihydropterin pyrophosphokinase. *Biochem.* 43:1469-1477

Blaszczyk J, Shi G, Li Y, Yan H, Ji X (2004b) Reaction trajectory of pyrophosphoryl transfer catalyzed by 6-hydroxymethyl-7,8-dihydropterin pyrophosphokinase. *Structure* 12:467-75

Blaszczyk J, Shi G, Yan H, Ji X (2000) Catalytic center assembly of HPPK as revealed by the crystal structure of a ternary complex at 1.25 A resolution. *Structure* 8:1049-1058

Bognar AL, Osborne C, Shane B, Singer SC, Ferone R (1985) Folylpoly-γ-glutamate synthetase-dihydrofolate synthetase. Cloning and high expression of the *Escherichia coli folC* gene and purification and properties of the gene product. *J. Biol. Chem.* 260:5625-5630

Bouma JE, Lenski RE (1988) Evolution of a bacteria/plasmid association. *Nature* 335:351-352

Brochet M, Couve E, Zouine M, Poyart C, Glaser P (2008) A naturally occurring gene amplification leading to sulfonamide and trimethoprim resistance in *Streptococcus agalactiae. J. Bacteriol.* 190:672-680

Brooks DR, Wang P, Read M, Watkins WM, Sims PF, Hyde JE (1994) Sequence variation of the hydroxymethyldihydropterin pyrophosphokinase: dihydropteroate synthase gene in lines of the human malaria parasite, *Plasmodium falciparum*, with differing resistance to sulfadoxine. *Eur. J. Biochem.* 224:397-405

Brown GM (1962) The biosynthesis of folic acid. II. Inhibition by sulfonamides. *J. Biol. Chem.* 237:536-540

Carter JY, Loolpapit MP, Lema OE, Tome JL, Nagelkerke NJ, Watkins WM (2005) Reduction of the efficacy of antifolate antimalarial therapy by folic acid supplementation. *Am. J. Trop. Med. Hyg.* 73:166-170

Castelli LA, Nguyen NP, Macreadie IG (2001) Sulfa drug screening in yeast: fifteen sulfa drugs compete with p-aminobenzoate in *Saccharomyces cerevisiae. FEMS Microbiol. Lett.* 199:181-184

Chin W, Contacos PG, Coatney GR, King HK (1966) The evaluation of sulfonamides, alone or in combination with pyrimethamine, in the treatment of multi-resistant falciparum malaria. *Am. J. Trop. Med. Hyg.* 15:823-829

Chio L-C, Bolyard LA, Nasr M, Queener SF (1996) Identification of a class of sulfonamides highly active against dihydropteroate synthase from *Toxoplasma gondii*, *Pneumocystis carinii*, and *Mycobacterium avium. Antimicrobial Agents Chemother.* 40:727-733

Chulay JD, Watkins WM, Sixsmith DG (1984) Synergistic antimalarial activity of pyrimethamine and sulfadoxine against *Plasmodium falciparum in vitro. Am. J. Trop. Med. Hyg.* 33:325-330

Clarke L, Rosowsky A, Waxman DJ (1987) Inhibition of human liver folylpolyglutamate synthetase by non-γ-glutamylatable antifolate analogs. *Mol. Pharmacol.* 31:122-127

Coats EA, Cordes HP, Kulkarni VM, Richter M, Schaper KJ, Wiese M, Seydel JK (1985) Multiple regression and principal component analysis of antibacterial activities of sulfones and sulfonamides in whole cell and cell-free systems of various DDS sensitive and resistant bacterial strains. Quantitative structure-activity relationships 4:99-109

Dallas WS, Gowen JE, Ray PH, Cox MJ, Dev IK (1992) Cloning, sequencing, and enhanced expression of the dihydropteroate synthase gene of *Escherichia coli* MC4100. *J. Bacteriol.* 174:5961-5970

De Benedetti PG, Iarossi D, Folli U, Frassineti C, Menziani MC, Cennamo C (1989a) Quantitative Structure activity relationships in dihydropteroate synthase inhibition by multisubstituted sulfones - Design and synthesis of some new derivatives with improved potency. *J. Med. Chem.* 32:2396-2399

De Benedetti PG, Iarossi D, Frassineti C, Menziani MC, Cocchi M, Cennamo C (1989b) Comparative QSAR analysis in dihydropteroate synthase inhibition by sulfones. Design and synthesis of some new derivatives with improved potency. *Prog. Clin. Biol. Res.* 291:345-348

De Benedetti PG, Iarossi D, Menziani C, Caiolfa V, Frassineti C, Cennamo C (1987) Quantitative structure-activity analysis in dihydropteroate synthase inhibition of sulfones. Comparison with sulfanilamides. *J. Med. Chem.* 30:459-464

Degowin RL, Powell RD (1964) Drug-resistant Falciparum malaria. J Lab Clin Med 64:851

Demanche C, Guillot J, Berthelemy M, Petitt T, Roux P, Wakefield AE (2002) Absence of mutations associated with sulfa resistance in Pneumocystis carinii dihydropteroate synthase gene from non-human primates. *Med. Mycol.* 40:315-318

DeMartino JK, Hwang I, Xu L, Wilson IA, Boger DL (2006) Discovery of a potent, nonpolyglutamatable inhibitor of glycinamide ribonucleotide transformylase. *J. Med. Chem.* 49:2998-3002

Dhople AM (1999) *In vitro* and *in vivo* activity of K-130, a dihydrofolate reductase inhibitor, against *Mycobacterium leprae*. *Arzneimittelforschung* 49:267-271

Dorsey G, Dokomajilar C, Kiggundu M, Staedke SG, Kamya MR, Rosenthal PJ (2004) Principal role of dihydropteroate synthase mutations in mediating resistance to sulfadoxine-pyrimethamine in single-drug and combination therapy of uncomplicated malaria in Uganda. *Am. J. Trop. Med. Hyg.* 71:758-763

Eagon RG, McManus AT (1989) Phosphanilic acid inhibits dihydropteroate synthase. *Antimicrobial. Agents Chemother.* 33:1936-1938

Enne VI, Livermore DM, Stephens P, Hall LM (2001) Persistence of sulphonamide resistance in *Escherichia coli* in the UK despite national prescribing restriction. Lancet 357:1325-8

Feldman HA (1966) Meningococcal disease, 1965. *JAMA* 196:391-393

Feldman HA (1986) The meningococcus: a twenty-year perspective. Rev Infect Dis 8:288-94

Fermer C, Kristiansen B-E, Skoeld O, Swedberg G (1995) Sulfonamide resistance in *Neisseria meningitidis* as defined by site-directed mutagenesis could have its origin in other species. *J. Bacteriol.* 177:4669-4675

Fermer C, Qvarnstrom Y, Swedberg G (1997) Adaptation to sulfonamide resistance in *Neisseria meningitidis* may have required compensatory changes to retain enzyme function. Kinetic analysis of dihydropteroate synthases from *Neisseria meningitidis*. Chemistry and Biology of Pteridines and Folates 1997, Proceedings of the International Symposium on Pteridines and Folates, 11th, Berchtesgaden, Germany, June 15-20, 1997:449-454

Fleming MP, Datta N, Gruneberg RN (1972) Trimethoprim resistance determined by R factors. *Br. Med. J.* 1:726-728

Gabelli SB, Bianchet MA, Xu W, Dunn CA, Niu Z-D, Amzel LM, Bessman MJ (2007) Structure and function of the *E. coli* dihydroneopterin triphosphate pyrophosphatase: A Nudix enzyme involved in folate biosynthesis. *Structure* 15:1014-1022

Garcon A, Levy C, Derrick JP (2006) Crystal structure of the bifunctional dihydroneopterin aldolase/6-hydroxymethyl-7,8-dihydropterin pyrophosphokinase from *Streptococcus pneumoniae*. *J. Mol. Biol.* 360:644-653

Giordanetto F, Fowler PW, Saqi M, Coveney PV (2005) Large scale molecular dynamics simulation of native and mutant dihydropteroate synthase-sulphanilamide complexes suggests the molecular basis for dihydropteroate synthase drug resistance. Philosophical Transactions of the Royal Society of London, Series A: Mathematical, Physical and Engineering Sci. 363:2055-2073

Goulding CW, Apostol MI, Sawaya MR, Phillips M, Parseghian A, Eisenberg D (2005) Regulation by oligomerization in a mycobacterial folate biosynthetic enzyme. *J. Mol. Biol.* 349:61-72

Grape M, Sundstrom L, Kronvall G (2003) Sulphonamide resistance gene sul3 found in *Escherichia coli* isolates from human sources. *J. Antimicrob. Chemother.* 52:1022-1024

Greenberg J, Richeson EM (1950) Potentiation of the antimalarial activity of sulfadiazine by 2,4-diamino-5-aryloxypyrimidines. *J. Pharmacol. Exp. Ther.* 99:444-449

Hampele IC, D'Arcy A, Dale GE, Kostrewa D, Nielsen J, Oefner C, Page MGP, Schonfeld H-J, Stuber D, Then RL (1997) Structure and function of the dihydropteroate synthase from *Staphylococcus aureus*. *J. Mol. Biol.* 268:21-30

Harinasuta T, Viravan C, Reid HA (1967) Sulphormethoxine in chloroquine-resistant falciparum malaria in Thailand. *Lancet* 1:1117-1119

Hauser PM, Macreadie IG (2006) Isolation of the *Pneumocystis carinii* dihydrofolate synthase gene and functional complementation in *Saccharomyces cerevisiae*. *FEMS Microbiol. Lett.* 256:244-250

Hauβmann C, Rohdich F, Schmidt E, Bacher A, Richter G (1998) Biosynthesis of pteridines in *Escherichia coli*. Structural and mechanistic similarity of dihydroneopterin-triphosphate epimerase and dihydroneopterin aldolase. *J. Biol. Chem.* 273:17418-17424

Helweg-Larsen J, Benfield TL, Eugen-Olsen J, Lundgren JD, Lundgren B (1999) Effects of mutations in *Pneumocystis carinii* dihydropteroate synthase gene on outcome of AIDS-associated *P. carinii* pneumonia. *Lancet* 354:1347-1351

Hennig M, D'Arcy A, Hampele IC, Page MG, Oefner C, Dale GE (1998) Crystal structure and reaction mechanism of 7,8-dihydroneopterin aldolase from *Staphylococcus aureus*. *Nat. Struct. Biol.* 5:357-362

Hennig M, Dale GE, D'Arcy A, Danel F, Fischer S, Gray CP, Jolidon S, Muller F, Page MG, Pattison P, Oefner C (1999) The structure and function of the 6-hydroxymethyl-7,8-dihydropterin pyrophosphokinase from *Haemophilus influenzae*. *J. Mol. Biol.* 287:211-9

Hill R, Goodwin M (1937) "Prontosil" in treatment of malaria, report of 100 cases. *South Med. J.* 30:1170-1172

Ho RI, Corman L, Morse SA, Schneider H (1975) Structure-activity of sulfones and sulfonamides on dihydropteroate synthetase from *Neisseria meningitidis*. *Antimicrob. Agents Chemother.* 7:758-763

Hopfinger AJ, Lopez de Compadre RL, Koehler MG, Emery S, Seydel JK (1987) An extended QSAR analysis of some 4-aminodiphenylsulfone antibacterial agents using molecular modeling and LFE-relationships. Quantitative Structure-Activity Relationships 6:111-117

Huang T, Barclay BJ, Kalman TI, von Borstel RC, Hastings PJ (1992) The phenotype of a dihydrofolate reductase mutant of *Saccharomyces cerevisiae. Gene* 121:167-171

Hurly MG (1959) Potentiation of pyrimethamine by sulphadiazine in human malaria. *Trans. R. Soc. Trop. Med. Hyg*. 53:412-413

Hyde RM, Paterson RA, Beddell CR, Champness JN, Stammers DK, Baker DJ, Goodford PJ, Kuyper LF, Ferone R, *et al.* (1983) The activity of sulfonamide-substituted benzylpyrimidines against dihydropteroate synthase, dihydrofolate reductase, and bacterial cell cultures. Chem. Biol. Pteridines, Proc. Int. Symp. Pteridines Folic Acid Deriv.: Chem., Biol. Clin. Aspects, 7th:505-509

Iliades P, Berglez J, Meshnick S, Macreadie I (2003) Promoter strength of folic acid synthesis genes affects sulfa drug resistance in *Saccharomyces cerevisiae. Microb. Drug Resist.* 9:249-255

Iliades P, Meshnick SR, Macreadie IG (2004a) Dihydropteroate synthase mutations in *Pneumocystis jiroveci* can affect sulfamethoxazole resistance in a *Saccharomyces cerevisiae* model. *Antimicrob. Agents Chemother.* 48:2617-2623

Iliades P, Meshnick SR, Macreadie IG (2005a) Analysis of *Pneumocystis jiroveci* DHPS alleles implicated in sulfamethoxazole resistance using an *Escherichia coli* model system. *Microbial Drug Resistance* (Larchmont, NY, United States) 11:1-8

Iliades P, Meshnick SR, Macreadie IG (2005b) Mutations in the *Pneumocystis jiroveci* DHPS gene confer cross-resistance to sulfa drugs. *Antimicrob Agents Chemother* 49:741-748

Iliades P, Walker DJ, Castelli L, Satchell J, Meshnick SR, Macreadie IG (2004b) Cloning of the *Pneumocystis jiroveci* trifunctional FAS gene and complementation of its DHPS activity in *Escherichia coli. Fungal Genet Biol.* 41:1053-1062

Kazanjian P, Locke AB, Hossler PA, Lane BR, Bartlett MS, Smith JW, Cannon M, Meshnick SR (1998) *Pneumocystis carinii* mutations associated with sulfa and sulfone prophylaxis failures in AIDS patients. *AIDS* 12:873-878

Koehler MG, Hopfinger AJ, Seydel JK (1988) A comparison of QSARs proposed for the inhibition of dihydropteroate synthase by substituted 4-aminodiphenylsulfones. *J. Mol. Structure: THEOCHEM* 179:319-332

Kovacs JA, Gill VJ, Meshnick S, Masur H (2001) New insights into transmission, diagnosis, and drug treatment of *Pneumocystis carinii* pneumonia. *JAMA* 286:2450-2460

Lane BR, Ast JC, Hossler PA, Mindell DP, Bartlett MS, Smith JW, Meshnick SR (1997) Dihydropteroate synthase polymorphisms in *Pneumocystis carinii. J. Infect. Dis.* 175: 482-485

Lawrence MC, Iliades P, Fernley RT, Berglez J, Pilling PA, Macreadie IG (2005) The three-dimensional structure of the bifunctional 6-hydroxymethyl-7,8-dihydropterin pyrophosphokinase/dihydropteroate synthase of *Saccharomyces cerevisiae. J. Mol. Biol.* 348: 655-670

Leil TA, Endo C, Adjei AA, Dy GK, Salavaggione OE, Reid JR, Ames MM, Adjei AA (2007) Identification and characterization of genetic variation in the folylpolyglutamate synthase gene. *Cancer Res.* 67:8772-8782

Lever OW, Jr., Bell LN, Hyman C, McGuire HM, Ferone R (1986a) Inhibitors of dihydropteroate synthase: substituent effects in the side-chain aromatic ring of 6-[[3-

(aryloxy)propyl]amino]-5-nitrosoisocytosines and synthesis and inhibitory potency of bridged 5-nitrosoisocytosine-p-aminobenzoic acid analogs. *J. Med. Chem.* 29:665-670

Lever OW, Jr., Bell LN, McGuire HM, Ferone R (1985) Monocyclic pteridine analogs. Inhibition of *Escherichia coli* dihydropteroate synthase by 6-amino-5-nitrosoisocytosines. *J. Med. Chem.* 28:1870-1874

Lever OW, Jr., Hyman C, Ray PH, Ferone R, Kelsey JE (1986b) A galactoside derivative of a nitrosoisocytosine inhibitor of dihydropteroate synthase: synthesis and biological evaluation. *J. Heterocyclic. Chem.* 23:629-631

Li G, Felczak K, Shi G, Yan H (2006) Mechanism of the conformational transitions in 6-hydroxymethyl-7,8-dihydropterin pyrophosphokinase as revealed by NMR spectroscopy. *Biochem.* 45:12573-12581

Li Y, Blaszczyk J, Wu Y, Shi G, Ji X, Yan H (2005) Is the critical role of loop 3 of Escherichia coli 6-hydroxymethyl-7,8-dihydropterin pyrophosphokinase in catalysis due to loop-3 residues arginine-84 and tryptophan-89? Site-directed mutagenesis, biochemical, and crystallographic studies. *Biochem.* 44:8590-8599

Lopez de Compadre RL, Pearlstein RA, Hopfinger AJ, Seydel JK (1987) A quantitative structure-activity relationship analysis of some 4-aminodiphenyl sulfone antibacterial agents using linear free energy and molecular modeling methods. *J. Med. Chem.* 30:900-906

Ma L, Borio L, Masur H, Kovacs JA (1999) *Pneumocystis carinii* dihydropteroate synthase but not dihydrofolate reductase gene mutations correlate with prior trimethoprim-sulfamethoxazole or dapsone use. *J. Infect. Dis.* 180:1969-1978

Ma L, Kovacs JA, Cargnel A, Valerio A, Fantoni G, Atzori C (2002) Mutations in the dihydropteroate synthase gene of human-derived *Pneumocystis carinii* isolates from Italy are infrequent but correlate with prior sulfa prophylaxis. *J. Infect Dis.* 185:1530-2

Martin JN, Rose DA, Hadley WK, Perdreau-Remington F, Lam PK, Gerberding JL (1999) Emergence of trimethoprim-sulfamethoxazole resistance in the AIDS era. *J. Infect Dis.* 180:1809-1818

Mathieu M, Debousker G, Vincent S, Viviani F, Bamas-Jacques N, Mikol V (2005) *Escherichia coli FolC* structure reveals an unexpected dihydrofolate binding site providing an attractive target for anti-microbial therapy. *J. Biol. Chem.* 280:18916-18922

McCullough JL, Maren TH (1973) Inhibition of dihydropteroate synthetase from *Escherichia coli* by sulfones and sulfonamides. *Antimicrob Agents Chemother.* 3:665-669

McGregor IA, Williams K, Goodwin LG (1963) Pyrimethamine and sulphadiazine in treatment of malaria. *Br. Med. J.* 2:728-729

Mei Q, Gurunathan S, Masur H, Kovacs JA (1998) Failure of co-trimoxazole in *Pneumocystis carinii* infection and mutations in dihydropteroate synthase gene. Lancet 351:1631-1632

Meneau I, Sanglard D, Bille J, Hauser PM (2004) *Pneumocystis jiroveci* dihydropteroate synthase polymorphisms confer resistance to sulfadoxine and sulfanilamide in *Saccharomyces cerevisiae. Antimicrob Agents Chemother* 48:2610-2616

Merali S, Zhang Y, Sloan D, Meshnick S (1990) Inhibition of *Pneumocystis carinii* dihydropteroate synthetase by sulfa drugs. Antimicrob Agents Chemother 34:1075-8

Meshnick SR (1999) Drug-resistant *Pneumocystis carinii. Lancet* 354:1318-1319

Michel R (1968) Comparative study of the association of sulfalene and pyrimethamine and of sulfalene alone in mass chemoprophylaxis of malaria. *Med. Trop.* (Mars) 28:488-494

Navin TR, Beard CB, Huang L, del Rio C, Lee S, Pieniazek NJ, Carter JL, Le T, Hightower A, Rimland D (2001) Effect of mutations in *Pneumocystis carinii* dihydropteroate synthase gene on outcome of *P carinii* pneumonia in patients with HIV-1: a prospective study. *Lancet* 358:545-549

Nichols BP, Guay GG (1989) Gene amplification contributes to sulfonamide resistance in *Escherichia coli. Antimicrob Agents Chemother* 33:2042-2048

Nopponpunth V, Sirawaraporn W, Greene PJ, Santi DV (1999) Cloning and expression of *Mycobacterium tuberculosis* and *Mycobacterium leprae* dihydropteroate synthase in Escherichia coli. *J. Bacteriol.* 181:6814-6821

Nzila AM, Nduati E, Mberu EK, Hopkins Sibley C, Monks SA, Winstanley PA, Watkins WM (2000) Molecular evidence of greater selective pressure for drug resistance exerted by the long-acting antifolate pyrimethamine/sulfadoxine compared with the shorter-acting chlorproguanil/dapsone on Kenyan *Plasmodium falciparum. J. Infect Dis.* 181: 2023-2028

Otzen T, Wempe EG, Kunz B, Bartels R, Lehwark-Yvetot G, Haensel W, Schaper K-J, Seydel JK (2004) Folate-synthesizing enzyme system as target for development of inhibitors and inhibitor combinations against *Candida albicans*-Synthesis and biological activity of new 2,4-diaminopyrimidines and 4'-substituted 4-aminodiphenyl sulfones. *J. Med. Chem.* 47:240-253

Patel O, Fernley R, Macreadie I (2003a) *Saccharomyces cerevisiae* expression vectors with thrombin-cleavable N- and C-terminal 6x(His) tags. *Biotechnol. Lett.* 25:331-334

Patel O, Karnik K, Macreadie IG (2004a) Over-production of dihydrofolate reductase leads to sulfa-dihydropteroate resistance in yeast. *FEMS Microbiol. Lett.* 236:301-5

Patel O, Satchell J, Baell J, Fernley R, Coloe P, Macreadie I (2003b) Inhibition studies of sulfonamide-containing folate analogs in yeast. *Microbial Drug Resistance* (Larchmont, NY, United States) 9:139-146

Patel OG, Mberu EK, Nzila AM, Macreadie IG (2004b) Sulfa drugs strike more than once. Trends Parasitol 20:1-3

Pato ML, Brown GM (1963) Mechanisms of resistance of *Escherichia coli* to sulfonamides. *Arch Biochem. Biophys.* 103:443-448

Pattishall KH, Acar J, Burchall JJ, Goldstein FW, Harvey RJ (1977) Two distinct types of trimethoprim-resistant dihydrofolate reductase specified by R-plasmids of different compatibility groups. *J. Biol. Chem.* 252:2319-2323

Peltola H (1983) Meningococcal disease: still with us. *Rev. Infect Dis.* 5:71-91

Peters W (1987) Chemotherapy and drug resistance in malaria, vol 1, 2nd edn. Academic Press, Inc., New York, NY

Poe M (1976) Antibacterial synergism: a proposal for chemotherapeutic potentiation between trimethoprim and sulfamethoxazole. *Science* (Washington, DC, United States) 194:533-535

Raadstroem P, Fermer C, Kristiansen B-E, Jenkins A, Skold O, Swedberg G (1992) Transformational exchanges in the dihydropteroate synthase gene of *Neisseria meningitidis*: a novel mechanism for acquisition of sulfonamide resistance. *J. Bacteriol.* 174: 6386-6393

Reeder JC, Rieckmann KH, Genton B, Lorry K, Wines B, Cowman AF (1996) Point mutations in the dihydrofolate reductase and dihydropteroate synthetase genes and *in vitro* susceptibility to pyrimethamine and cycloguanil of *Plasmodium falciparum* isolates from Papua New Guinea. *Am. J. Trop. Med. Hyg.* 55:209-213

Rieckmann KH, Brewer GJ, Powell RD (1968) Effects of diphenylsulphone (dapsone) against *Plasmodium vivax* of South West Pacific origin. *Trans R Soc. Trop. Med. Hyg.* 62:649-53

Roland S, Ferone R, Harvey RJ, Styles VL, Morrison RW (1979) The characteristics and significance of sulfonamides as substrates for *Escherichia coli* dihydropteroate synthase. *J. Biol. Chem.* 254:10337-10345

Rosowsky A, Forsch RA, Reich VE, Freisheim JH, Moran RG (1992) Side chain modified 5-deazafolate and 5-deazatetrahydrofolate analogues as mammalian folylpolyglutamate synthetase and glycinamide ribonucleotide formyltransferase inhibitors: synthesis and *in vitro* biological evaluation. *J. Med. Chem.* 35:1578-1588

Salcedo E, Cortese JF, Plowe CV, Sims PF, Hyde JE (2001) A bifunctional dihydrofolate synthetase-folylpolyglutamate synthetase in *Plasmodium falciparum* identified by functional complementation in yeast and bacteria. *Mol. Biochem. Parasitol.* 112:239-252

Sanders WJ, Nienaber VL, Lerner CG, McCall JO, Merrick SM, Swanson SJ, Harlan JE, Stoll VS, Stamper GF, Betz SF, Condroski KR, Meadows RP, Severin JM, Walter KA, Magdalinos P, Jakob CG, Wagner R, Beutel BA (2004) Discovery of potent inhibitors of dihydroneopterin aldolase using CrystaLEAD high-throughput X-ray crystallographic screening and structure-directed lead optimization. *J. Med. Chem.* 47:1709-1718

Schrag SJ, Perrot V, Levin BR (1997) Adaptation to the fitness costs of antibiotic resistance in *Escherichia coli*. *Proc. Biol. Sci.* 264:1287-91

Sheehy TW, Reba RC, Neff TA, Gaintner JR, Tigertt WD (1967) Supplemental sulfone (dapsone) therapy. Use in treatment of chloroquine-resistant falciparum malaria. *Arch Intern. Med.* 119:561-566

Sheng Y, Sun X, Shen Y, Bognar AL, Baker EN, Smith CA (2000) Structural and functional similarities in the ADP-forming amide bond ligase superfamily: implications for a substrate-induced conformational change in folylpolyglutamate synthetase. *J. Mol. Biol.* 302: 425-438

Shi G, Blaszczyk J, Ji X, Yan H (2001) Bisubstrate analogue inhibitors of 6-hydroxymethyl-7,8-dihydropterin pyrophosphokinase: Synthesis and biochemical and crystallographic studies. *J. Med. Chem.* 44:1364-1371

Skold O (1976) R-factor-mediated resistance to sulfonamides by a plasmid-borne, drug-resistant dihydropteroate synthase. *Antimicrob Agents Chemother* 9:49-54

Snow RW, Guerra CA, Noor AM, Myint HY, Hay SI (2005) The global distribution of clinical episodes of *Plasmodium falciparum* malaria. *Nature* 434:214-7

Stammers DK, Achari A, Somers DO, Bryant PK, Rosemond J, Scott DL, Champness JN (1999) 2.0 A X-ray structure of the ternary complex of 7,8-dihydro-6-hydroxymethylpterinpyrophosphokinase from *Escherichia coli* with ATP and a substrate analogue. *FEBS Lett.* 456:49-53

Sun X, Bognar AL, Baker EN, Smith CA (1998) Structural homologies with ATP- and folate-binding enzymes in the crystal structure of folylpolyglutamate synthetase. *Proc. Natl. Acad. Sci. USA* 95:6647-6652

Sun X, Cross JA, Bognar AL, Baker EN, Smith CA (2001) Folate-binding triggers the activation of folylpolyglutamate synthetase. *J. Mol. Biol.* 310:1067-1078

Swarbrick JD, Buyya S, Gunawardana D, Gayler KR, McLennan AG, Gooley PR (2005) Structure and substrate-binding mechanism of human Ap4A hydrolase. *J. Biol. Chem.* 280:8471-8481

Swedberg G, Skoeld O (1980) Characterization of different plasmid-borne dihydropteroate synthases mediating bacterial resistance in sulfonamides. *J. Bacteriol.* 142:1-7

Takahashi T, Hosoya N, Endo T, Nakamura T, Sakashita H, Kimura K, Ohnishi K, Nakamura Y, Iwamoto A (2000) Relationship between mutations in dihydropteroate synthase of *Pneumocystis carinii* f. sp. hominis isolates in Japan and resistance to sulfonamide therapy. *J. Clin. Microbiol.* 38:3161-3164

Tapsall J, Merlino J (2007) Towards an integrated approach to the problem of antimicrobial reistance in Australia. *Microbiol. Australia*, 28:152-153

Then RL (1993) History and future of antimicrobial diaminopyrimidines. *J. Chemother.* 5:361-368

Triglia T, Menting JGT, Wilson C, Cowman AF (1997) Mutations in dihydropteroate synthase are responsible for sulfone and sulfonamide resistance in *Plasmodium falciparum*. *Proc. Natl. Acad. Sci. USA* 94:13944-13949

van Hensbroek MB, Morris-Jones S, Meisner S, Jaffar S, Bayo L, Dackour R, Phillips C, Greenwood BM (1995) Iron, but not folic acid, combined with effective antimalarial therapy promotes haematological recovery in African children after acute falciparum malaria. *Trans R Soc. Trop. Med. Hyg.* 89:672-676

Vinnicombe HG, Derrick JP (1999) Dihydropteroate synthase from *Streptococcus pneumoniae*:Characterization of substrate binding order and sulfonamide inhibition. *Biochemical and Biophysical Research Communications* 258:752-757

Visconti E, Ortona E, Mencarini P, Margutti P, Marinaci S, Zolfo M, Siracusano A, Tamburrini E (2001) Mutations in dihydropteroate synthase gene of *Pneumocystis carinii* in HIV patients with *Pneumocystis carinii* pneumonia. *Int. J. Antimicrob Agents* 18:547-551

Volpe F, Ballantine SP, Delves CJ (1993) The multifunctional folic acid synthesis fas gene of *Pneumocystis carinii* encodes dihydroneopterin aldolase, hydroxymethyldihydropterin pyrophosphokinase and dihydropteroate synthase. *Eur. J. Biochem.* 216:449-458

Walter RD, Koenigk E (1974) Biosynthesis of folic acid compounds in Plasmodia. Purification and properties of the 7,8-dihydropteroate-synthesizing enzyme from *Plasmodium chabaudi*. Hoppe-Seyler's Zeitschrift fuer Physiologische Chemie 355:431-437

Walzer PD, Foy J, Steele P, Kim CK, White M, Klein RS, Otter BA, Allegra C (1992) Activities of antifolate, antiviral, and other drugs in an immunosuppressed rat model of *Pneumocystis carinii* pneumonia. *Antimicrob Agents Chemother* 36:1935-1942

Wang P, Brooks DR, Sims PF, Hyde JE (1995) A mutation-specific PCR system to detect sequence variation in the dihydropteroate synthetase gene of *Plasmodium falciparum*. *Mol. Biochem. Parasitol.* 71:115-125

Wang P, Lee CS, Bayoumi R, Djimde A, Doumbo O, Swedberg G, Dao LD, Mshinda H, Tanner M, Watkins WM, Sims PF, Hyde JE (1997) Resistance to antifolates in *Plasmodium falciparum* monitored by sequence analysis of dihydropteroate synthetase and dihydrofolate reductase alleles in a large number of field samples of diverse origins. *Mol. Biochem. Parasitol.* 89:161-177

Wang P, Wang Q, Sims PF, Hyde JE (2007) Characterisation of exogenous folate transport in *Plasmodium falciparum. Mol. Biochem. Parasitol.* 154:40-51

Watkins WM, Mosobo M (1993) Treatment of *Plasmodium falciparum* malaria with pyrimethamine-sulfadoxine: selective pressure for resistance is a function of long elimination half-life. *Trans R Soc. Trop. Med. Hyg.* 87:75-78

Wettergren Y, Odin E, Nilsson S, Willen R, Carlsson G, Gustavsson B (2005) Low expression of reduced folate carrier-1 and folylpolyglutamate synthase correlates with lack of a deleted in colorectal carcinoma mRNA splice variant in normal-appearing mucosa of colorectal carcinoma patients. *Cancer Detect Prev.* 29:348-355

White PJ, Woods DD (1965a) Biochemical properties of staphylococci sensitive and resistant to sulphonamides. *J. Gen. Microbiol.* 40:255-271

White PJ, Woods DD (1965b) The synthesis of p-aminobenzoic acid and folic acid by staphylococci sensitive and resistant to sulphonamides. *J. Gen. Microbiol.* 40:243-253

Wiese M, Schmalz D, Seydel JK (1996) New antifolate 4,4'-diaminodiphenyl sulfone substituted 2,4-diamino-5-benzylpyrimidines. Proof of their dual mode of action and autosynergism. Archiv der Pharmazie (Weinheim, Germany) 329:161-168

Wiese M, Seydel JK, Pieper H, Krueger G, Noll KR, Keck J (1987) Multiple regression analysis of antimalarial activities of sulfones and sulfonamides in cell-free systems and principal component analysis to compare with antibacterial activities. Quantitative Structure-Activity Relationships 6:164-172

Williams DL, Spring L, Harris E, Roche P, Gillis TP (2000) Dihydropteroate synthase of *Mycobacterium leprae* and dapsone resistance. *Antimicrob Agents Chemother* 44:1530-1537

Wise EM, Jr., Abou-Donia MM (1975) Sulfonamide resistance mechanism in *Escherichia coli.* R plasmids can determine sulfonamide-resistant dihydropteroate synthases. *Proc. Natl. Acad. Sci. USA* 72:2621-2625

Woods DD (1940) The relationship of p-aminobenzoic acid to the mechanism of action of sulphanilamide. *Br. J. Exp. Pathol.* 21:74-90

World Health Organization (2003) Immunization, Vaccines and Biologicals: Pneumococcal vaccines

World Health Organization EaoCD, Control Sa (2003) Control of epidemic meningococcal disease. WHO practical guidelines. 2nd edition

Xiao B, Shi G, Chen X, Yan H, Ji X (1999) Crystal structure of 6-hydroxymethyl-7,8-dihydropterin pyrophosphokinase, a potential target for the development of novel antimicrobial agents. *Structure* 7:489-496

Xiao B, Shi G, Gao J, Blaszczyk J, Liu Q, Ji X, Yan H (2001) Unusual conformational changes in 6-hydroxymethyl-7,8-dihydropterin pyrophosphokinase as revealed by X-ray crystallography and NMR. *J. Biol. Chem.* 276:40274-81

Xie L, Smith JA, Gross SS (1998) GTP Cyclohydrolase I Inhibition by the Prototypic Inhibitor 2,4-Diamino-6-Hydroxypyrimidine. Mechanisms and unanticipated role of GTP cyclohydrolase I feedback regulatory protein. *J. Biol. Chem.* 273:21091-21098

Young PG, Smith CA, Sun X, Baker EN, Metcalf P (2006) Purification, crystallization and preliminary X-ray analysis of *Mycobacterium tuberculosis* folylpolyglutamate synthase (MtbFPGS). Acta Crystallographica Section F 62:579-582

Zhang Y, Merali S, Meshnick SR (1992) p-Aminobenzoic acid transport by normal and Plasmodium falciparum-infected erythrocytes. *Mol. Biochem. Parasitol.* 52:185-94

Zhang Y, Meshnick SR (1991) Inhibition of *Plasmodium falciparum* dihydropteroate synthetase and growth *in vitro* by sulfa drugs. *Antimicrob Agents Chemother* 35:267-71

Zimmerman M, Tolman RL, Morman H, Graham DW, Rogers EF (1977) Inhibitors of folate biosynthesis. 1. Inhibition of dihydroneopterin aldolase by pteridine derivatives. *J. Med. Chem.* 20:1213-1215

Reviewed by Dr Ross Fernley, CSIRO Molecular and Health Technologies, Parkville, Victoria, Australia

In: New Developments in Medicinal Chemistry
Editors: Marta P. Ortega and Irene C. Gil
ISBN 978-1-60456-810-3

Chapter 2

STUDIES ON ANTI-CANCER AGENTS: PHENOLIC COMPOUNDS AND THEIR PHARMACOLOGICAL ACTIVITY

Maria Dolors Pujol ***and Isabel Sánchez***
Laboratori de Química Farmacèutica (Unitat associada al CSIC),
Facultat de Farmàcia, Universitat de Barcelona, Barcelona, Spain

INTRODUCTION

Polyphenols constitute the most abundant group of antioxidants of normal human food (tea, red wine, grapes, olive oil, chocolate, broccoli, cherries, pomegranates, peanuts, berries and other fruits or vegetables including Ginkgo *biloba*) that protect against oxidative stress and their associated pathologies such as inflammation, cancer and coronary heart diseases. [1] The presence of phenol functions in their structures confers stability against oxidation. Also the biological properties are related to the phenol groups and their disposition in the structure. These compounds that occurs naturally in various food and beverages of plant origin, were also named Vitamin P (more than 1500 compounds) and their main beneficial biological effects are: [2, 3] a) The diminution of reactive oxygen species related with the inflammation process, with the immune system by the recruitment of leucocytes and the blood homeostasis. b) Inhibition of growth of several tumors. Thus, following the American Cancer Society dietary guidelines [4] of "five or more pieces of fresh fruit and vegetables per day" to help prevent cancer and anti-inflammatory diseases. Also fruit and vegetable juices or tea might provide substances that help prevent cancer. c) Reduction of inflammatory effects such as coronary diseases related to the oxidation of LDL (light density lipoprotein). d) Treatment of skin aging in humans (Figure 1).

In general, the pure natural flavonoids show low bioavailability and this is a problem for comparing abilities in various *in vivo* and *in vitro* models. [5] Polyphenols absorbed the free-radical generated in the body by cells because of the oxidative stress. These free-radicals

* Av. Diagonal 643. 08028-Barcelona (Spain). E-mail: mdpujol@ub.edu.

cause diseases and premature aging. [6] Free radicals are considered chemical structures with incomplete electron shells which make them more chemically reactive than those with complete electron shells. Exposure to various environmental factors, including tobacco smoke and different radiation types, can also lead to free radical formation. The most common form of free radicals is oxygen which leads to the superoxide. When the oxygen becomes free-radical may cause damage to the DNA and other molecules. Over time, such damage may become irreversible and lead to disease including cancer.

Polyphenols can be grouped and classified according the number of carbon atoms and the type and number of phenol groups: [7]

- Simple phenols C6
- Benzoic acids and derivatives C6-C1
- Aketofenones and phenylacetic acids C6-C2
- Cinnamic acids and derivatives C6-C3
- Cumarines and derivatives C6-C3
- Flavonoids and derivatives C6-C3-C6 (Considered the main polyphenol group)
- Benzophenones and stilbenes C6-C1-C6
- Xanthones C6-C2-C6

Flavonoids and simple phenolic acids received a singular attention such as natural antioxidants attributable to their intrinsic reducing properties.

Flavonoid structures vary according to the substitution of the pyran ring, which is fused with an aromatic ring and further substituted with another aromatic ring at the C-2-position (flavonoid) or C-3 (isoflavonoid).

Flavonoids are considered a subclass of the polyphenols and can be sub-divided according to the modification of their structure: connection of the B ring to the C ring, the oxidation state of the B ring, and the number and position of the hydroxyl groups. The flavones (650 structures) and flavonoles (more than 1000 structures) are the most present compounds in the natural products. By migration of the ring C of the position 2 to 3 was formed the isoflavonoid group (more than 230 structures) (Figure 2). These compounds are very important for the vegetables that contain them due to its role of protection against the pathogenic agents.

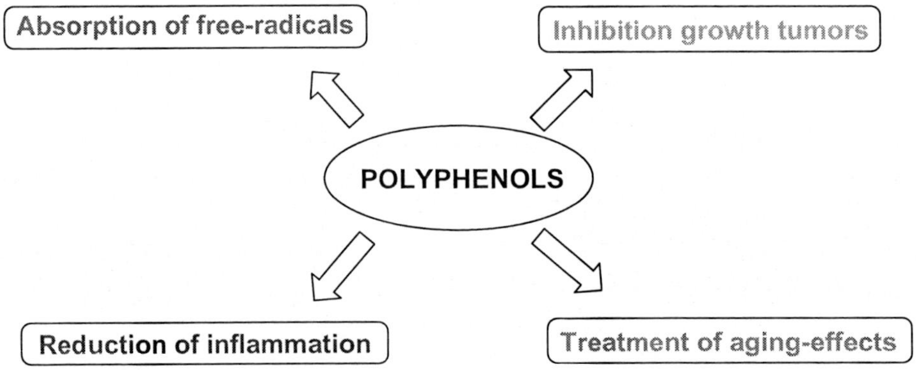

Figure 1. Biological effects of polyphenols.

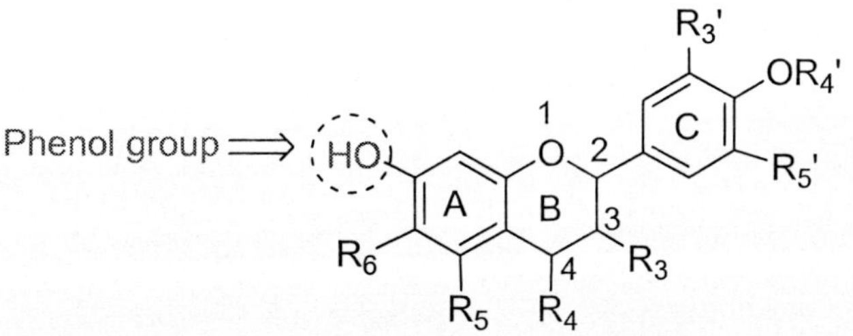

Flavonoid types	B-C ring connection (position on B ring)	B-ring insaturation	B-functional groups	Flavonoids*
Flavanols	2	None	3-OH	(+)-Catechin (+)-Gallocatechin
			3-*O*-gallate	(+)-Epigallocatechin
Flavanones	2	None	4-Oxo	Eriodictyol Hesperetin
Flavones	2	2-3 double bond	4-Oxo	Apigenin
Isoflavones	3	2-3 double bond	4-Oxo	Genistein Daitzein
Flavonols	2	2-3 double bond	3-Hydroxy, 4-Oxo	Quercetin Myricetin
Anthocyanidins	2	1-2, 3-4 double bond	3-Hydroxy	Cyanidin Pelargonidin Petunidin Peonidin

* Flavonoids reported in the USDA Database for the Flavonoid Content of Selected Foods. [7-9]

Figure 2. Flavonoid types.

Black tea was the major flavonoid source, predominantly of flavan-3-ols, representing 70% total intake. Apple was the main quercetin source and onion also was a prominent source. Apple, apricot and grapes were the major sources of epicatechin and catechin. Flavonoid consumption profiles and flavonid sources varied according the age and the habits of the people worldwide.

A study of research suggests that flavonoids may inhibit breast cancer development by inhibition of estrogen production, inhibition of cell proliferation and decreasing the production of reactive oxygen species. Among the clinical effects (Figure 3) is interesting to note: [10]

Several isolated flavonoids possess one or more of the above indicated biological activities, but it is known that a combination of flavonoids (natural foods) have best anti-inflammatory activity and are of interest for the human health.

Isoflavones are similar in structure and also in activity to the 17-β-estradiol representative of human estrogens and the activity led to the denomination of phytoestrogens. These compounds bind selectively the estrogenic receptor β while several flavonoids exhibit antiestrogenic activity.

Anti-inflammatory	Reduce the mobility of leucocytes
	Reduce the activation of the complement
	Reduce the cyclo-oxygenase levels
Anti-cancer	Protection of DNA against oxidation
	Modification of bases
Antiviral	Reduction of effects VIH
Anti-trombogenesis	Inhibition of platelet aggregation
Cardiovascular protection	Inhibition of oxidation of LDL
	Reduce the coronary health
	Reduce the cholesterol levels
Anti-osteoporosis	Protection of osseous density
Estrogenic properties	Mimic estrogens in mammals
Anti-aging	Activating endogenous defense systems
Anti-apoptosis [11-12]	Activating endogenous mechanisms
Immunomodulation	Interaction with immune response

Figure 3. Clinical effects of polyphenols.

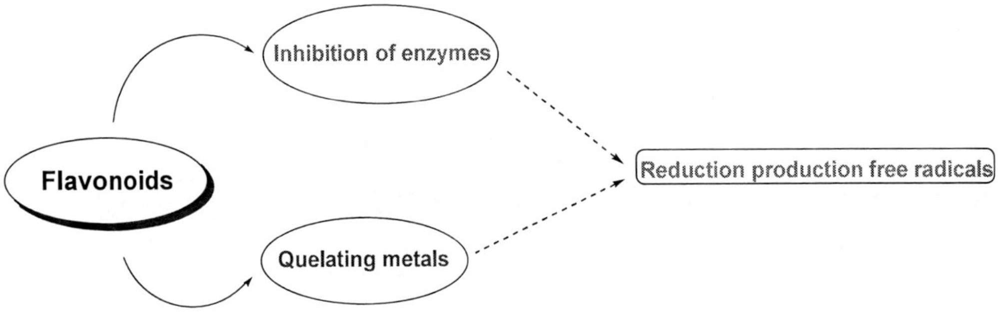

Figure 4. Reduction of production of free radicals by flavonoids.

Several studies related to the polyphenols have been demonstrated *in vitro* the tumor prevention and also the inhibition of cancer cell proliferation, [13] and other authors such as Lin, J. and col. [14] have been reported recently the results of a work on cytostatic properties of polyphenols *in vivo* realized with 878 cases, concluding that the flavonoid intake was not quantitatively associated with colorectal cancer.

The antioxidant activity of the polyphenol group is due to the reduction of production of free radicals by two possible mechanisms: a) directly inhibition of enzymes or b) by quelation of the transition metals implicated on the generation of free radicals (free radicals were associated with pathogenesis of various disorders such as cancer, cardiovascular diseases, neurovegetative disorders and are implicated in the aging) (Figure 4). Moreover, flavonoids are capable to reduce the oxygen reactive species (ORS), highly oxidized.

The oxidative stress and the lipid peroxidation are the cause of several chronic diseases including cardiovascular diseases, cancer, dementia and others. Polyphenols possess antioxidant properties, and can prevent the oxidative damage by inhibition of the generation of reactive species by sequestration of free radicals o by increasing the endogen protectors.

The phenolic compounds are found in many vegetables and fruits. The fruits content quercetin, kanferol, hesperetin and cinnamic acids. Beverages like tea and wine content procianidines and catechins. A varied diet of natural foods containing 800 mg/day of polyphenols is considered interesting for a good health. Polyphenols were found to be pharmacological active as preventive or therapeutic agents for several diseases, but their bioavailability depends on several factors. The intake of natural foods containing polyphenols are best than the intake of dietary supplementation.

Most studies in a variety of animal models have shown that various polyphenols have *in vitro* and *in vivo* chemopreventive properties against several cancers.

Quercetin, silymarin, rutin, rosmarinic acid, tea flavonoids, hydroxytyrosol, ellagic acid, apigenin, cosmociin and narciclasine are natural polyphenol compounds with several therapeutic indications which will be described in this work.

QUERCETIN

Quercetin (3,3',4',5,7-pentahydroxyflavone) is a common flavonoide found in the current diet containing fruits and vegetables. [15]

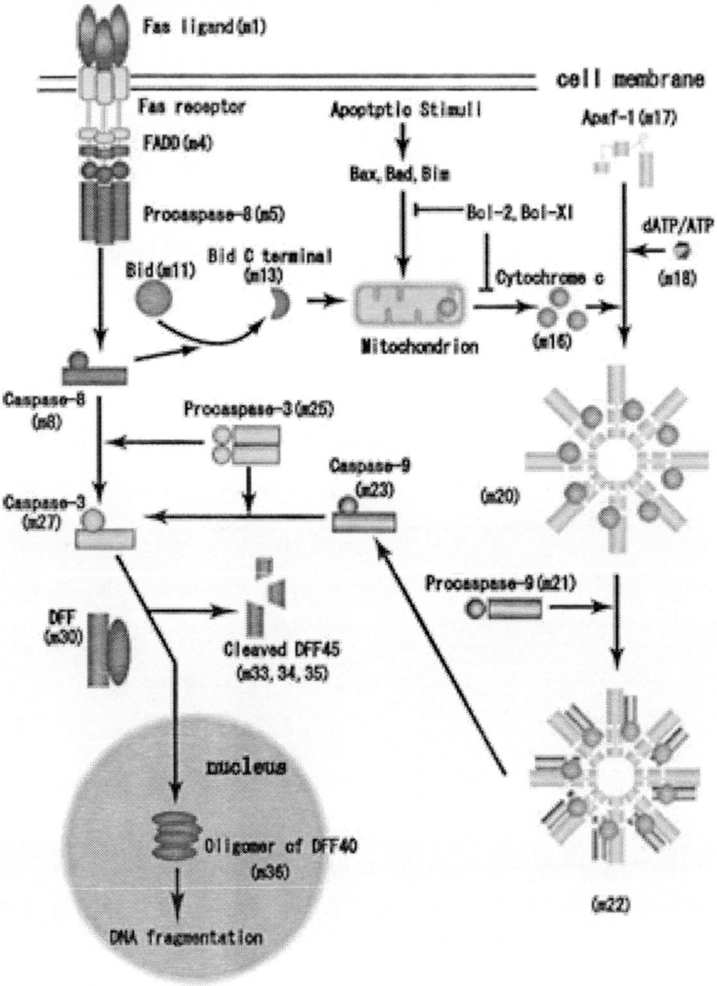

Figure 5. Apoptosis mechanism. [27]

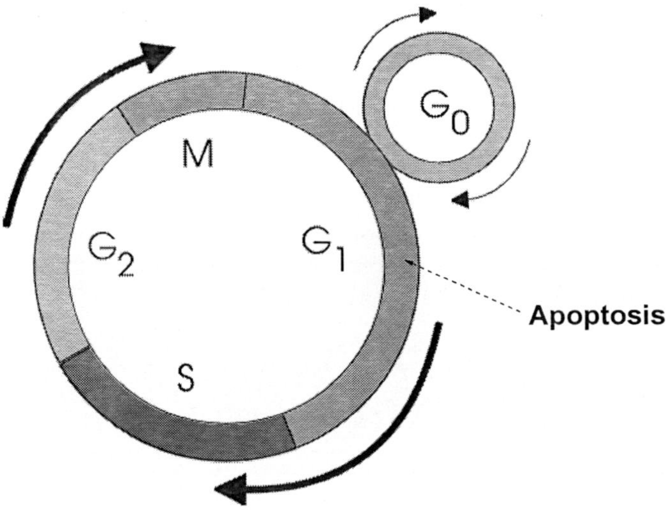

Figure 6. The cycle cellular phases.

The raw and red onions contain more quercetin than other kind of onions, while raw spinachs present more quantity of quercetin than apples and apples more than raspberries or strawberries (Table 2). Quercetin may induce apoptosis by direct activation of caspase cascade implicated in the mitochondrial apoptosis pathway in leukemia, [16] breast, [17] lung, [18] hepatoma, [19] oral [20] and colon cancer cell lines. [21] Quercetin activates caspase-3 and caspase-9 but no caspase-8 and releases cytochrome *C* in HL-60 cells (Figure 5), [22] blocks the phase G_1 of the cellular cycle (Figure 6) in human gastric, nasopharyngeal and hepatic cancer cells. [23] Moreover quercetin produces DNA fragmentations in hepatocarcinoma cell lines, [24] leukemia [25] and other tumor cells.

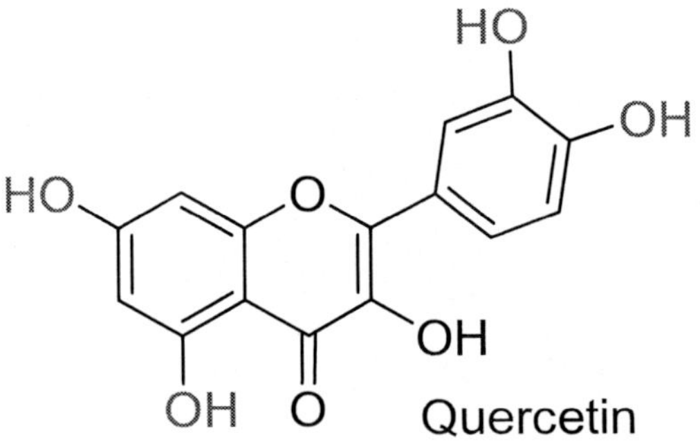

Table 1. Flavonoid (quercetin) content of onions

Description	n	Min*	Max*	References
Onions, cooked, boiled	14	8.70	31.00	31, 32
Onions, raw	294	1.50	41.00	33
Onions, red, raw	58	0.00	75.55	33, 34
Onions, spring, raw	3	6.71	18.00	31, 35
Onions, white, sweet	78	0.00	63.4	36

*Min and max units = mg/100 g, edible portion. n = number of assays.

Table 2. Flavonoid (quercetin) content of diver vegetables

Description	n	Min*	Max*	References
Apples, raw, with skin	11	1.00	14.00	35, 37
Apples, raw, with skin	24	0.00	2.00	34
Raspberries, raw	4	0.50	0.96	34
Spinach, raw	6	0.00	27.22	35, 38
Strawberries, raw	10	0.47	0.86	34, 39

*Min and max units = mg/100 g, edible portion. n = number of assays.

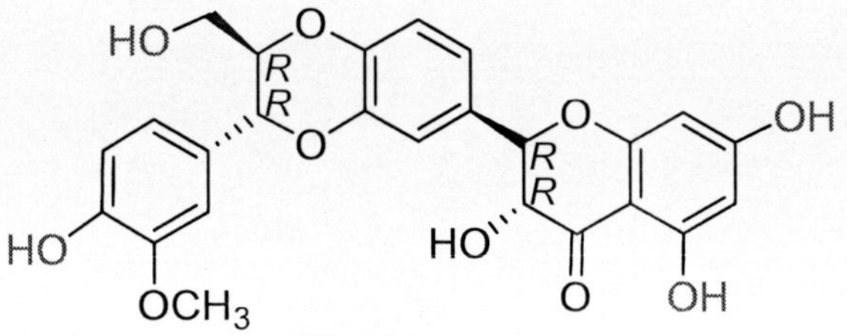

Figure 7. Silybin structure.

Quercetin concentrations that significantly decrease growth of HepG2 cells have also similar effects on other cancer cell lines, remaining demonstrated the inhibitory effect of this flavonoide in the cellular tumor growth and their chemopreventive or therapeutic effects. [26]

In general, it was found two different pathways for the initiation of the apoptosis: one of this is the extrinsic pathway involved the death receptor signaling and the other is the intrinsic pathway involved the mitochondrial cascades. [28-29]

The work of Yang et al. [30] provides evidence that quercetin-glucuronide induced apoptosis through intrinsic pathway involved mitochondrial cascades in NCI-H209 cells and causing a cell cycle arrest at G_2/M phase.

SILYMARIN

Seeds or the milk thistle extracts of *Silybum marianum* L. Gaertn. (Asteraceae) have been shown to have antiproliferative effects in several tumor types. Several preparations have been used for centuries for the treatment of various liver disorders such as chronic hepatitis and cirrhosis, *Amanita phalloides* mushroom poisoning, cytoprotection for chemical substances and prophylaxis against several chemotherapeutic side-effects. [40-43]

The extracts from *Silybum* contain a mixture of natural polyphenolic compounds. These flavonolignans [44] possessing antioxidant properties are grouped under the name of silymarin and silybin (also named silybine, silybin A or silybinin) (Figure 7) is the most abundant and also most active component. Other active constituents are: silychristin, silydianin, and isosilybin. [45] Really, they are known the silybins A and B, the salychristins A and B and the isosilybins A and B considered isomers. [46-48]

Silybin induces p21/Cip1 and p27/Kip1 and G1 arrest in prostate cancer cells irrespective of p53 status. The inhibition of cell cycle regulatory proteins plays a fundamental role in the mechanism of action of silybin and this is suitable for combined use with other chemotherapeutic agents possessing other action mechanism. Silymarin protects the liver by inhibition of lipid peroxidation and prevents liver damage caused by several drugs, toxins and alcohol. Pharmacological studies have been show that the silymarin mixture (Figure 8) is not toxic and possessess an interesting anti-atherosclerotic activity by inhibition of adhesion of molecules. Silymarin isolated from artichokes, has been shown to prevent tumor promotion in animal experimentation.

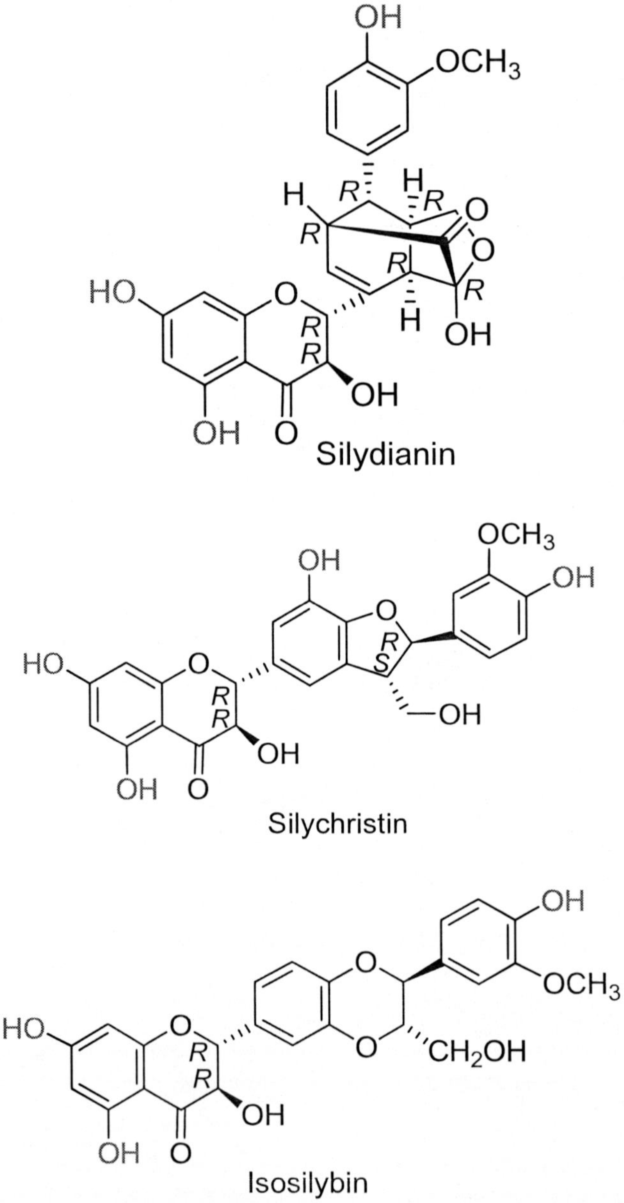

Figure 8. Components of Silymarin.

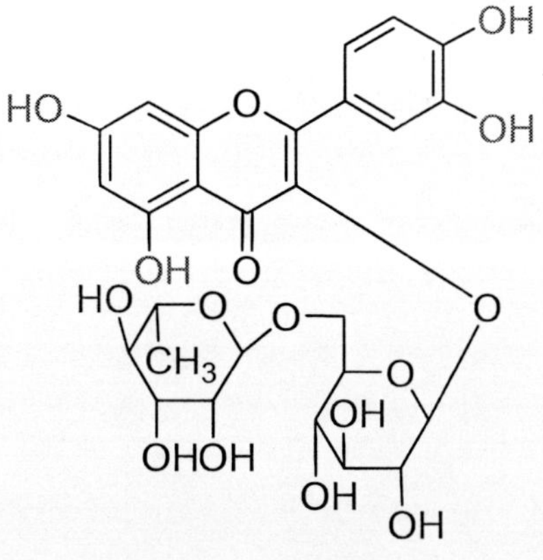

Figure 9. Rutin structure.

Rutin

Rutin (3-[[6-*O*-(6-deoxy-α-*L*-mannopyranosyl)-β-*D*-glucopyranosyl]oxy]-2-(3,4-dihydroxy phenyl)-5,7-dihydroxy-4*H*-1-benzopyran-4-one) also named rutoside, sophorin or quercetin-3-rutinoside (quercetin-glucoside and rutinose) is a flavonoid found in the fruit of Fava *D'Anta* tree (Dimorphandra, from Brazil), leaves of Rheum *species* and other species (specially fruits like apricots, cherries, prunes and all citric fruits) with a wide range of pharmacological activities. [49-50]

Structurally rutin is a flavonol glucoside comprised of the flavonol backbone (aglycon) and the rutinose as a glycon (Figure 9). It possesses antioxidant activity and can play a role in the inhibition of some tumours. Actually, the main pharmaceutical application of the rutin is related to the effect over the capillaries and the blood due to their ability to strengthen and modulate the permeability of the walls of the blood vessels including delicate capillaries. Rutin give nutritional support to the circulatory systems including the vessels in eyes (glaucoma, cataracts, etc.). [51-52] Rutin works with vitamin C and assists in reducing pain and intraocular pressure. Rutin strengthens the blood sanguineous vessels (treatment of hemorrhoids and Meniere's disease), and therefore can reduce the symptoms of haemophilia. [51] It also may help to prevent a common unpleasant-looking venous edema of the inferior extremities due to the poor blood circulation. [53-54] Rutin can also reduce the cytotoxicity of oxidized cholesterol and lower the risk of heart disease. Moreover rutin alone has been shown to protect LDL (Low-density lipoprotein) against oxidation.

Rosmarinic Acid

Rosmarinic acid (3,4-dihydroxycinnamic acid, (*R*)-1-carboxy-2-(3,4-dihydrxyphenyl) ethyl ester) is a component of Rosmarinus *officinalis,* Salvia *officinalis,* Prunella *vulgaris* and other plant leaves. [55] The rosmarinic acid by hydrolysis give the cafeic acid also with structure of acid-phenol (Figure 10). Recent rapports suggest protective effects of rosmarinic acid in Alzheimer disease. [56] Rosmarinic acid has also anti-inflammatory, anti-infective, antioxidant activity and it is capable of inducing apoptosis of Jurkat and peripheral T-cells in an Lack-dependent manner. [57]

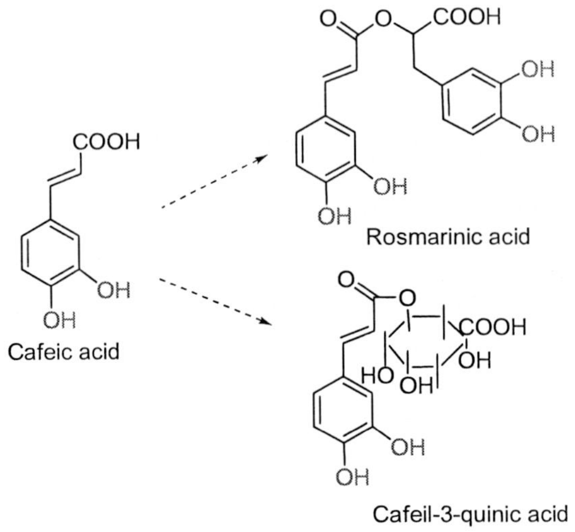

Figure 10. Rosmarinic acid structure.

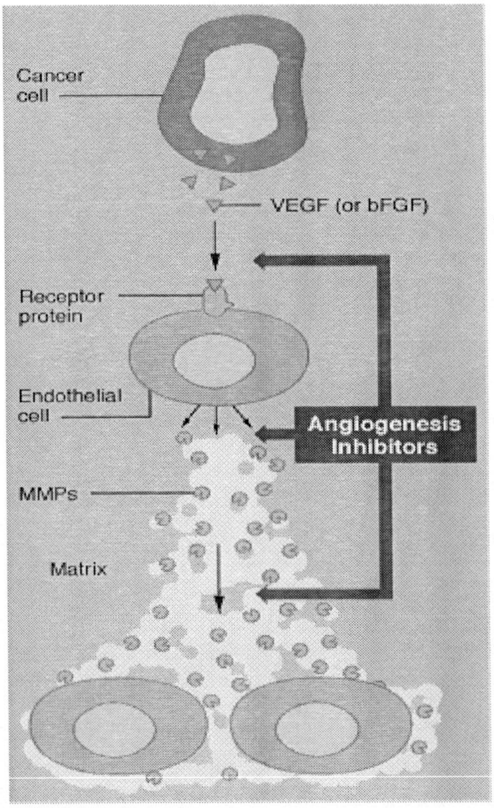

Figure 11. Angiogenesis process.

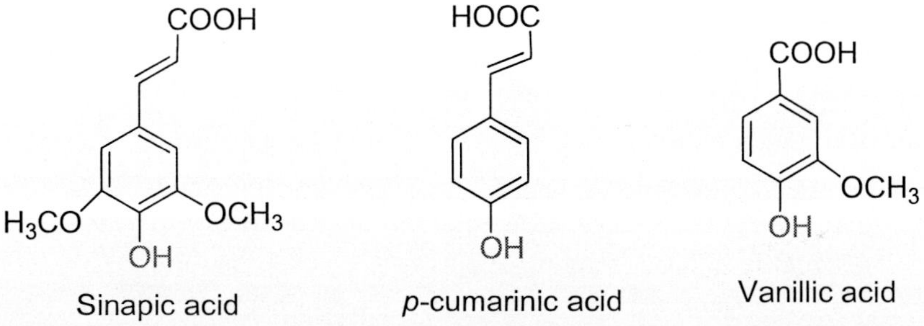

Figure 12. Sinapic, *p*-cumarinic and vanillic acid structures.

The rosmarinic acid was able to induce apoptosis in three different cell lines: erythroleukemia (K562); papillary (NPA) and anaplastic (ARO) thyroid cancer. [58] The rosmarinic acid also inhibited several important steps of angiogenesis (Figure 11) such as proliferation, migration, adhesion and tube formation on human umbilical vein endothelial cells (HUVEC). Moreover showed the capacity of reduce intracellular reactive oxygen species (ROS). [59] These studies suggested that the antiangiogenic activity might be related to its anti-oxidative activity.

Related to the cafeic acid structure it found the sinapic, *p*-cumarinic and vanillic acid. Sinapic acid (4-hydroxy-3,5-dimethoxycinnamic acid) is a phenolic compound presents in the beer and in the rice. This carboxylic acid is the responsible of the potential colon and breast tumor-suppressive properties of rice. An extraction of brown rice with ethyl acetate presents eight carboxylic acids among them: caffeic acid, *p*-coumarinic acid and vanillic acid were also components of these nutrients (Figure 12).

These carboxylic acid show cancer chemopreventive properties under conditions of frequently intake rice as part of the diet. [60]

TEA FLAVONOIDS

One of the richest sources for polyphenols is from the tea. Tea made from the leaves of the Camelia *sinensis* is a popular beverage with important antioxidant properties. The preparation of tea modifies the composition of flavonoides (Figure 14). The most important change was the reduction of flavanols levels and the increase in the concentration of thearubigins (see Table 3). These changes of composition are due to the oxidative process that occurs during production of black and oolong tea, the flavan-3-ols were converted to the thearubigins. The decaffeination and other manipulation process such as the preparation of instant and ready-to-drink teas, according with the literature data, decrease the flavonol concentrations (Table 3). The environment, the type of tea, and other parameters also affect the concentration and proportion of the different flavonoids of the tea. [7]

The cancer-preventive activity of tea components has been studied extensively the last decade at the cellular level. A tea extract containing Polyphenon E (PPE) and the most abundant, most active and most studied component (-)-epigallocatechin gallate (EGCG) decrease the tumor multiplicity and cell proliferation. Diet appears to be a better route administration for PPE than drinking fluid. Studies in cell line led to the proposal of many mechanisms on the preventive cancer action of EGCG. The autoxidation of EGCG observed in cell cultures may produce activities demonstrated in many animal models including cancer of the skin, lung, mammary gland, oral cavity, stomach, liver, pancreas, small intestine, colon, bladder, and prostate. [61] Several studies have shown that EGCG induces apoptosis in cancer cells, but the molecular mechanism is still largely unknown. [62] Potential mechanisms (Figure 13) have been suggested to include inhibition of enzymes such as cyclooxygenase (COX) and lipoxygenase, inhibition of activator protein-1, inhibition of angiogenesis, activation of p53 tumour suppressor protein and inhibition of telomerase activity. EGCG are the major catechin of the green tea and has been studied more often than other catechins. [63] Recently was reported that EGCG and ECG are cell growth suppressors, and ECG showed better modulation the cell growth arrest in several cancer cells. [64]

Table 3. Transformation of components during tea processing [7]

Tea	Total flavonols*	Total flavan-3-ols**	Thearubigins***
Green	5.2	132.1	1.1
Green decaffeinate	4.8	55.8	8.8
Black	3.9	34.3	73.4
Black, ready-to-drink	4.4	2.7	49.0
Oolong (red tea)	2.7	51.6	-

7. Data from Beecher, G. R. J. Nutr. 2003, 3248S-3254S (from Anonymous 2003 USDA database for the flavonoid content of selected foods. Nutrient data laboratory. http://www.nal.usda.gov/fnic/foodcomp/Data/Flav/flav.html). Last update 2003.

* Total flavonols: kaempferol, myricetin and quercetin. In most teas, quercetin is the prominent flavonol.

**Total flavan-3-ols: (+)-catechin, (-)-epicatechin, (-)-epicatechin-3-gallate, (-)-epigallocatechin, (-)-epigallocatechin-3-gallate, and (-)-gallocatechin. (-)-Epigallocatechin-3-gallate is the most abundant flavonol.

*** Thearubigins: expressed on basis of gallic acid.

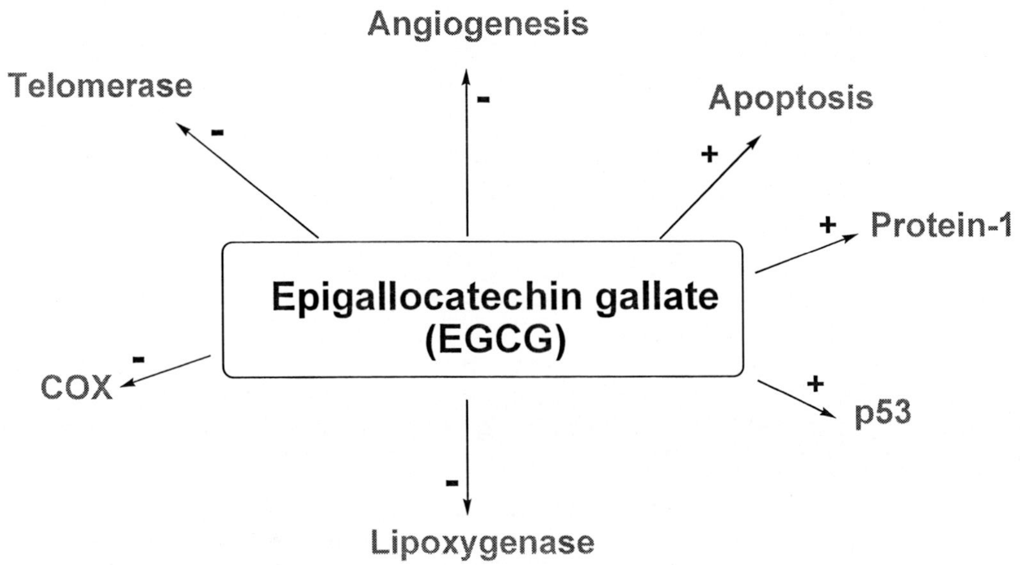

Figure 13. Biological activities of EGCG.

Based on several studies, the polyphenols of green tea appear shown chemopreventive properties against prostate cancer. The beverage of tea, specifically green tea, has gained considerable attention for the prevention and treatment of prostate cancer. In the studies of antitumor activity of the tea components such as EGCG or EGC at the cellular level, was found that the apoptosis induction and the cell arrest at the G1 cellular cycle phase are the main mechanisms found. The minor component epigallocatechin (EGC) suppresses cyclin D1 expression in head and neck squamous cell carcinoma (HNSCC) cell lines. [65]

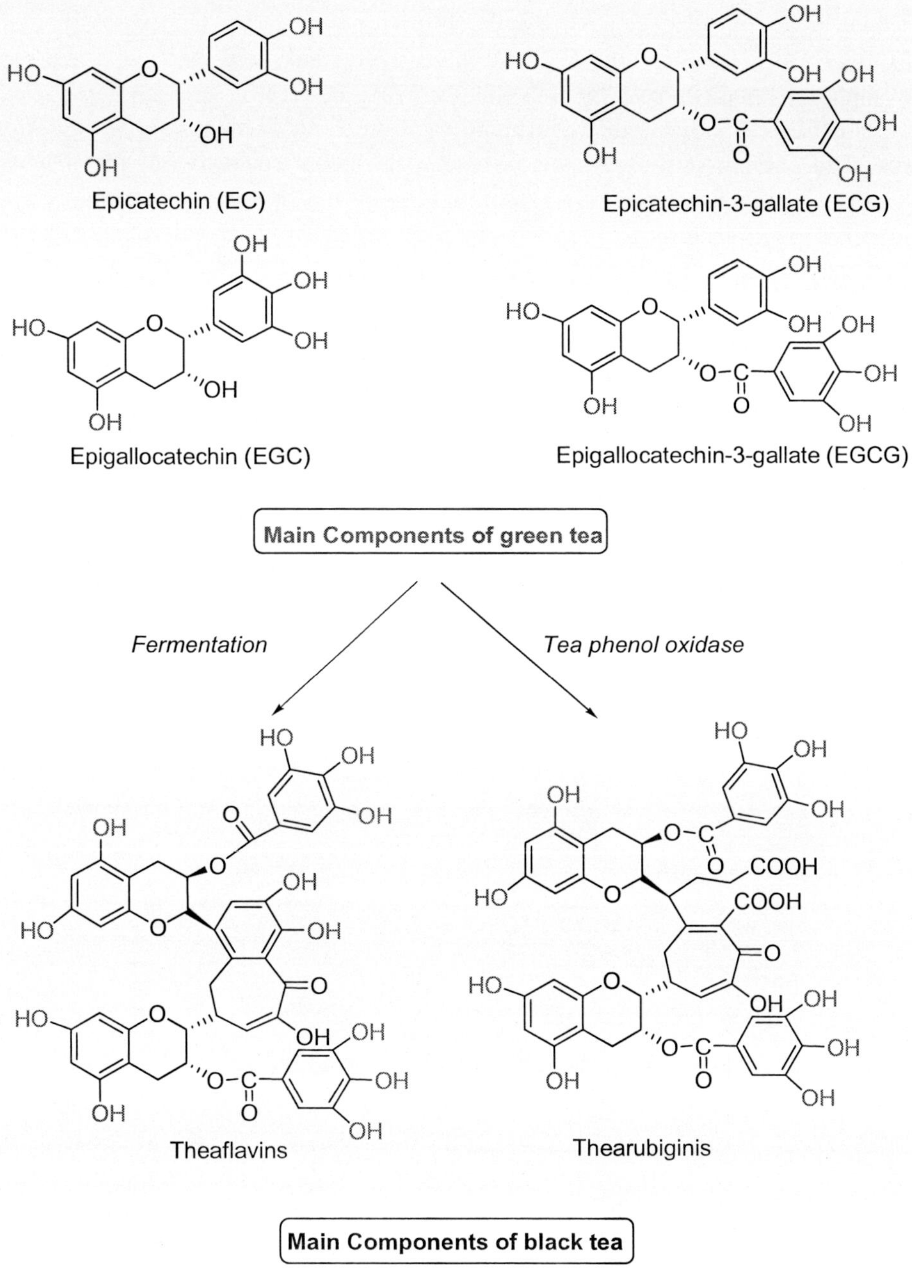

Figure 14. Components of green and black tea. [69]

Tea constituents may be used for the prevention of cancer at selected organs if sufficient concentration of the antioxidant components can be delivered until this organs. [66]

Other authors reported an interesting work over the treatment of the sarcoma S180 (in mice) combining radiotherapy and isolated tea polyphenols. [67]

In summary, green tea polyphenols inhibit angiogenesis and metastasis. Specifically, EGCG regulates expression of VEGF, matrix metalloproteinases and EGFR. The catechin gallate (GC), a minor polyphenol in green tea induced apoptosis and the cytotoxicity was not due to oxidative stress. [68]

After the treatment in liver cancer cells with tea polyphenols, the expressions of cyclin D1 and CDK4 was decreased, and the arrest in G1 cycle phase was induced. Tea polyphenols possibly suppressed the expressions of cyclin D1. [70]

Several evidences indicate that consumption of green tea is interesting for the prevention of cancer. The components of green tea reduce the carcinogenesis but the mechanism is unknown today, while topoisomerase I, protein kinases, matrix metalloproteinases and others are considered possible targets. [71] The bioavailability of these polyphenols to different organ sites may contribute to the different preventive efficacy of polyphenols against urinary bladder and mammary cancers. In general, these polyphenols have a minimal preventive effect of breast cancers. [72]

Tea polyphenols decreased ROS (reactive oxygen species) generation and down regulated MMP-9 expression. Moreover, epicatechin (EC), epigallocatechin (EGC) and other polyphenols inhibited invasion of tumor cells by their antioxidative capacity. [73]

HYDROXYTYROSOL

The hydroxytyrosol (2-(3,4-dihydroxyphenyl)ethanol (DPE)), an olive oil polyphenol confers anticarcinogenic activity which may be due to their antioxidant properties but also to their ability to reduce the bioavailability of food carcinogens and to inhibit their possible metabolic activation. [74]

The main glycoside of the olives is oleuropein, and hydroxytyrosol is a degradation product of oleuropein in the virgin olive oil, considered a major antioxidant compound (Figure 15). [75]

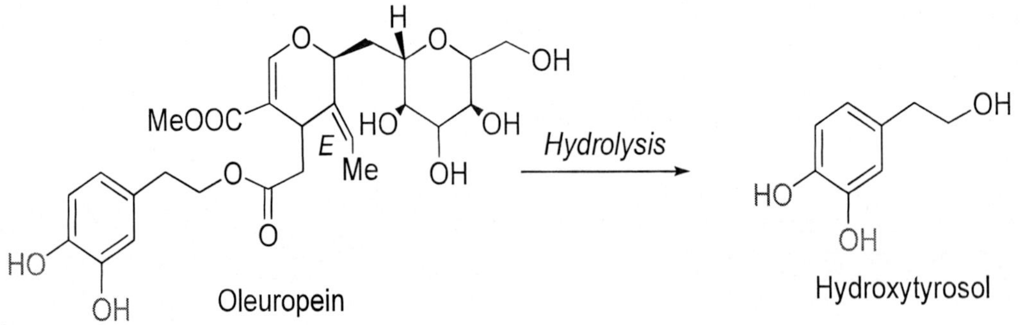

Figure 15. Europein and hydroxytyrosol structures.

The Mediterranean diet that contains a high consumption of olive oil might confer a potential protection against breast cancer. The same work indicates that the olive oil reverses acquired autoresistance to monoclonal antibody trastuzumab (Herceptin) in HER-2-overexpressing breast cancer cells and completely recovered the sensitivity. [76] Also was demonstrated that the incidence of other cancers including colon cancer is lower in Mediterranean countries compared with the North of Europe. [77] *In vitro* studies confirm that hydroxytyrosol is capable of inhibiting several stages in colon carcinogenesis (HT-29 cells line).

Scientific data suggests that the beneficial effects of the olive oil are linked to the antioxidant phenolic components such as hydroxytyrosol. *In vitro* the hydroxytyrosol showed HepG2 cells cytotoxicity and similar to other natural polyphenols demonstrate prevention of LDL oxidation. [78]

The hydroxytyrosol from olive oil caused growth arrest and apoptosis in human colon carcinoma HT-29 cells *in vitro*. Their mechanism is an overexpression of the pro-apoptotic factor CHOP/CADD153 and activation of the Jun-NH"-terminal kinase/activator protein-1. A recent work with phosphatases inhibitors concludes that the hydroxytyrosol activated PP2A, which has a key initiating role of various pathways that lead to apoptosis in colon cancer cells. [79]

Hydroxytyrosol may exerts a protective activity against cancer by arresting the cell cycle G0/G1 phase and the possibility of inducing apoptosis in tumor cells. [80]

Ellagic Acid

Ellagic acid (EA), presents in the muscadine grapes, black raspberries, pomegranate, strawberries, cranberries, nuts blackberries and other foods, was an example of the phenolic compounds and possesses several biological activities. Ellagic acid is formed in several fruits from two molecules of gallic acid as an ellagitannins or as a glucoside (Figure 16). This compound shows low biodisponibility and different this one depending on if it is presented in pure or combined form. In general, the ellagic acid joint molecules of sugar in their natural form. Ellagic acid shows antiangiogenic properties and the capacity of restraining tumour growth. The inhibition of the angiogenesis process blocks the formation of new blood vessels and reduces the expansion of tumors. Several authors announce that the ellagic acid cause also selectively apoptosis (natural cell death) in cancer cells. [81] Of another side, prevent the binding of chemical carcinogens including aromatic hydrocarbons, *N*-nitrosamines and aromatic amines to DNA and protect the p53 gene from free-radical damage. Moreover, ellagic acid has been found to promote the mental and cardiovascular health, and favored the vitality and the longevity. [82] Biological studies suggest that a possible mechanism by which ellagic acid arrests the growth of cancer cells is by forming adducts with DNA. A work over the study of enzymatic inhibition of ellagic acid shows a potent inhibition of DNA topoisomerases *in vitro*. [83]

A first study of the pomegranate juice, containing polyphenols like ellagic acid, in patients with recurrent prostate cancer shows statically significant effects on PSADT (pretreatment prostate-specific antigen), with antiproliferative and proapoptotic properties. A reduced risk of cancer is associated with the intake of polyphenol-rich diet that includes vegetables and fruits. The pomegranate (Punica *granatum* L.) fruit has been used in the old cultures for its medicinal purposes. This fruit contents different polyphenols, including ellagic acid in free and bound forms (as ellagitannins and ellagic acid glycosides), gallotannins, anthocyanins and other flavonoids such as quercetin. Ellagic acid and quercetin have been showed *in vitro* and *in vivo* anticarcinogenic properties. These antioxidant components decrease proliferation cellular and induce apoptosis via inhibition of NF-kB activity. [84]

Besides positive effects against the cancer, these compounds can allow to design strategies of prevention that it should benefit to the whole population in general.

Another compound strictly related to the gallic acid is the tannic acid, formed by the combination of gallic acid and glucose linked by esters bonds (Figure 17). Tannic acid is naturally occurring polyphenol present in several vegetables and fruits that has numerous pharmacological applications in human health such as antioxidant, radical scavenging and antitumor effects. Tannic acid presents a potent inhibition of epidermal growth factor receptor tyrosine kinase (EGFr tyrosine kinase), this inhibition was competitive with respect to ATP and non-competitive with respect to peptide substrate. [85]

Nepka and col. have been studied the chemopreventive effect of tannic acid in low-dose for comparing with the tannic acid intake in the diet, and the results confirmed the chemopreventive activity of tannic acid in hepatoma development. [86]

Ellagic acid, tannic acid and caffeic acid are well considered promising antitumor agents. Caffeic acid was the most effective inhibitor of the tumor promotion, when these polyphenols were investigated in the promotional phase of carcinogenesis. [87]

Topical application of tannic acid inhibits the cutaneous carcinogenesis and the immunosuppressant induced by UV-B irradiation. [88]

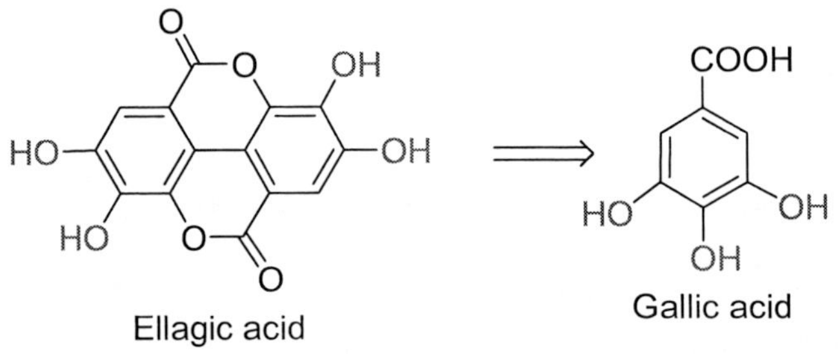

Figure 16. Ellagic acid structure.

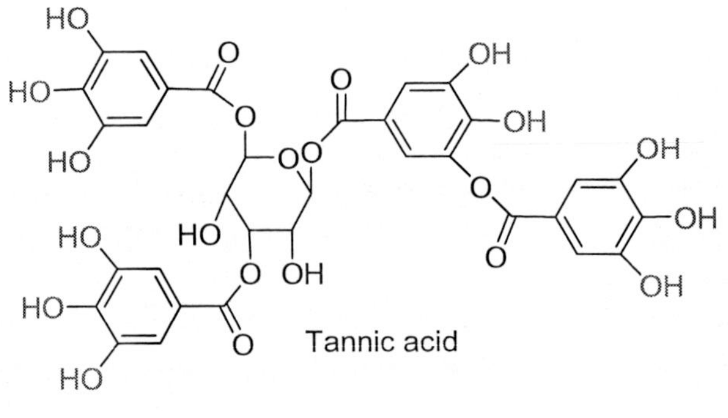

Figure 17. Tannic acid structure.

APIGENIN

Apigenin (4',5,7-trihydroxyflavone) (Figure 18), a natural flavonoid found in a variety of fruits and vegetables, has been shown to possess multi-biological functions especially related to the anti-cancer activity and promote cell cycle arrest and apoptosis in several cancer cell lines. Also was implicated as chemopreventive against prostate and breast cancers. Apigenin and genistein, both flavonoids have been shown to possess cancer-protective and antiproliferative effects on prostate (DU-145) and breast (MDA-MB-231) cancer cells expressing only estrogen receptor. It was concretely found that these compounds present activity through caspase-3-activation. [89]

Apigenin was also showed anti-hepatoma activity on 3 selected human hepatoma cells, such as Hep-G2, Hep-3B, and PLC/PRF/5 cells. These activity was similarly in potency to the 5-fluorouracil against Hep-G2 cells, showing IC50 value of 8.02 ±1.30 μg/mL.90 After several studies for investigate the cellular mechanism of action, it was found that apigenin induce apoptosis in Hep G2 cells. Moreover, the accumulation of p53 was increased and also the induction of p21 expression was implicated in the apoptosis process.

In vivo study using oral intake of apigenin against prostate cancer showed increase in the protein expression of several factors such as WAF1/p21, KIP1/p27, INK4a/p16; down-modulation of the protein expression of cyclins D1, D2 and E and also resulted in induction of apoptosis which was correlated with serum and tumor apigenin levels. [91]

Recently, apigenin inhibits growth of pancreatic cancer cells through suppression of cyclin B-associated cdc2 activity and arrests the cellular cycle in a G2/M phase and may be of interest for the prevention or treatment of pancreatic cancer. [92]

The flavonoid apigenin has been implicated as a modulator of MDR in colon HCT-15 cancer cells. [93]

Chan and col. [94] have been prepared a series of apigenin derivatives possessing a poly(ethylene glycol) such as substituent a the C-4' position and apigenin dimers linked with poly(ethylene glycol) spacers of different lengths (Figure 19). The dimmer with a spacer of four ethylene glycol units exhibited the most dramatic reversal activity. Increased drug retention in resistant cancer cells and enhanced chemosensitivity of several antitumor agents like paclitaxel, doxorubicin, vincristine, and others in breast cancer and leukemia cells *in vitro*, resulting in reduction of IC50 by 5-50 times.

Apigenin has been inhibited the proliferation and invasion of breast cancer cell MCF-7 and MDA-MB-435S. [95]

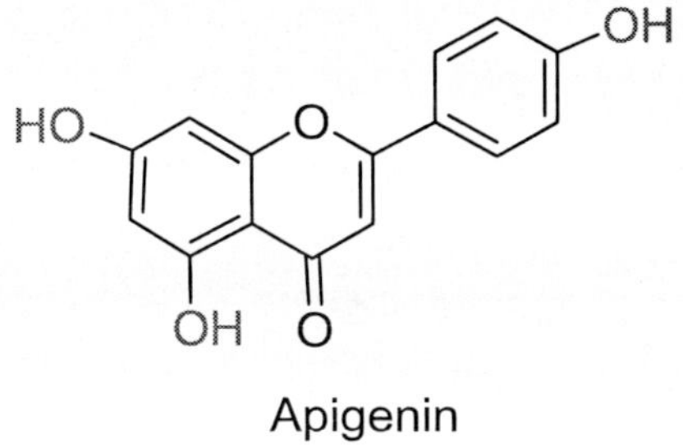

Figure 18. Apigenin structure.

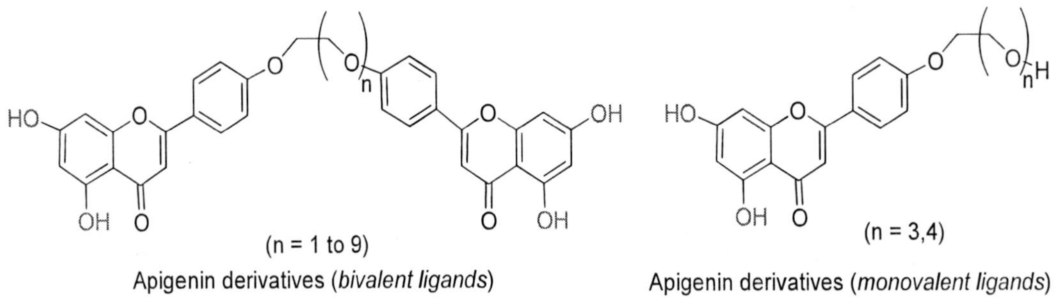

Figure 19. Apigenin derivatives.

The cosmociin, cosmoside, apigetrin, cosmetin, cosmosiine, cosmosioside or apigenin-7-O-glucoside (7-(β-D-glucopyranosyloxy)-5-hydroxy-2-(4-hydroxyphenyl)-4*H*-1-benzopyran-4-one) was extracted from Chamomille (Matricaria *chamomilla*) or S. swainsonii subspecies *swainsonii* between other plants. [96]

The anticancer properties of aqueous and methanolic extracts of chamomile against various human cancer cell lines was evaluated, and was observed apoptosis in cancer cells lines while the normal cells was not affected . The major component of the chamomile extracts was the apigenin-7-*O*-glucoside. The same studies indicate that the apigenin glucoside was less cytotoxic than the corresponding aglycon, apigenin. *In vivo* tests suggest that the deconjugation of glycosides occurs *in vivo* in the small intestine. [97]

Cosmociin and Narciclasine

In the most frequently occurring compounds in the CHMIS-C (Comprehensive herbal medicine information system for cancer) amongst the most potent anticancer agents that have a GI50 value < 1 mM was found Cosmociin (Figue 20) NSC = 407303 (76 nM) [98] and Narciclasine (Figue 21) NSC = 266535 (45 nM). [98]

The narciclasine or lycoricidinol (3,4,4a,5-tetrahydro-2,3,4,7-tetrahydroxy-(2*S*,3*R*,4*S*,4a*R*)-[1,3]dioxolo[4,5-*j*]phenanthridin-6(2*H*)-one) isolated from the bulbs of Hymenocallis *littoralis* and other plants of the amaryllidaceous. [99] Recently, Kiss, R. and col. reported that the narciclasine induces marked apoptosis-mediated cytotoxic effects in human cancer cells of MCF-7 breast and PC-3 prostate carcinoma cells. [100] Narciclasine was considered highly selective for the cancer cells and their sensitivity to normal human fibroblasts was very low probably due to the absence of death pathway activation.

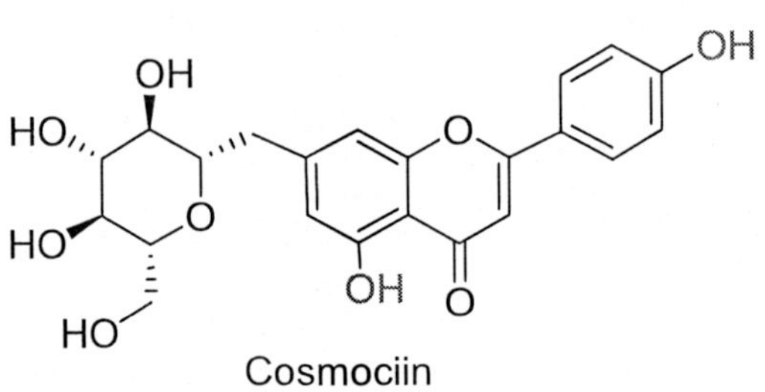

Figure 20. Cosmociin structure.

Narciclasine NSC = 266535 (45 nM). [98]

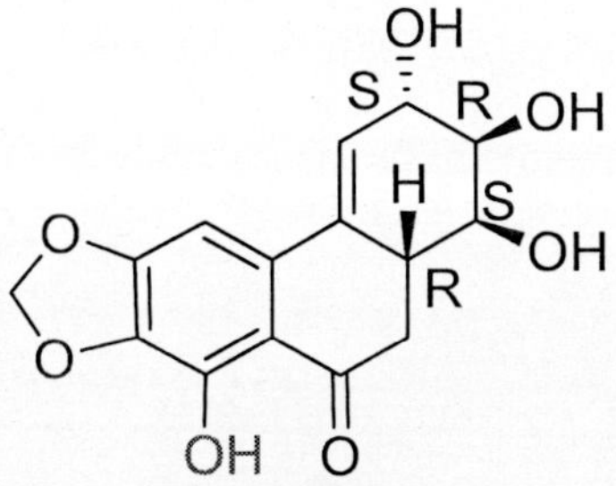

Figure 21. Narciclasine structure.

Polyphenols are bioactive compounds found in foods or dietary supplements which may be beneficial to health. Several studies [101-105] show a relation between intake of polyphenols and reduced risk of certain cancers. Thus cancer risk was lower in people who ate larger amounts of polyphenols compared with those who intake lower amounts. Polyphenols are substances that may protect cells from the damage caused by free-radicals. In general, polyphenols have the chemical capacity of interact and stabilize free-radicals.

REFERENCES

[1] Harborne, J. B. *Prog. Clin. Biol. Res.* 1986, *213*, 15-24.

[2] Tapiero, H.; Tew, K. D.; Ba, G. N.; Mathé, G. *Biomed. Pharmacother.* 2002, *56*, 200-207.

[3] Yang, C. Y.; Landau, J. M.; Huang, M. T.; Newmark, H. L. *Annu. Rev. Nutr.* 2001, *21*, 381-406.

[4] American Cancer Society, website: http://www.cancer.org.

[5] Maestri, D M.; Nepote, V; Lamarque, A. L.; Zygadlo, J. A. *Phytochemistry* 2006, 15-35.

[6] Joseph, J. A.; Shukitt-Hale, B.; Lau, F. C. *Ann. N. Y. Acad. Sci.* 2007, 470-485.

[7] Beecher, G. R. *J. Nutr.* 2003, 3248S-3254S.

[8] Anonymous. USDA database for the flavonoid content of selected foods 2003. www.nal.usda.gov/fnic/foodcomp/Data/Flav/flav.html

[9] http://www.ars.usda.gov/nutrientdata or contact the Nutrient Data Laboratory, 10300Baltimore Avenue, Bldg. 005, Rm. 107, BARC-WEST, Beltsville, MD 20705. Tel. No. 301-504-0630, e-mail: ndlinfo@ba.ars.usda.gov.

[10] Grippo, A. A.; Capps, K.; Rougeau, B.; Gurley, B. J. *An. Pharmacother.* 2007, *41*, 1375-1381.

[11] Birt, D.; Hendrich, S.; Wang, W. *Pharmacol. Ther.* 2001, *90*, 157-177.

[12] Gupta, S.; Afaq, F.; Mukhtar, H. *Oncogene* 2002, *21*, 3727-3738.

[13] Kuo, S. M.; Morehouse, H. F.; Lin, C. P. *Cancer Lett.* 1997, *116*, 131-137.

[14] Lin, J.; Zhang, S. M.; Wu, K.; Willett, C. W.; Fuchs, C. S.; Giovannucci, E. *Am. J. Epidemiol.* 2006, *164*, 644-651.

[15] Yang, C.; Landau, J.; Huang, M.; Newmark, H. *Annu. Rev. Nutr.* 2001, *21*, 381-397.
[16] Shen, S.; Chen, Y.; Hsu, F.; Lee, W. *J. Cell. Biochem.* 2003, *89*, 1044-1055.
[17] Singhal, R.; Yeh, Y. A.; Praja, N.; Olah, E.; Sledge, G. J.; Weber, G. *Biochem. Biophys. Res. Commun.* 1995, *208*, 425-431.
[18] Nguyen, T.; Tran, E.; Nguyen, T.; Do, P.; Huynh, T.; Huynh, H. *Carcinogenesis* 2004, *25*, 647-659.
[19] Chi, Y. S.; Jong, H. G.; Son, K. H.; Chang, H, W. ; Kang, S. S.; Kim, H. P. *Biochem. Pharmacol.* 2001, *62*, 1185-1191.
[20] Ong, C.; Tran, E.; Nguyen, T.; Ong, C.; Lee, J.; Ng, C.; Leong, C.; Huynh, H. *Oncol. Rep.* 2004, *11*, 727-733.
[21] van Erk, M.; Roepman, P.; van der Lende, T.; Stierum, R.; Aarts, J.; van Bladeren, P.; van Ommen, B. *Eur. J. Nutr.* 2005, *44*, 143-156.
[22] Wang, I.; Lin-Shiau, S.; Lin, J. *Eur. J. Cancer* 1999, *35*, 1517-1525.
[23] Ramos, S.; Alia, M.; Bravo, L.; Goya, L. *J. Agric. Food Chem.* 2005, *53*, 1271-1280.
[24] Shi, M.; Wang, F.; Wu, Z. *World J. Gastroenterol.* 2003, *9*, 73-78.
[25] Csokay, B.; Prajda, N.; Weber, G.; Olah, E. *Life Sci.* 1997, *60*, 2157-2163.
[26] Granados-Serrano, A. B.; Martín, M. A.; Bravo, L.; Goya, L.; Ramos, S. *J. Nutr.* 2006, *136*, 2715-2721.
[27] Image of: Matsuno,H.; Tanaka, Y.; Aoshima, H.; Doi, A.; Matsui, M.; Miyano, S. *In Silico Biology* 2003, *3*, 32.Ferreira, C. G.; Epping, M.; Kruyt, F. A.; Giaccone, G. *Clin. Cancer Res.* 2002, *8*, 2024-2034.
[29] Henry-Mowatt, J.; Dive, c.; Martinou, J. C.; James, D. *Oncogene* 2004, *23*, 2850-2860.
[30] a) Yang, J. H.; Hsia, T. C.; Kuo, H. M.; Lee Chao, P. D.; Chou. C. C.; Wei, Y. H.; Chung, J. G. *Drug Metabolism Disposition* 2006, *34*, 296-304. b) Choi, J. A.; Kim, J. Y.; Lee, J. Y.; Kang, C. M.; Kwon, H. J.; Yoo, Y. D.; Kim, T. W.; Lee, Y. S.; Lee, S. J. *Int. J. Oncol.* 2001, *19*, 837-844.
[31] Crozier, A.; Lean, M. E. J.; McDonald, M. S.; Black, C. *J. Agric. Food Chem.* 1997, *45*, 590-595.
[32] Ewald, C.; Fjelkner-Modig, S.; Johansson, K.; Sjöholm, I.; Åkesson B. *Food Chem.* 1999, *64,* 231-235.
[33] Lugasi, A.; Hovari, J. *Acta Alimentaria 2000, 29,* 345-352.
[34] Justesen, U.; Knuthsen, P.; Leth, T. *J. Chromatogr. A* 1998, *799,* 101-110.
[35] Hertog, M. G. L., Hollman, P. C. H., and Katan, M. B. *J. Agric. Food Chem.* 1992, *40,* 2379-2383.
[36] Bilyk, A.; Cooper, P. L.; Sapers, G. M. *J. Agric. Food Chem.* 1984, *32,* 274-276.
[37] Patil, B. S., Pike, L. M.; Hamilton, B. K. *New Phytol.* 1995, *130,* 340-355.
[38] Chu, Y-H.; Chang, C-L.; Hsu, H-F. *J. Sci. Food Agric.* 2000, *80,* 561-566.
[39] Häkkinen, S. H.; Kärenlampi, S. O.; Mykkänen, H. M.; Törrönen, A. R. *J. Agric. Food Chem.* 2000, *48,* 2960-2965.
[40] National Cancer Institute-Clinical Trials. Phase II. Randomized pilot study of silymarin in treating patients with leukemia who are receiving hepatotoxic chemotherapy. 13 july 2005. (http://www.clinicaltrials.gov/ct/show/NCT00055718).
[41] Flaig, T. W.; Su, L.-J.; Harrison, G.; Agarwal, R.; Glode, L. M. *Int. J. Cancer* 2007, *120*, 2028-2033.
[42] Kien, V.; Walterová, D. *Biomed Papers* 2005, *149*, 29-41.
[43] Saller, R.; Meier, R.; Brignoli, R. *Drugs* 2001, *61*, 2035-2063.

[44] Kim, N.; Graf, T. N.; Sparacino, C. M.; Wani, M. C.; Wall, M. E. *Org. Biomol. Chem.* 2003, *1*, 1684-1689.
[45] Kvasnicka, F.; Biba, B.; Sevcik, R. *J. Chromatogr.* 2003, *990*, 239-245.
[46] Smith, W. A.; Lauren, D. R.; Burgess, E. J.; Perry, N. B.; Martin, R. J. *Planta Med.* 2005, *71*, 877-880.
[47] Lee, D. Y. W.; Liu, Y. *J. Nat. Prod.* 2003, *66*, 1171-1174.
[48] Shibano, M.; Lin, A.-S.; Itokawa, H.; Lee, K.-H. *J. Nat. Prod.* 2007, *70*, 1424-1428.
[49] Anon: Natural Medicines Comprehensive Database, 4th ed. Therapeutic Research Faculty, Stockton, CA. 2002.
[50] Milde, J.; Elstner, E. F.; Grabmann, J. *Phytomedicine* 2004, *11*, 105-113.
[51] Wijayanegara, H.; Mose, J.C.; Achmad, L. *J. Int. Med. Res.* 1992, *20*, 54-60.
[52] De Jongste, A. B.; Jonker, J. J.; Huisman, M. V. *Thromb. Haemost.* 1989, *62*, 826-829.
[53] Bergqvist, D.; Hallbook, T.; Lindblad, B. *VASA*, 1981, *10*, 253-260.
[54] Mann, R.J. *Br. J. Clin. Pract.* 1981, *35*, 79-81.
[55] Al-Sereiti, M. R.; Abu-Amer, K. M.; Sen, P. *Indian J. Exp. Biol.* 1999, *37*, 124-130.
[56] Iuvone, T.; De Filippis, D.; Esposito,G.; D'Amico, A.; Izzo, A. A. *J. Pharmacol. Exp. Ther.* 2006, *317*, 1143-1149.
[57] Hur, Y-G.; Yun, Y.; Won, J. *J. Immunol.* 2004, *172*, 79-87.
[58] Salimei, P. S.; Marfe, G.; Di Renzo, L.; Di Stefano, C.; Giganti, M. G.; Filomeni, G.; Ciriolo, M. R. *Front. Biosc.* 2007, *12*, 1308-1317.
[59] Huang, S.-S.; Zheng, R.-L. *Cancer Lett.* 2006, *239*, 271-280.
[60] Hudson, E. A.; Dinh, P. A.; Simmonds, M. S. J. Gescher, A. *Cancer Epidem. Biomark. Prev.* 2000, *9*, 1163-1170.
[61] Hao, X.; Bose, M.; Lambert, M.; Joshua, D.; Ju, J.; Lu, G.; Lee, M.-J.; Park, S.; Husain, A.; Wang, S.; Sun, Y.; Yang, C. S. *Nutr. Cancer* 2007, *59*, 62-69.
[62] Yang, C. S.; Lambert, J. A. D.; Joshua, D.; Ju, J.; Lu, G.; Sang, S. *Toxicol. Applic. Pharmacol.* 2007, *224*, 265-273.
[63] Lim, Y. C.; Lee, S.-H.; Song, M. Y.; Yamaguchi, K.; Yoon, J.-H.; Choi, E. C.; Baek, S. J. *Eur. J. Cancer* 2006, *42*, 3260-3266.
[64] Ravindranath, M. H.; Saravanan, T. S.; Monteclaro, C. C. *Evid. Based Complement Alternat. Med.* 2006, *3*, 237-247.
[65] Siddiqui, I. A.; Saleem, M.; Adhami, V. M.; Asim, M.; Mukhtar, H. *Acta Pharmacol. Sin.* 2007, *28*, 1392-1408.
[66] Wang, H.; Yao, L.; Wang, G.; Li, Q.; Hu, L. *Dulixue Zazhi* 2006, *20*, 109-110 (AN 2007: 917338).
[67] Khan, N.; Mukhtar, H. *Life Sciences* 2007, *81*, 519-533.
[68] Shankar, S.; Ganapathy, S.; Srivastava, R. K. *Front. Biosc.* 2007, *12*, 4881-4899.
[69] a) Hertog, M. G. L., Hollman, P. C. H.; van de Putte, B. *J. Agric. Food Chem.* 1993, *41*, 1242-1246. b) Uniformed Services University of the Health Sciences (2007, December-9). Green Tea Shown To Possess Antitumor Effect In Breast Cancer. *ScienceDaily*. Retrieved January 28, 2008, from http://www.sciencedaily.com /releases/2007/12/071208171411.htm
[70] Li, X.; Tong, L.; Li, D. *Zhongguo Gonggong Weisheng* 2006, *22*, 327-328. (AN 2007: 681160).
[71] Lubet, R. A.; Yang, C. S.; Lee, M.-J.; Hara, Y.; Kapetanovic, I. M.; Crowell, J. A.; Steele, V. E.; Juliana, M. M.; Grubbs, C. J. *Mol. Cancer Therap.* 2007, *6*, 2022-2028.

[72] Chen, L.; Zhang, H. Y. *Molecules* 2007, *12*, 946–957.
[73] Guenther, S.; Ruhe, C.; Derikito, M. G.; Boese, G.; Sauer, H.; Wartenberg, M. *Cancer Lett.* 2007, *250*, 25-35.
[74] Manna, C.; Galletti, P.; Cucciolla, V.; Moltedo, O.; Leone, A.; Zappia, V. *J. Nutr.* 1997, *127*, 286-292.
[75] Manna, C.; Della Ragione, F.; Cucciolla, V.; Borriello, A.; D'Angelo, S.; Galletti, P.; Zappia, V. *Adv. Exp. Med. Biol.* 1999, *472*, 115-130.
[76] Menéndez, J. A.; Vázquez-Martín, A.; Colomer, R.; Brunet, J.; Carrasco-Pancorbo, A.; García-Villalba, R.; Fernández-Gutiérrez, A.; Segura-Carretero, A. *BMC Cancer* 2007, 7, 80-99.
[77] Gill; C. I. R.; Boyd, A.; McDermott, E.; McCann, M.; Servili, M.; Selvaggini, R.; Taticchi, A.; Esposto, S.; Montedoro, G.; McGlynn, H.; Rowland, I. *Int. J. Cancer* 2005, *117*, 1-7.
[78] Goya, L.; Mateos, R.; Bravo, L. *Eur. J. Nutr.* 2007, *46*, 70-78.
[79] Guichard, C.; Pedruzzi, E.; Fay, M.; Marie, J.-C.; Braut-Boucher, F.; Daniel, F.; Grodet, A.; Gougerot-Pocidalo, M.-A.; Chastre, E.; Kotelevets, L.; Lizard, G.; Vandewalle, A.; Driss, F.; Ogier-Denis, E. *Carcinogenesis* 2006, *27*, 1812-1827.
[80] Fabiani, R.; De Bartolomeo, A.; Rosignoli, P.; Servili, M.; Montedoro, G. F.; Morozzi, G. *Eur. J. Cancer Prev.* 2002, *11*, 351-358.
[81] Constantinou, A.; Stoner, G. D.; Young-Han, M. R. G.; Runyan, C.; Moon, R. *Nutr. Cancer* 1995, *23*, 121-130.
[82] Wang, B. H.; Lu, Z. X.; Polya, G. M. *Planta Med.* 1998, *64*, 195-199.
[83] Narayanan, B. A.; Geoffroy, O.; Willingham, M. C.; Re, G. G.; Nixon, D. W. *Cancer Lett.* 1999, *136*, 215-221.
[84] Pantuck, A. J.; Leppert, J. T.; Zomorodian, N.; Aronson, W.; Hong, J.; Barnard, R. J.; Seeram, N.; Liker, H.; Wang, H.; Elashoff, R.; Heber, D.; Aviram, M.; Ignaro, L.; Belldegrun, A. *Clin. Cancer Res.* 2006, *12*, 4018-4026.
[85] Yang, E. B.; Wei, L.; Zhang, K.; Chen, Y. Z.; Chen, W. N. *J. Biochem.* 2006, *139*, 495-502.
[86] Nepka, C.; Sivridis, E.; Antonoglou, O.; Kortsaris, A.; Georgellis, A.; Taitzoglou, I.; Hytiroglou, P.; Papadimitriou, C.; Zintzaras, L.; Kouretas, D. *Cancer Lett.* 1999, *141*, 57-62.
[87] Kaul, A.; Khanduja, K. L. *Nutr. Cancer* 1998, *32*, 81-85.
[88] Gensler, H. L.; Gerrish, K. E.; Willians, T.; Rao, G.; Kittelson*, J. Nutr. Cancer* 1994, *22*, 121-130.
[89] Mak, P.; Leung, Y.-K.; Tang, W.-Y.; Harwood, C.; Ho, S. -M. *Neoplasia* 2006, *8*, 896-904.
[90] Chiang, L.-C.; Ng, L. T.; Lin, I. C.; Kuo, P. L.; Lin, C.-C. *Cancer Lett.* 2006, *237*, 207-214.
[91] Shukla, S.; Gupta, S. *Mol. Cancer Therapeut.* 2006, *5*, 843-852.
[92] Ujiki, M. B.; Ding, X. Z.; Salabat, M. R.; Bentrem, D. J.; Golkar, L.; Milam, B.; Talamonti, M. S.; Bell, R. H.; Iwamura, T.; Adrian, T. E. *Molecular Cancer* 2006, 576-582.
[93] Di Pietro, A.; Conseil, G.; Pérez-Victoria, J. M.; Dayan, G.; Baubichon-Cortay, H.; Trompier, D.; Steinfels, E.; Jault, J. M.; de Wet, H.; Maitrejean, M.; Comte, G.;

Boumendjel, A.; Mariotte, A. M.; Dumontet, C.; McIntosh, D. B.; Goffeau, A.; Castanys, S.; Gamarro, F.; Barron, D. *Cell. Mol. Sci.* 2002, *59*, 307-322.
[94] Chan, K.-F.; Zhao, Y.; Burkett, B. A.; Wong, I. L. K.; Chow, L. M. C.; Chan, T. H. *J. Med. Chem.* 2006, *49*, 6742-6759.
[95] Sheng, X.; Qing, S.; Ji, X.; Ba, M.; Li, L.; Nanfang, H. *Zhongua Shiyan Waike Zazhi* 2006, *23*, 629-634.
[96] Skaltsa, H.; Georgakopoulos, P.; Lazari, D.; Karioti, A.; Heilmann, J.; Sticher, O.; Constantinidis, T. *Biochem. Syst. Ecol.* 2007, *35*, 317-320.
[97] Srivastava, J. K.; Gupta, S. *J. Agricult. Food Chem.* 2007, *55*, 9470-9478.
[98] Fang, X.; Shao, L.; Zhang, H.; Wang, S. *J. Med. Chem.* 2005, *48*, 1481-1488.
[99] Pettit, G. R.; Melody, N. *J. Nat. Prod.* 2005, *68*, 207-211.
[100] Dumont, P.; Ingrassia, L.; Rouzeau, S.; Ribaucour, F.; Thomas, S.; Roland, I.; Darro, F.; Lefranc, F.; Kiss, R. *Neoplasia* 2007, *9*, 766-776.
[101] Steinmetz, K. A.; Kushi, L. H.; Bostick, R. M., Folsom, A. R.; Potter, J. D. *American Journal of Epidemiology* 1994, *139*, 1-15.
[102] Setiawan, V. W.; Yu, G. P.; Lu, Q. Y. *Asian Pacific Journal of Cancer Prevention* 2005, *6*, 387-395.
[103] Hsing, A. W.; Chokkalingam, A. P.; Gao, Y. T. *Journal of the National Cancer Institute* 2002, *94*, 1648-1651.
[104] Chan, J. M.; Wang, F.; Holly, E. A. *Cancer Epidemiology Biomarkers and Prevention* 2005, *14*, 2093-2097.
[105] Challier, B.; Perarnau, J. M.; Viel, J. F. *European Journal of Epidemiology* 1998, *14*, 737-747.

In: New Developments in Medicinal Chemistry
Editors: Marta P. Ortega and Irene C. Gil
ISBN 978-1-60456-810-3

Chapter 3

CYTOTOXIC ANTICANCER DRUGS FROM MEDICINAL PLANTS*

Anh-Tho Nguyen* and Pierre Duez
Laboratory of Pharmacognosy, Bromatology and Human Nutrition,
Institute of Pharmacy CP 205-9,
Université Libre de Bruxelles, B-1050 Brussels, Belgium

ABSTRACT

Although there have been large improvements in cancer treatment over the last twenty years, the lack of cancer chemotherapeutic drugs is still a major cause of death in this century. Medicinal plants play an important role in the treatment of cancer by offering unique active drugs or their templates for clinical uses, as exemplified by paclitaxel (Taxol®), vinca alkaloids (vincristine, vinblastine), and flavopiridol. The strategies for developing anticancer agents from medicinal plants have changed in the last decade for a number of reasons, including advances in technology, changes in the plant selection mode and biological activity testing. This review reflects and discusses the newest methods applied for searching anticancer agents from medicinal plants including plant selection and extraction, the active principle isolation, structure elucidation and biological testing. This review reflects and discusses the newest methods applied for searching anticancer agents from medicinal plants including plant selection and extraction, the active principle isolation, structure elucidation and biological testing.

Keywords: *Medicinal Plants; Plant chemotherapeutic agents; Extraction; Isolation; Structure elucidation; Biological testing.*

* A version of this chapter was also published in *Phytochemistry Research Progress*, edited by Takumi Matsumoto published by Nova Science Publishers, Inc. It was submitted for appropriate modifications in an effort to encourage wider dissemination of research.

* Corresponding author. Email: anguyen@vub.ac.be; Tel.: +32-2-477 48 59; Fax: +32-2-477 48 55.

1. Role of Medicinal Plants in Cancer Treatment

Plants have been used in traditional medicine for thousands of years and continue to play an important role in heath care (Pezzuto, 1997). These plants, used in traditional medicine, are the so-called "medicinal plants". Today, about 80% of the world population live in developing countries and mostly rely on plant products for their primary health care. For the remaining 20% inhabitants, more than 25% of their pharmaceuticals have been directly derived from plant products. As recently summarized, about 120 drugs currently in use with a large number of therapeutic activities are obtained from plants. These include for example, steroids, cardiotonic glycosides (*Digitalis* glycosides), analgesics and antitussives (opium alkaloids), cholinergics (physostigmine, pilocarpine), antimalarials (*Cinchona* alkaloids), antigout (colchicine), anesthetic (cocaine), skeletal muscle relaxant (tubocurarine), and anticancer agents (see below) (Cragg et al., 1996; Pezzuto, 1997; Mans et al., 2000).

Plants have a long history of use for the treatment of cancer (Harwell, 1982; Graham et al., 2000; Mans et al., 2000) although many claims of efficacy of the treatment should be viewed with some skepticism; in fact cancer, as a specific disease entity, is likely to be poorly defined in terms of folklore and traditional medicine (Cragg et al., 1994). Medicinal plants contribute to the modern treatment of cancer by either providing the active substances or a template for synthesis or synthetic modification resulting in more effective anticancer agents.

2. Cytotoxic-Anticancer Agents Isolated from Medicinal Plants

Phytochemical investigation of plants, which have a history in traditional use to treat conditions consistent with cancer symptomatology, has indeed often resulted in the isolation of active principles with anticancer activity. The resins from the roots of *Podophyllum hexandrum* Royle and *Podophyllum peltatum* L. (Berberidaceae) have been used in India and China for more than 2000 years in the treatment of various cancers *e.g.* sarcoma, adenocarcinoma and melanoma (Hartwell, 1982; Dewick, 2002; Balachandran, 2005). Phytochemical investigation of these roots led to the isolation of lignans, podophyllotoxin (1), α and β-peltatin (2 and 3) (Figure 1) having antitumor properties (Hartwell, 1982; Dewick, 2002). Podophyllotoxin and peltatins were found to be unsuitable for clinical use due to toxic side-effects. However, the semi-synthetic derivatives etoposide (4) and teniposide (5) (Figure 1), which are still obtained from natural podophyllotoxin, have proved to be excellent anticancer drugs. Etoposide and the water-soluble etopophos (etoposide 4'-phosphate) (6) (Figure 1) are very effective anticancer agents used in the treatment of small cell lung cancer, testicular cancer and lymphomas, usually in combination with other anticancer drugs. Teniposide has similar anticancer properties but is not as widely used as etoposide. It is also used in paediatric neuroblastoma (Wang, 1998; Dewick, 2003). Podophyllotoxin was found to inhibit microtubule assembly (Desbene et al., 2002). Epotoside and teniposide, however, were identified as DNA topoisomerase II inhibitors (Hande, 1998).

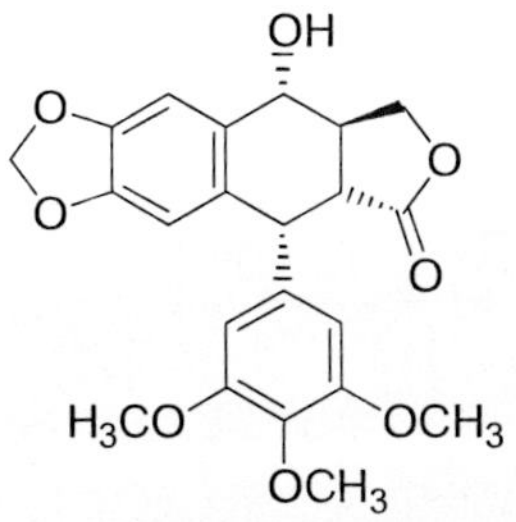

1 Podophyllotoxin

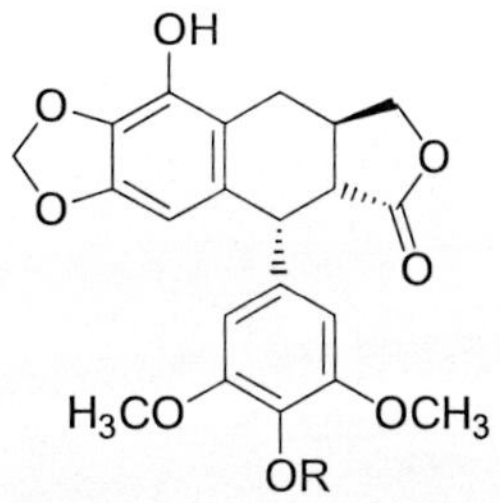

2 *β*-peltatin, R = CH_3
3 *α*-peltatin, R = H

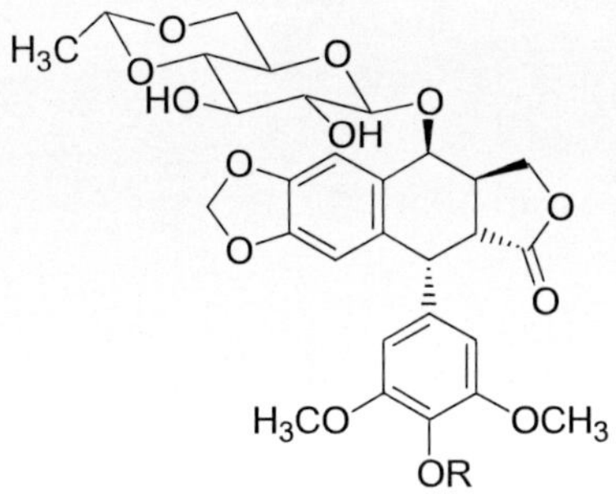

4 Etoposide, R = H
6 Etopophos, R = P

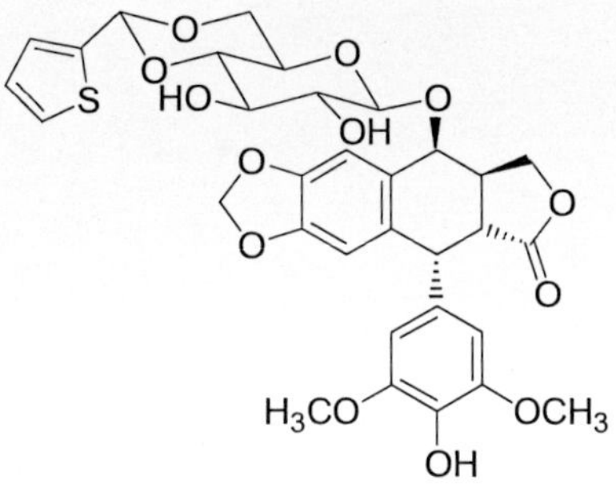

5 Teniposide

Figure 1. Structures of lignans.

Since about AD 180, *Solanum dulcamara* L. (Solanaceae) has been used to treat cancers in China and Chile and its use has been described for many centuries (Hartwell, 1982). The active tumor-inhibitory principle from this plant has been identified as a steroidal alkaloid glycoside *β*-solamarine (7) (Figure 2). Various lichens *e.g. Cetraria* and *Usnea* species also have a long historical use for cancer treatment in Chile, in India and in Argentina (Hartwell, 1982). They are all rich in usnic acid (8) (Figure 2), the compound is recognized as a tumor inhibitor (Dewick, 2002).

The most successful study of higher plant used for cancer treatment led to the discovery of an effective anticancer drug complex diterpene, paclitaxel (9) (Taxol®) (Figure 3) from *Taxus* species *e.g. Taxus brevifolia* Nutt. and *Taxus baccata* L. (Taxaceae). The barks of *Taxus brevifolia*, indeed, have been used for many years in America to treat skin cancer (Gramham, 2000); the leaves of *Taxus baccata* have also been used in India to treat what was believed to be cancer (Cragg et al., 2002; Newman et al., 2000; Hertwell, 1982). Taxol® was first isolated and identified by Wani and coworkers from the toxic plant, *Taxus brevifolia* in 1971.

It is an important new anticancer agent, with a broad spectrum of activity against some cancers (breast, ovarian, non-small-cell lung, small-cell lung, head and neck) which do not respond to other agents. The discovery of taxol has spurred the isolation of many additional taxonoids and by this time more than 400 analogs have been discovered (Cragg et al., 1996; Mukherjee et al., 2001; Dewick, 2003). Taxol was isolated in a very small yield (about 0.02%) from the bark of *Taxus brevifolia*. One centenary large tree, therefore, provides only one treatment for one cancer patient.

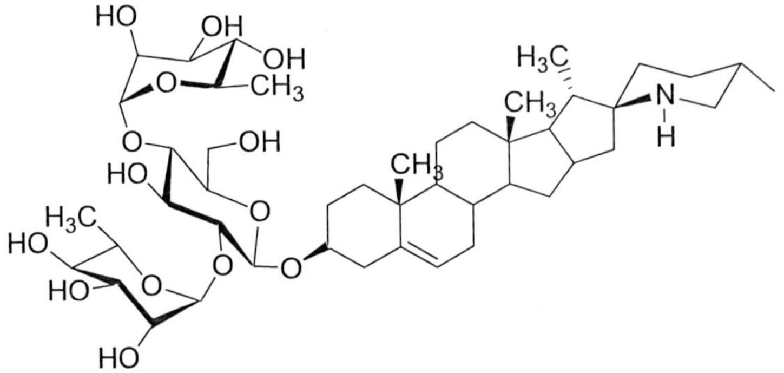

7 *β*-solamarine

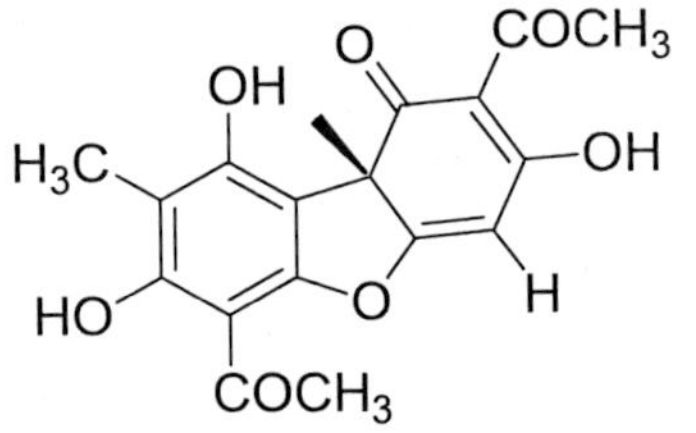

8 Usnic acid

Figure 2. Structures of *β*-solamarine and usnic acid.

To solve this problem, taxol could be semi-synthesized from baccatin III (10) (Figure 3) or from 10-deacetyl baccatin III (11) (Figure 3), which are readily extracted from the leaves and the twigs of *Taxus baccata*. From these precursors, not only taxol but also taxotere (Docetaxel® (12)) (Figure 3) could be obtained. Up to 2% taxol with regard to the dried plant material (*Taxus brevifolia, Taxus baccata*) could be obtained by semi-synthesis which made the treatment available to many more patients (Dewick, 2003). Taxol and docetaxel both exhibit a unique mechanism of action. They promote polymerization of tubulin and stabilize the structure of intracellular microtubules. This process effectively inhibits the normal dynamic reorganization of the microtubules that is necessary for interphase and mitotic functions (Attard, 2005).

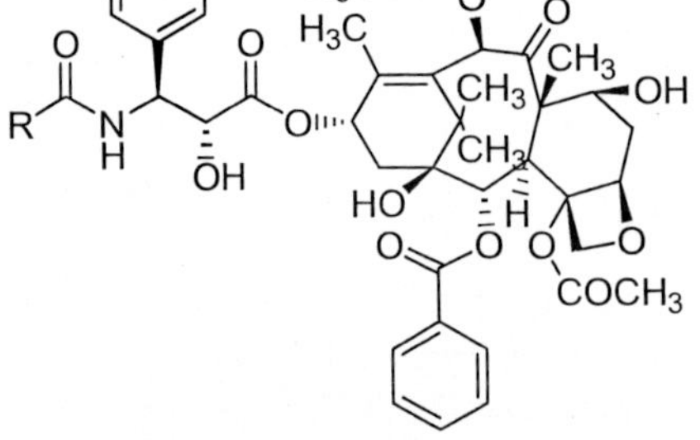

9 Taxol, R = C_6H_5
12 Docetaxel, R = $COC(CH_3)_3$

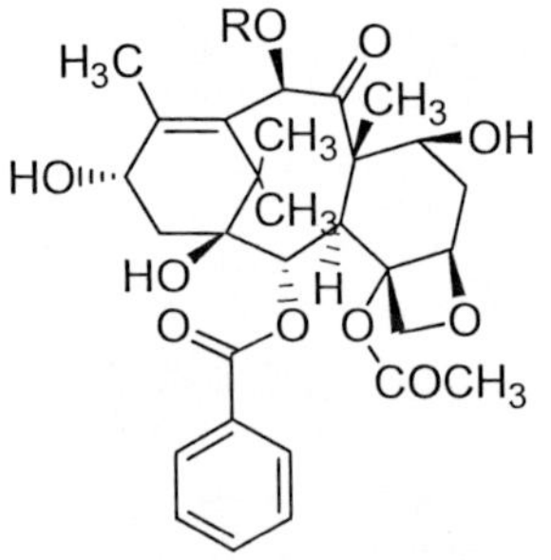

10 Baccatin III, R = $COCH_3$
11 10-Deacetylbaccatin III, R = H

Figure 3. Taxol derivatives.

By contract, the most well-known *Catharanthus* alkaloids, vinblastine (13) and vincristine (14) (Figure 4) were isolated from *Catharanthus roseus* (L.) G. Don. (Apocynaceae), a plant known in folk medicine, not as a cure for cancer, but in the treatment of hematuria, diabetes mellitus and menstrual disorders (Dewick, 2002; IMM, 1990). Vinblastine and vincristine were first discovered in 1960 during the investigation of the plant for hypoglycaemic agents, but no hypoglycaemic activity was detected. These alkaloids showed excellent antileukaemic properties and are now extracted commercially from the genus *Catharanthus*.

Although only slightly different in the structure, vincristine is clinically more important than vinblastine and is especially useful in the treatment of childhood leukemia. Structure modifications of vinblastine afforded two new drugs, vindesine (15) and vinfosiltine (16) (Figure 4).

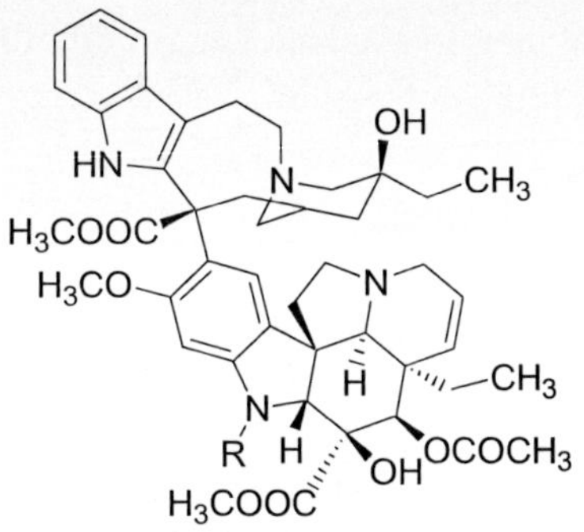

13 Vinblastine, R = CH_3
14 Vincristine, R = CHO

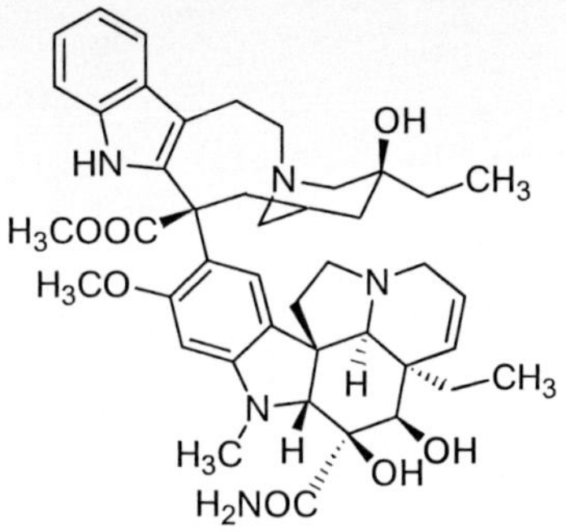

15 Vindesine

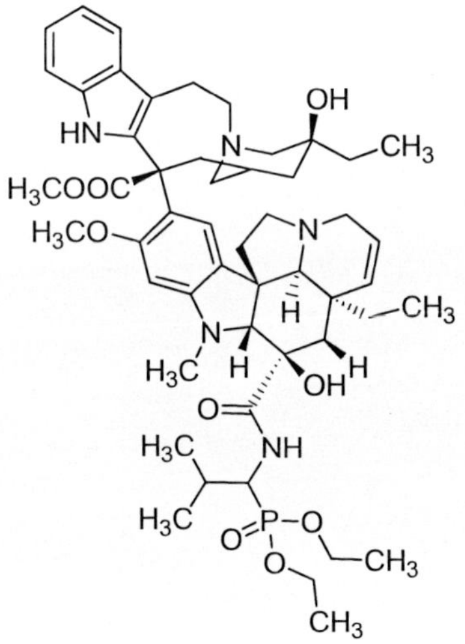

16 Vinfosiltine

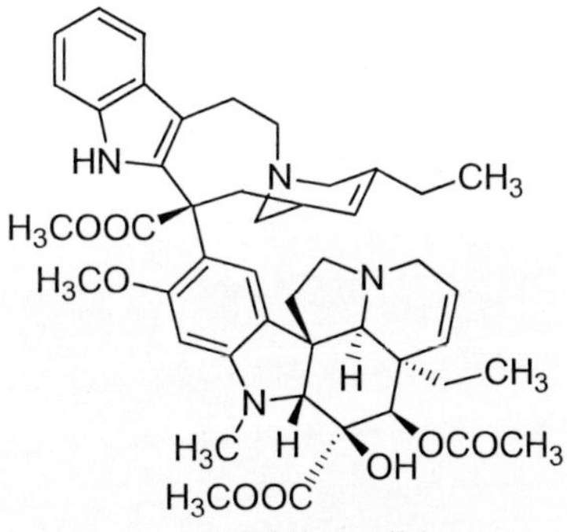

17 Vinorelbine

Figure 4. Structures of Catharanthus alkaloids.

While vindesine is used to treat acute lymphoid leukaemia in children, hepatocellular cancer, and non-small cell lung cancer, vinfosiltine is useful for the treatment of advanced malignant melanoma and advanced breast cancer (Dewick, 2002; Wang, 1998). A more recent hemi-synthesis from vinblastine yielded vinorelbine (17) (Figure 4), which has broader anticancer activity and lower side-effects than vinblastine. Vinorelbine gained approval in France for the treatment of non-small cell lung and breast cancers in 1989, and in 1995 it gained approval in the USA for the treatment of advanced non-small cell lung cancer (Wang, 1998). Vinblastine and vincristine are minor constituents from *Catharanthus* genus so the isolation process is a very tedious process. The chemical synthesis of these alkaloids, hence, has attracted considerable attention and some dimetric structures of Catharanthus alkaloids have been achieved (Dewick, 2002). *Catharanthus* alkaloids prevent microtubule assembly by binding to a specific binding site on tubulin. They represent a new class of naturally occurring oncolytic agents extensively used in the cancer chemotherapy (Attard, 2005). Flavopiridol (18) (Figure 5) is among the most exciting plant-based agent currently in clinical trial for cancer treatment. This original structure is totally synthetic; based on a template natural product rohitukine (19) (Figure 5).

The molecule was first isolated from the leaves and stems of Amoora rohituka Wight and Arn and later from Dysoxylum binectariferum (Roxb.) Hook. f. ex Bedd. (Maliaceae) (Harmon et al., 1979; Naik et al., 1988). Flavopiridol was found to have a low toxicity and to be effective in a variety of solid and haematological malignancies. It is useful for the treatment of colorectal, prostate, non-small cell lung, and renal cell carcinoma, as well as non-Hodgkin's lymphoma and chronic lymphocytic leukemia in combination with paclitaxel or cisplastin (Newman et al., 2000; Mans et al., 2000; Kim, 2004). Flavopiridol inhibits tumor growth by suppressing cyclin-dependent kinases (CDKs). It is now in Phase II clinical trial against a broad range of tumors (Flinn et al., 2005).

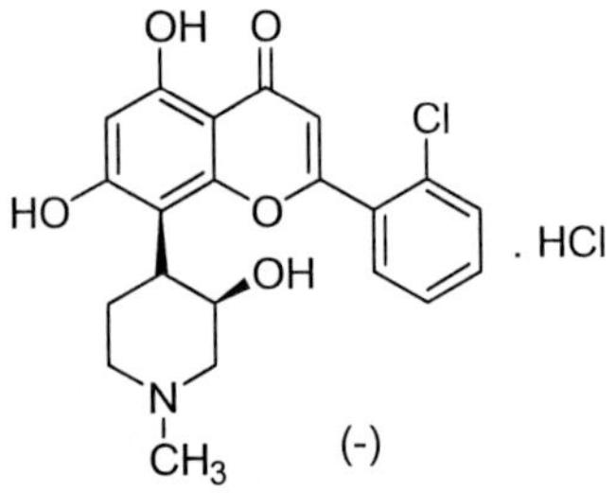

18 Flavopiridol

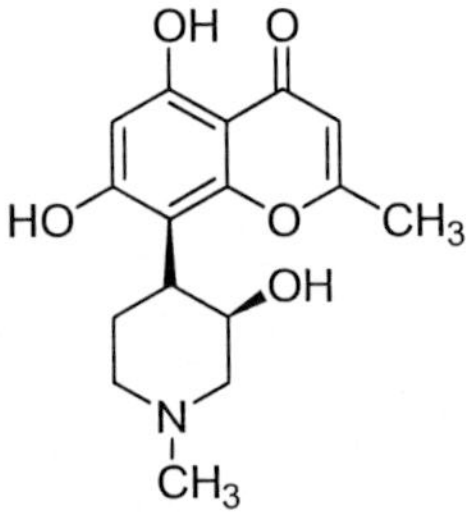

19 Rohitukine

Figure 5. Structure of flavone derivatives,

3. The Search for Cytotoxic Anticancer from Medicinal Plants

As already mentioned, medicinal plants have a great potential for the discovery of promising new anticancer drugs. On the earth, there are about 800,000 plant species in which the angiosperms represent from 300,000 to 500,000 species. Until now, however, less than 10% of the higher plants have received attention to detect anticancer constituents. This means that the majority of important plant cancer chemotherapeutic drugs still probably awaits discovery (Hamburger and Hostettmann, 1991; Pettit, 1994, Mukherjee et al., 2001). Methods of investigation for cytotoxic anticancer from medicinal plants are described hereby.

3.1. Plant Collection

Various methods may be employed for the discovery of plant-derived anticancer drugs. Scientists claiming to have interest in anticancer drug discovery have used four approaches for the selection of medicinal plants that may contain new biological agents *e.g.* random, taxonomic, phytochemical and ethnomedical approaches (Cordell et al., 1991; Kim et al., 2002).

Random plant collection will acquire biodiversity in order to broaden the chemodiversity of available structures. A random collection of 10,000 different plants is estimated to represent 50,000-100,000 different natural product structures (Cordell, 1995). In the more developed countries, many large-scale research institutions investigate randomly acquired plant by high-throughput screening methods. Biological assays are the fundamental frameworks of such discovery programs. Since 1955, the United States National Cancer Institute (NCI) has developed a random plant selection screening program. Considering that the novel compounds may be found anywhere in the plant kingdom, more than 114,000 plant extracts – seeds, leaves, roots, etc. - representing 12,000 to 13,000 species so far were tested for antitumor activity by NCI. Taxol, the major clinically active agents, have emerged from this program (Cragg et al., 1996; Mukherjee et al., 2001).

In the taxonomic approach, medicinal plants of a certain genus or interesting family are collected in different locations and then evaluated. The knowledge about the phytochemical composition of related plants will be a clue to study the presence of comparable structures with the hope that they may improve the chemotherapeutic index. By this way, the research is likely to identify new analogs rather than novel lead compounds.

In the phytochemical approach, particular types of compounds *e.g.* indole alkaloids, sesquiterpenes… are regarded as being of biological interest and plants likely to present related compounds are collected and evaluated. For the most part, this approach is hardly reliable and leads to low success rate. Serendipity may indeed come into play, and straightforward isolation procedures may lead to the discovery of extremely valuable therapeutic agents. Nonetheless, most drug discovery programs, including anticancer drug discovery program can hardly accept to await for a serendipitous occurrence.

In the ethnomedical approach, credence is given to folkloric claims of efficacy of the plants medicinally used, and, on the basis of this information, the plants are collected and evaluated.

For each collection, the most reasonable method of evaluating plant material is a range of bioassays. Active leads are submitted to bio-guided fractionation leading to the active compounds. Using this method, active principles contained in the crude plant samples can be identified in a reasonable time. The keystones for plant collection are *e.g.* the NAPRALERT database (Loub et al., 1985; Graham et al., 2000), the herbalist information, the traditional use of plants for the treatment of "cancer" symptoms or of parasitic diseases and the "toxic" plants (Farnsworth, 1994; Iwu, 2002).

3.2. Isolation Procedure

Three methods are usually applied for the isolation of the active principles from medicinal plants.

The classical method, yielding pharmacologically active, pure constituents from the plant, has always been a long and very tedious process, requiring manpower and financial resources. Combination of various chromatographic methods is applied and the knowledge of phytochemistry is highly requested for this process. When the pure compounds are obtained, they are then evaluated for their bioactivity and the lead compounds will be selected for further characterization.

The bioassay-guided fractionation is a much more efficient way which leads from the intact plant to its pure bioactive compounds (Figure 6). Following this procedure, crude plant extracts, which showed highly activity on a given bioassay (most often *in vitro*), are fractionated with the help of different chromatographic methods. The bioassay serves as a guide during the fractionation process until the pure compounds are obtained. Structure elucidation of the isolated compounds is then performed at the end of the isolation process and their biological activity can be further evaluated. Utilization of this method allows to shorten what however remains a long and tedious isolation procedure. It has been used for more than two decades and is much valued in many laboratories. Nevertheless, known natural products which are already tested for the bioactivity of interest may be unnecessarily isolated by this process and so the need arises for dereplication, a method that contributes to reduce this phenomenon.

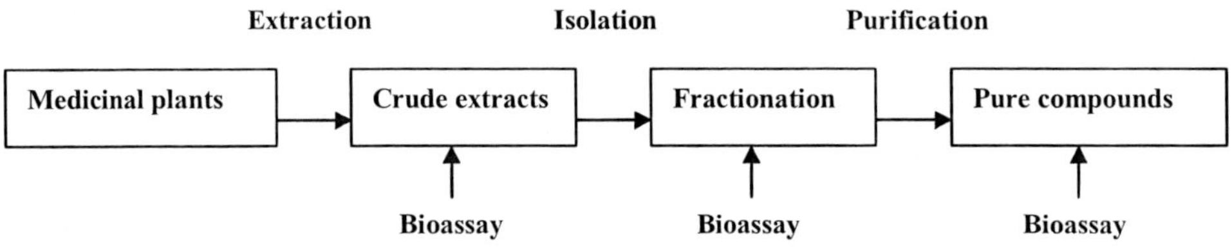

Figure 6. Procedure of bio-guided fractionation to obtain active principles from medicinal plants.

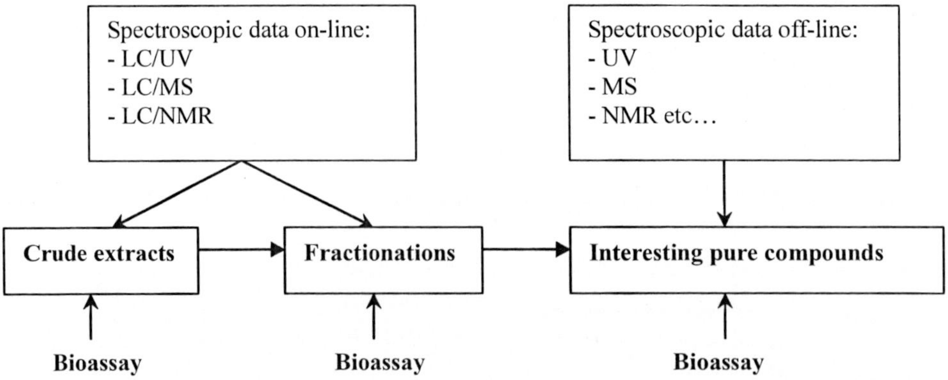

Figure 7. The use of LC-hyphenated techniques in combination with bio-guided fractionation for obtaining interesting active principles from medicinal plants.

An efficient way to perform dereplication is the combination of bio-guided fractionation with quasi-simultaneous structure elucidation of active principles in extracts or fractions (Figure 7) as recently proposed by Professor Hostettmann (Hostettmann et al, 2003). Whereas the classical and the bio-guided fractionation methods mainly use TLC and specific spray reagents to follow the chemical constituents of crude extracts, this method utilizes HPLC for separation of crude extracts or fraction constituents.

The introduction of hyphenated techniques related to high performance liquid chromatography in the past 20 years has provided powerful tools *e.g.* LC/UV-DAD, LC/MS, LC/NMR…

Combination of high separation efficacy on HPLC with the different detectors has make possible the acquisition of on-line complementary spectroscopic data on a LC peak of interest within a complex mixture. This type of approach has shortened the isolation process, given preliminary structure information and rapidly brought interesting results. Structures of isolated interesting compounds will then be classically examined by off-line spectroscopy *e.g.* UV, MS, NMR.

3.3. Structure Elucidation

Elucidating the structure of bioactive compounds from medicinal plants has always been and remains a field of interest to pharmacists, biologists or chemists. Most papers published in the principal journals of natural product chemistry present, in whole or in part, structural elucidation of analogs or new compounds. Due to the diversity of structures obtained from the isolation processes, their complete elucidation remains a considerable task despite of many analytical advances. Therefore numerous techniques *e.g.* UV, MS, 1- and 2-D NMR spectroscopy, optical rotation … have been employed to permit solving structures. What has changed nowadays results from technical evolution, software revolution, the automation level of instrumentation, the sophistication of the experiments to be performed, and the precise integration of data into coherent structures (Cordell, 1995; Vlietinck, 2000).

The first question is to define the general characteristics of the isolated compounds. The phytochemical isolation often leads to suspect in which chemical family the compounds isolated belong *e.g.* alkaloids, flavonoids or triterpens etc. This will much facilitate the structure elucidation process. However, in the frame of bio-guided fractionation or the use of LC-hyphenated techniques the exact type of compound isolated is often not known. In this context, information on the compounds isolated from the same plant, the same genus or the same family, *e.g.* spectroscopic data, molecular weight, solubility, polarity, and structures, will be very useful. There is no special recipe to elucidate structures of plant-derived anticancer compounds and most anticancer agents or candidates isolated from plants have very complex structures.

NMR spectroscopy is certainly the most powerful tool for structural investigations. When the pure compound is obtained, it is customary to record a proton spectrum. ^{1}H NMR spectrum indicates the number of protons and is much useful to predict aromatic-related and sugar protons. With the present availability of 2-D NMR experiments such as COSY, HMQC, HMBC, the spin-spin analysis on ^{1}H NMR spectrum has lost its previous importance. These more recent techniques give much information in terms of connectivities. While inter-proton connectivities are obtained with the COSY experiment, carbon and proton are linked in

HMQC and HMBC experiments. The HMQC indicates C-H connectivities over one bond and HMBC gives connectivities over two or three bonds. Due to low abundance of the ^{13}C isotope, the ^{13}C NMR spectra (broad-band, DEPT 135 and 90) are more difficult to obtain but are indeed extremely useful in structure elucidation. With regard to stereochemistry, NOESY experiment is frequently employed to detect H-H interactions. INADEQUATE, which indicates one bond C-C coupling, can be considered as a possible alternative to X-ray to unravel complicated carbon networks but it is not so often used because it is a "voracious" experiment which requires about a millimole to be performed (typically, 200 to 500 mg of pure compounds).

Of course, NMR spectroscopy is not the only physical science which has been used to resolve many structures. Mass spectroscopy now contributes a tremendous impact on the way in which natural product chemistry is conducted. With highly polar and non-volatile molecules *e.g.* polysaccharides, sugar esters, glycans…, obtaining a molecular ion has frequently been a challenge. Thanks to electrospray, fast atom bombardment and matrix-assisted laser desorption, these compounds are now accessible in mass spectroscopy. High resolution mass spectrometry tends to replace elemental analysis as far as composition is concerned. It is not, however, proof of sample purity. A main problem in mass spectrometry remains the obtention of a molecular ion but chemical ionization techniques have been developed since the last two decades. Using these methods, the molecular ion of labile compounds can be recorded. MS is a destructive technique but thankfully requires only minute amounts of compounds.

In spite of these advanced spectroscopic techniques, structure elucidation of plant-derived substances still requires time and experience. That is why computer-based structure elucidation has been developed in very recent years. Three major components of such programs *e.g.* spectrum interpretation, structure generation, and spectrum prediction are all very useful for structural investigations. Spectrum interpretation is the process by which spectral data are reduced to structural inferences. Structure generation generates all molecules compatible with these inferred structures. Structure prediction evaluates the relative probability of the different inferred structures to be correct. Examples of the latter systems include CHEMICS® utilizing ^{1}H, ^{13}C, 2-D NMR, IR and MS spectral properties and SESAMI® using 1-D and 2-D NMR spectra. Other structural elucidation systems *e.g.* ACCESS®, CSEARCH®, EPIOS® and SpecSolv® use a ^{13}C NMR interpretative library from assigned ^{13}C NMR databases to interpretate and generate defined structures (Vlietinck, 2000).

3.4. Biological Testing

In the search for drugs used in the treatment of cancer, we always meet the terms *e.g.* “cytotoxic compounds”, “tumor inhibitors” and “anticancer drugs” which are used to describe the activities of the isolated compounds. Cytotoxic drugs kill cells and when their activity can transfer to *in vivo*, they inhibit tumor aggression. Anticancer drugs are tumor inhibitors.

The most efficient and practical method for anticancer drugs discovery is bioactivity-guided fractionation. The plant extracts and substances demonstrating a positive response in bioassay systems are considered as “active leads”. When a number of active leads are identified, the most promising materials are selected for further fractionation. Each fraction is monitored in the bioassay system for its potential to mediate a positive response and this process continues until pure active compounds are obtained. When structures of active isolates are achieved, more biological evaluation procedures are performed and, on the basis of accumulated data, the compounds may be selected for advanced biological testing. Only candidates showing acceptable therapeutic index in animal experiments can enter clinical trials (Pezzuto, 1997).

3.4.1. In Vitro Cytotoxicity Tests

Although cytotoxicity is neither necessary nor sufficient for anticancer activity, it is an activity consistent with antitumor activity as it is sensitive to every mechanism required for cell survival. In conjunction with the results of other biological assays, the results thus will help in deciding which materials are subjected to the fractionation process. Cytotoxicity could also be considered as a primary screen for mechanism-based assay. For many years, the lead compounds for anticancer chemotherapy have been selected on the basis of cytotoxicity, frequently via *in vitro* tests (Cordell et al., 1991). A number of cytotoxic test systems has been developed with attempt to have results interrelated with *in vivo* efficacy studies. These include cytotoxicity assays with cell cultures (MTT, LDH, neutral red, ATP content assays), the potato disc assay, and the brine shrimp assay (Cordell, 1995; Weyermann et al., 2005).

3.4.1.1. Cell Culture

The cytotoxic assays with cell cultures usually require specialized facilities.

MTT assay measures the metabolic activity of viable cell mitochondria. Only metabolically active cells can reduce 3-(4,5-dimethylthiazol-2-yl)-2,5-diphenyltetrazolium bromide (MTT) to a blue violet formazan, which is quantified by spectrometry (Mosmann, 1983).

An other frequently used parameter for cell death is the integrity of the cell membrane, which can be measured by the cytoplasmic enzyme activity released by damaged cells. Lactate dehydrogenase (LDH) is a stable cytoplasmic enzyme present in all cells. When plasma membrane is damaged, LDH is rapidly released into the cell culture supernatant. It is determined by an enzymatic test (Korzeniewski and Callewaert, 1983). NAD^+ is reduced to $NADH/H^+$ by LDH-catalyzed conversion of lactate to pyruvate, then the catalyst (diaphorase) transfers H/H^+ from $NADH/H^+$ to the tetrazolium salt 2-(4-iodophenyl)-3-(4-nitrophenyl)-5-phenyltetraolium chloride (INT), which is reduced to a red formazan (Decker and Lohmann-Matthes, 1988; Nachlas et al., 1960).

Neutral red assay is based on the uptake and subsequent lysosomal accumulation of the supervital dye neutral red (3-amino-*m*-dimethylamino-2-methyl-phenazine hydrochloride). Quantification of the dye extracted from the cells has been shown to be linear with living cells number, both by direct cell counts and by protein determination of cell population (Borenfreund and Puerner, 1985).

ATP (adenosine triphosphate) present in all metabolically active cells can be determined by a bioluminescent measurement. This method uses an enzyme, luciferase, which catalyses the formation of light from ATP and luciferin. The emitted light intensity is linearly related to the ATP concentration and then to the number of living cells (Crouch et al., 1993).

3.4.1.2. Other Tests

The potato disc assay and the brine shrimp assay have been proposed by Professor J. L. McLaughlin's group at Purdue University as low cost substitutes for cytotoxicity assays with cell cultures (Hamburger and Hostettmann, 1991).

The potato disc assay (crown gall tumor inhibition) is a test system used as a pre-screen for antitumor activity. Crown gall is a neoplastic disease of plants induced by specific strains of *Agrobacterium tumorigenes* through the transfer of tumor-inducing plasmids from the bacterium into plant cells. For the assay, potato slices are inoculated in Petri dishes with *Agrobacterium tumorigenes* in a growth medium containing samples to be tested. After inoculation for 21 days, the number of crown galls is counted and percentage inhibition calculated by comparison with control (Ferrigni et al., 1982).

In contract with crown gall tumor assay, the brine shrimp assay does not require sterile conditions. Eggs of brine shrimp are readily available and larvae hatch rapidly when placed into artificial sea water. This assay is based on the premise that bioactive compounds are toxic at higher doses and that lethality in a simple organism might be used as a mean of monitoring activity-directed fractionation (Meyer et al., 1982). Inhibition of crown gall tumor and of brine shrimp growth was shown to correlate to *in vitro* cytotoxicity on leukemia cells (Anderson et al., 1991).

3.4.2. Mechanism-Based Assays

The activity of many drugs in use in cancer chemotherapy can be ascribed to inhibition of nucleic acid synthesis but their mechanisms of action largely differ. With increasing understanding of cell biology and molecular biology, determining the mechanism of action became very important. Mechanism-based assays are now used in the field of anticancer drug discovery, *e.g.* tests for detecting inhibitors of topoisomerase I and II, tubulin polymerization and inhibitors of protein kinase C.

Drugs inhibiting topoisomerases have recently received a great attention. Topoisomerases are enzymes involved in DNA replication by their ability to break and reseal the DNA strands, and are classified as type I or II according to their ability to cleave one or both strands of DNA, respectively (Mukherjee et al., 2001). The mechanism of action of tubulin-binding drugs has been extensively investigated. Soluble tubulin exists in the cell as heterodimers (α and β tubulins). During the polymerization, these heterodimers link together to form protofilaments; thirteen of these protofilaments organize in a hollow cylinder to make up the backbone of the microtube. Microtubes are important in the movement of organelles during interphase and during mitosis. They form the mitotic spindle that brings daughter chromosomes to separate at the poles of the dividing cell. Drugs which interfere with microtubule function lead to failure of alignment of the daughter chromosomes and the cell can not pass through the mitosis checkpoint. The mitosis is stopped and apoptosis follows (Attard et al., 2005).

Interfering with the transduction of signals from cell membrane receptors to the nucleus or to targets like the phospholipase C, phosphatidylinositol kinases (PIKs) and protein kinases C (PKCs) are potentially useful targets for anticancer therapeutic intervention. Flavopiridol that was presented earlier is a protein kinase inhibitor and its activity is strongest on cyclin dependent kinases (CDKs) (Sedlacek, 2001).

Many other mechanism-based assays are being developed which strongly increase the chances of finding original active compounds.

3.4.3. In Vivo Study

There is no substitute method in cases where entirety of living animal is required. From 1960-1982, all compounds and extracts entering the NCI drug discovery program were first tested for *in vivo* anticancer activity, mainly using the L1210 and P388 mouse leukemia models. However, the detected antitumor agents were effective on these fast-growing tumors often with very little activity against slow-growing tumors. In addition, the slow-growing tumors grow and divide more slowly than rapidly proliferating normal human tissues such as bone marrow and gastrointestinal epithelium. These tissues thus suffer toxic effect of the administered drugs (Dewick, 2002; Cragg, 1998). In an attempt to overcome this problem, the NCI developed an alternative *in vitro* method for primary screening. Sixty human cancer cell lines representative for nine cancer types *e.g.* leukemia, lung, colon, central nervous system, melanoma, ovarian, renal, prostate, and breast, were extensively used. It also gradually introduced and developed techniques to understand the mechanism of action of selected compounds tested on animal models. Although, xenografts in athymic mice were developed and used as the representative models for assessing *in vivo* antitumor activity (Povlsen, 1980), the use of multiple *in vivo* animal models for screening remains largely unpractical, given the costly requirements for adequate testing capacities and specific tumor type representation.

Promising cytotoxic compounds are usually selected by a range of standard experimental neoplasms, and then considered for preclinical toxicology studies if the results are satisfactory. At this step, a large amount of material will be required and large scale extraction and isolation may be necessary. However, only very few compounds will reach clinical trials and bring their benefit to cure cancer patients.

CONCLUSION

The question is which anticancer products from medicinal plants should be considered worthy of development as a drug for further studies. As nature is quite imaginative in designing compounds, many if not all, structures of plant-derived anticancer drugs are complicated and, very often, the total synthesis of these compounds is not feasible. Success so far in finding alternative drugs from medicinal plants for the treatment of cancer made all scientists proud and encourages the collaboration between botanists, phytochemists and pharmacologists. Medicinal plants certainly remain an untapped reservoir of potential anticancer agents to exploit. A number of plants extracts, fractions and pure compounds standing in phytochemical laboratories have to be tested and there are many interesting activities still to be discovered.

REFERENCES

Anderson JE, Goetz GE, McLaughlin JL, Suffness M. A blind comparison of simple bend-top bioassay and human tumor cell cytotoxicities and antitumor prescreens. *Phytochem. Anal.* 1991 ; 2: 107-11.

Attard G, Greystoke A, Kaye S, De Bono J. Update on tubulin-binding agents. *Pathol. Biol.* 2006; 54(2):72-84.

Balachandran P, Govindarajan R. Cancer-an ayurvedic perspective. *Pharmacol. Res.* 2005; 51: 19-30.

Borenfreund E, Puerner JA. Toxicity determined *in vitro* by morphological alterations and neutral red absorption. *Toxicol. Lett.* 1985; 24 (2): 119-24.

Cordell GA, Beecher CWW, Pezzuto JM. Can ethnopharmacology contribute to the development of new anticancer drugs? *J. Ethnopharmacol.* 1991; 32: 117-33.

Cordell GA. Changing strategies in natural products chemistry. *Phytochemistry* 1995; 40(6): 1585-612.

Crouch SP, Kozlowski R, Slater KJ, Fletcher J. The use of ATP bioluminescence as a measure of cell proliferation and cytotoxicity. *J. Immunol. Methods* 1993; 160 (1): 81-88.

Cragg GM, Boyd MR, Cardellina II JH, Newman DJ, Snader KM, Cloud TG. Ethnobotany and drug discovery: the experience of the US National Cancer Institute. In: *Ethnobotany and the search for new drugs*. Edited by Chadwick DJ, Marsh J. Ciba foundation symposium. Wiley, Chichester, UK 1994; 185: 178-96.

Cragg GM, Simon JE, Jato JG, Snader KM. Drug discovery and development at the national cancer institute: Potential for new pharmaceutical crops. In: *Potential for new pharmaceutical crops.* Edited by J. Janick. ASHS press, Arlington, USA 1996, p. 554-60.

Cragg GM. Paclitaxel (Taxol®): A success story with valuable lessons for natural product drug discovery and development. *Med. Res. Rev.* 1998; 18(5): 315-31.

Cragg GM, Newman DJ. Drugs from nature: past achievements, future prospects. In: *Ethnomedicine and drug discovery.* Edited by Iwu MM and Wootton JC. Elsevier science B.V, Amsterdam, The Netherlands 2002, p. 23-37.

Decker T, Lohmann-Matthes ML. A quick and simple method for the quatitation of lactate dehydrogenase release in measurements of cellular cytotoxicicty and tumor necrosis factor (TNF) activity. *J. Immunol. Methods* 1988; 115 (1): 61-69.

Desbene S, Giorgi-Renault S. Drugs that inhibit tubulin polymerization: the particular case of podophyllotoxin and analogues. *Curr. Med. Chem. Anti-Canc. Agents* 2002; 2(1): 71-90.

Dewick PM. Tumour inhibitors from plants. In: *Pharmacognosy*. Edited by Evans WC. Saunders WB, London, UK 2002, p. 394-406.

Dewick PM. *Medicinal natural products – a biosynthetic approach.* John Wiley and Sons Ltd., London, UK 2003, p. 205.

Farnsworth NR. *Ethnopharmacology and drug development.* Ciba foundation symposium 1994; 185: 42-59.

Flinn IW, Byrd JC, Bartlett N, Kipps T, Gribben J, Thomas D, Larson RA, Rai K, Petric R, Ramon-Suerez J, Gabrilove J, Grever MR. Flavopridol administered as a 24-hour continuous infusion in chronic lymphocyte leukaemia lacks clinical activity. *Leukemia Res.* 2005; 29(11): 1253-57.

Ferrigni NR, Putnam JE, Anderson JE, Jacobsen LB, Nichols DE, Moore DS, McLaughlin JL. Powell RG, Smith CRJ. *J. Nat. Prod.* 1982; 45 (6): 679-86.

Jaracz S, Chen J, Kuznetsova LV, Ojima I. Recent advances in tumor-targeting anticancer drug conjugates. *Bioorg. and Med. Chem.* 2005; 14: 5043-54.

Graham JG, Quinn ML, Fabricant DS, Farnsworth NR. Plants used against cancer – an extension of the work of Jonathan Hartwell. *J. Ethopharmacol.* 2000; 73: 347-77.

Hamburger M, Hostettmann K. Bioactivity in plants: the link between phytochemistry and medicine. *Phytochemistry* 1991; 30(12): 3864-74.

Hande KR. Etoposide: four decades of development of a topoisomerase II inhibitor. Eur J Cancer 1998; 34(10): 1514-21.

Harmon AD, Weiss U, Silverton JV. The structure of rohitukine, the main alkaloid of *Amoora rohituka* (syn. Aphanamixis polystachya) (Maliaceae). *Tetrahydron* 1979; 8: 721-24.

Harwell JL. *Plants used against cancer*. Quarterman, Lawrence, USA 1982.

Hostettmann K, Wolfender JL, Marston. Hyphenated HPLC techniques for screening and quality control. Society for medicinal plant research: 50 years 1953-2003, a jubilee edition. Wissenschaftliche Verlagsgesellschaft. Stuugart, Germany 2003. p. 99-108.

Institute of Medica Materia (IMM). *Medicinal plants from Vietnam*. Hanoi: Science and Technique press 1990, p. 107.

Iwu MM. Introduction: therapeutic agents from ethnomedicine. In: *Ethnomedicine and drug discovery*. Edited by Iwu MM and Wootton JC. Amsterdam, Elsevier 2002, p. 1-22.

Kim J, Park EJ. Cytotoxic anticancer candidates from natural products. *Curr. Med. Chem – Anticancer agents* 2002; 2: 485-537.

Kim JC, Saha D, Cao Q, Choy H. Enhancement of radiation effects by combined docetaxel and flavopiridol treatment in lung cancer cells. *Radiother. oncol.* 2004; 71: 213-21.

Korzeniewski C, Callewaert DM. An enzyme release assay for natural cytotoxicity. *J. Immuol. Methods* 1983; 64 (3): 313-20.

Loud WD, Farnsworth NR, Soejarto DD, Quinn ML. NAPRALERT: Computer-handling of natural products research data. *J. Chem. Inf. Comput. Sci.* 1985; 25(2): 99-103.

Mans DRA, Rocha AB, Schwartsmann G. Anti-cancer drug discovery and development in Brazil: Targeted plant collection as a rational strategy to acquire candidate anti-cancer compounds. *The oncologists* 2000; 5: 185-98.

Meyer BN, Ferrigni NR, Putnam JE, Jacobsen LB, Nichols DE, McLaughlin JL. Brine shrimp: A convenient general bioassay for active plant constituents. *Planta Med.* 1982; 45 (1): 31-34.

Mosmann T. Rapid colorimetric assay for cellular growth and survival: application to proliferation and cytotoxicity assay. *J. Immunol. Methods* 1983; 65 (1): 55-63.

Mukherjee AK, Basu S, Sarkar N, Ghosh AC. Advances in cancer therapy with plant based natural products. *Curr. Med. Chem.* 2001; 8: 1467-86.

Nachlas MM, Margulies SI, Goldberg JD, Seligman AM. The determination of lactic dehydrogenase with tetrazolium salts. *Anal. Biochem.* 1960; 1: 317-26.

Naik RG, Kattige SL, Bhat SV, Alreja B, De Sousa NJ, Rupp RH. An anti-inflammatory cum immunomodulatory piperidinylbenzopyranone from *Dysoxylum binectariferum*: isolation, structure and total synthesis. *Tetrahedron* 1988; 44 (7): 2081-86.

Newman DJ, Cragg GM, Snader KM. The influence of natural products upon drug discovery. *Nat. Prod. Rep.* 2000; 17: 215-34.

Pettit GR. Marine animal and terrestrial plant anticancer constituents. *Pure Appl. Chem.* 1994; 66(10/11): 2271-81.

Pezzuto JM. Plant-derived anticancer agents. *Biochem. Pharmacol.* 1997; 53: 121-33.

Povlsen CO. Heterotransplants of human tumors in nude mice. *Antibiot. Chemother* 1980; 28: 15-20.

Sedlacek HH. Mechanism of action of flavopiridol. *Crit. Rev. Oncol./Hematol.* 2001; 38: 139-70.

Vlietinck AJ. The future of phytochemistry in the new millennium. In: 2000 years of natural products research past, present and future. Edited by Luijendijk TJC. *Phytoconsult* 2000, p. 215-22.

Wadler S, Schwartz EL. New advanced in interferon therapy of cancer. *The oncologist* 1997; 2: 254-67.

Wang HK. Plant-derived anticancer agents currently in clinical use or in clinical trials. *IDrugs* 1998; 1(1) 92-102.

Weyermann J, Lochmann D, Zimmer A. A practical note on the use of cytotoxicity assay. *Int. J. Pharm.* 2005; 288 (2): 369-76.

In: New Developments in Medicinal Chemistry
Editors: Marta P. Ortega and Irene C. Gil ISBN 978-1-60456-810-3

Chapter 4

Emerging Applications of Quantum Dots in Medicinal Chemistry and Drug Development*

Ian D. Tomlinson, Michael R. Warnement and Sandra J. Rosenthal
Dept. of Chemistry, Vanderbilt University, Nashville, TN, USA

Abstract

Quantum dots have increasingly been incorporated into a wide variety of biological assays as improved fluorescent probes. Their photophysical properties permit the investigation of cellular processes and biological phenomena with unprecedented spatial resolution and temporal longevity. Consequently, quantum dots are poised to facilitate advances in future drug development applications. Multiplexed detection in whole cell assay format may ultimately provide added insight into the extremely complex biochemical mechanisms involved in drug receptor interactions. This article provides a detailed discussion of biological applications which have incorporated quantum dot detection, with a particular emphasis on their possible integration into drug discovery and medicinal chemistry applications.

Introduction

Traditionally drug discovery has focused on a single biological target in the development of new drugs and, while this methodology has been productive, there has been a steady decline in new drugs reaching the marketplace over the last two decades. The introduction of techniques, such as combinatorial chemistry and high throughput screening have increased the number of lead compounds, but these techniques address the issue of quantity rather than

* A version of this chapter was also published in *Quantum Dots: Research, Technology and Applications*, edited by Randolf W. Knoss published by Nova Science Publishers, Inc. It was submitted for appropriate modifications in an effort to encourage wider dissemination of research.

quality. Many of these compounds have subsequently failed due to toxicity or poor bioavailability. Critics have claimed that these screening techniques are not a true representation of the complex chain of events that occur when a drug binds to its target. This target based approach focuses on a single gene or the molecular mechanism of a key event in a disease process. Consequently, these approaches cannot probe the complex interplay between bioavailability, binding at the desired target and expression of other systems or the activation of alternate biochemical pathways. To overcome the limitations of the target based approach for drug discovery a number of researchers have proposed alternate approaches based on whole cell assays. These may be more realistic and enable the research community to gain valuable insights into the biochemical mechanisms of drug receptor interactions and, by using whole cells, biological complexity may be introduced into the drug discovery process. This should reveal both the intra- and intercellular processes occurring in a specific disease state.

Improved fluorescent detection, incorporating instrumentation such as the Flexstation^{TM} fluorescent plate reader, flow cytometry or laser scanning cytometry, has resulted in the development of assays which can ultimately be adapted to screen a wide variety of fluorescently tagged drug candidates in a whole cell assay format. Unfortunately, this approach has limited utility due to the properties of current organic fluorophores, which include photobleaching (many organic fluorophores bleach within seconds), chemical degradation, toxicity, broad emission spectra and low quantum yields. Consequently cellular imaging has remained a largely descriptive tool and, until recently, has only been amenable to small scale experimental samples. However this situation is likely to change with the development of new fluorophores such as quantum dots.

Quantum dots are nanometer sized crystals that offer the promise to solve some of these issues in drug development, while simultaneously being useful for developing deeper insights into long term drug-receptor interactions and downstream biochemical changes. Their unique photophysical properties make them ideal for the development of new diagnostic tools with a much higher signal to noise ratio than can ever be achieved with a conventional fluorophore, and they may be utilized in biological screens that require longer periods of illumination. Since many drugs interact with unknown receptors or enzymes, fluorescent nanocrystals may find utility in the purification process to identify new drug receptors and enzymatic targets. In addition to drug development, cellular imaging, and diagnosis applications, quantum dots may also have the potential as nanoscale drug-delivery devices or nano vectors. Given the multivalent nature of their surfaces, this alternate approach has focused on incorporating both targeting molecules and therapeutic molecules on the quantum dot surface to permit selective treatment of diseased cells and tissue.

Several unique photophysical properties make quantum dots useful as fluorescent markers in a wide range of biological applications, and their first reported use as imaging agents occurred in 1998.[1,2] Subsequently they have become a tool in many sandwich based assay platforms, cellular imaging and *in vivo* imaging protocols. A review of the literature shows that quantum dots have been utilized in a wide variety of biological applications. Of these, perhaps the most important from a medicinal chemistry and drug development perspective are antibody, peptide and small molecule conjugated quantum dots. To understand the importance of quantum dots in drug development it is necessary to understand both their physical properties as well as the current state of the art applications which employ them. As such, this review will initially provide an introduction to quantum dots and the physical underpinnings of their unique photophysical properties. We will then focus our discussion on surface modifications which have been employed to facilitate biological compatibility and ensure a specific interaction with an intended biological target. A review of current biological applications is then presented with a focus on antibody, peptide and small molecule quantum dot conjugates. We conclude by highlighting future directions for the incorporating quantum dots in medicinal chemistry and drug development applications.

AN INTRODUCTION TO QUANTUM DOTS

The photophysical properties of quantum dots arise from their small size, causing them to behave quite differently than bulk semiconductor material. A cursory examination of solid state physics indicates that the electrons in any semiconductor material are partitioned into either a ground state valence band or a higher energy conduction band. Absorbing a photon of sufficient energy can cause promotion of an electron in the valence band to the conduction band, leaving behind a 'hole' in the valence band. Due to numerous attractive and repulsive forces, this electron-hole pair, or exciton as it is commonly referred, seeks to be separated by a finite distance, known as the Bohr exciton radius. Nanoparticles, however, typically have radii on the order of 1-4 nm which is smaller than the typical Bhor exciton radius. Consequently, the exciton formed upon absorption of a photon cannot achieve its desired separation, resulting in an effect known as quantum confinement. The size-tunable properties of quantum dots are all due to this quantum mechanical quantum confinement effect, hence the name "quantum dots". Smaller particles are increasingly confined and give rise to higher energy electronic transitions and, consequently, a blue-shifted fluorescent emission. Additionally, optimization of this fluorescent emission can be achieved by wrapping this initial nanoparticle in a shell of a different semiconductor material. A wide variety of quantum dots synthesized in this manner have been reported in the literature[3-7], while the most extensively studied of these are quantum dots that have a cadmium selenide or cadmium telluride core encapsulated in a cadmium doped zinc sulfide shell (this review will focus on these quantum dots). The shell acts to passivate potential trap states on the surface of the core, enabling the quantum confinement of an electron hole pair, resulting in an increased likelihood of fluorescent emission.

Consequently, the fluorescent properties of these core shell quantum dots offer several improvements compared to organic fluorophores. Most notably, they are highly fluorescent and their quantum yields are significantly greater than fluorescent dyes (commercially available quantum dots have quantum yields in excess of 80-90% which is due in part to the cadmium dopant in the shell).[8-10] Additionally, the size dependant extinction coefficients of quantum dots are exceptionally large (on the order of $1x10^6$) [11,12] which means imaging applications based on quantum dots can utilize low powered excitation sources. This combination of extremely high extinction coefficients and quantum yields leads to exceptional brightness. Fluorescent dyes are also susceptible to photobleaching and metabolic and chemical degradation, while quantum dots have demonstrated excellent photostability in biological environments. The absorption spectrum of quantum dots (Figure 1A) is continuous in nature, so a single excitation source may be used to excite a wide range of colors.[13]

Additionally they have size tunable narrow emission spectra, commonly less than 30 nm full width at half maximum[14-17] (Figure 1B), which enables the development of multiplexing experiments where several distinct biological targets may be monitored simultaneously.

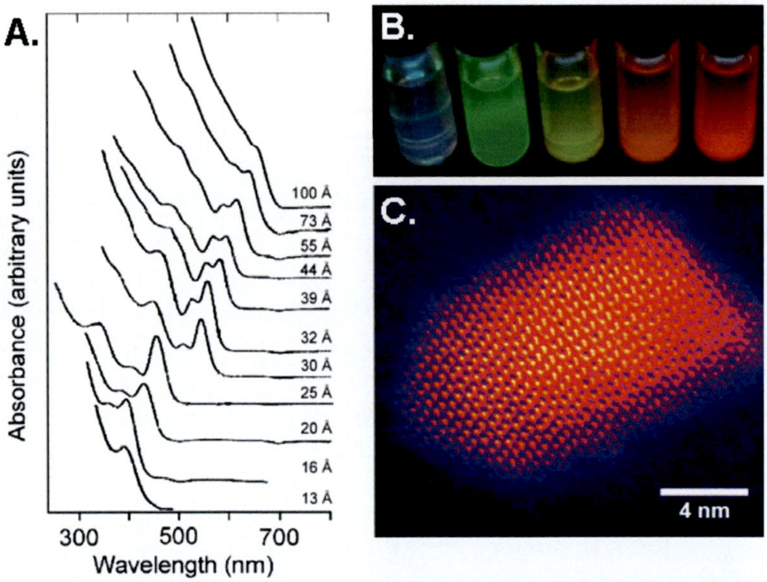

Figure 1. Photophysical properties and characterization of quantum dots. (A) Representative absorption spectra for various sizes of CdSe quantum dot nanocrystals. This tunable absorption is the result of quantum confinement effects upon absorption of a photon. Since this absorption is continuous above the band gap for all nanocrystals, a single excitation source (i.e. 350 nm) can be used to excite all sizes of nanocrystals. (B) UV illumination illustrates that the fluorescent emission from cadmium selenide nanocrystals can be tuned across the visible spectrum, and can even be extended into the near-IR or UV by varying composition. (C) High resolution atomic number contrast scanning transmission electron (Z-STEM) micrograph illustrating the atomic structure of an individual CdSe/ZnSe/Cd quantum dot.[8] Adapted from Rosenthal et. al. *Surface Science Reports*.[9]

Quantum Dot Surface Chemistry

Quantum dot synthesis generally utilizes a coordinating solvent such as trioctyl phosphine oxide (TOPO) to control growth rate, and are consequently coated with this ligand after synthesis.[18] TOPO coated dots are stable and highly fluorescent in organic solvents, but they have low solubility and quantum yields in aqueous solution. To make them compatible with an aqueous environment it is necessary to modify the surface chemistry.[19] This may be achieved by replacing the TOPO with a water soluble ligand such as a thiolated acid.[20] Examples of this approach include mercaptoacetic acid[2], mercaptopropionic acid[21], dihydrolipoic acid (DHLA)[22,23], and DL-cysteine.[24] As these thiols are not covalently bound to the surfaces of dots, an equilibrium exists in aqueous solution with dynamic ligand exchange. This may be problematic if there are other species present in the buffer that can displace the thiol. To overcome the limitations of thiolated ligands a number of alternative surface chemistries have been explored. These include displacing the TOPO with organic phosphine oligomer heterofunctional ligands[25], displacing the TOPO with peptides[26], encapsulating the TOPO coated quantum dots in a water soluble polymer[27,28] or cross linking the mecapto acids with the amino acid lysine.[29] Derivatives of TOPO that can be chemically modified to form a water soluble polymer after the quantum dot synthesis have also been reported.[30] To be useful for biological applications, additional surface modifications must facilitate improved colloidal stability, low nonspecific adsorption to cellular components and photostability in a wide range of buffers and pHs, regardless of salt concentration.

Colloidal stability and photostability in a wide range of buffers can be achieved by a number of capping methodologies where the dot is coated in materials such as amphillic polymers[31-33], multifunctional polymers[34] dendrimers[35,36] or creating a silicon dioxide layer around the dot.[1,19,37-40] Low nonspecific adsorption may be achieved by adding inert polymers to the surface such as polyethylene glycol[41], sugars, polysacarides,[45] proteins,[42] or encapsulating the quantum dots in micells.[43-46] The photostability of quantum dots has been studied in numerous buffers, and it has been shown that quantum dots will degrade when stored in buffers at low pH (pH~5) for a period of several days. For this reason quantum dots are usually supplied and stored in borate buffer at pH 8.4 until required. This degradation has been shown to be size dependant and occurs quicker at higher dilutions.[47] Additionally, the photostability and colloidal stability of quantum dots is dependant upon their surface coatings. Quantum dots with thio glycerol on their surfaces had better colloidal and photostability in 5 M sodium chloride solution than mercaptoacetic acid coated quantum dots, which aggregate after 30 minutes.[48] Quantum dots coated in polyethylene glycol, however, have even greater colloidal stability in high salt buffers.[49] A reduction in quantum yield has been observed in commonly used buffers such as Tris and PBS, but this is a slow process, requiring a period of many hours or days before it becomes significant, which is not likely to be problematic during the time frame of the average experiment.

BIOACTIVATION OF QUANTUM DOTS

Initial attempts to introduce biological activity onto quantum dots were first reported in 1998. One involved conjugating the quantum dots to transferin which were subsequently used for cellular imaging[2]; while in the same issue of *Science*, Bruchez et al. used phalloidin, an actin binding molecule, to image cells.[1] Following this pioneering work a great deal of effort has been expended to make the surfaces of quantum dots compatible with biologically active molecules. There are now many different strategies to conjugate biologically active molecules to quantum dots. These include conventional methods such as covalently attaching a reactive functionality on the ligand to the surface coating of the dot via a reactive ester, or to maleimide. The biologically active ligand may also be bound to the surface of the dot via electrostatic interactions or acid base interaction such as a thiol interacting with the zinc sulfide coat. Quantum dots have been coated in proteins such as avidin and streptavidin and biotinylated ligands may be conjugated to these protein modified quantum dots.[50]

Using these approaches a diverse range of biologically active molecules have been attached to quantum dots. Ligands that have been conjugated to quantum dots include antibodies[51], peptides[52], proteins[53-57], RNA[58], DNA[59-68], PNA[69], cytokines[70], viruses[71] and small molecules such as drugs and neurotransmitters.[72, 137] The biological properties of the quantum dot may be altered by changing the surface coating and enabling the development of cell penetrating quantum dots; a variety of different coatings have been reported including cationic cell penetrating peptides and nanogels.[2,73-78] Quantum dots have been utilized to label proteins and receptors in the cells membrane[79] and nuclear targeting peptides have also been attached to quantum dots to image cell nuclei.[80]

Peptides that are cleaved by proteolytic enzymes have been attached to the surfaces of quantum dots enabling the development of fluorescent probes for the expression of protease enzymes.[81-83] One such quantum dot assay for enzymatic activity was developed by Xu et al. and uses a FRET based system for the detection of β-lactamase activity. In this assay, a β-lactam was derivatized with biotin and the fluorescent dye Cy5, as illustrated in Figure 2. Cy5 acted as a fluorescence acceptor and quenched the fluorescent emission of the quantum dot. However, the presence of β-lactamase initiates cleavage of this lactam linker, permitting the Cy5 to dissociate from the dot's surface and restore the fluorescent quantum dot emission.[84]

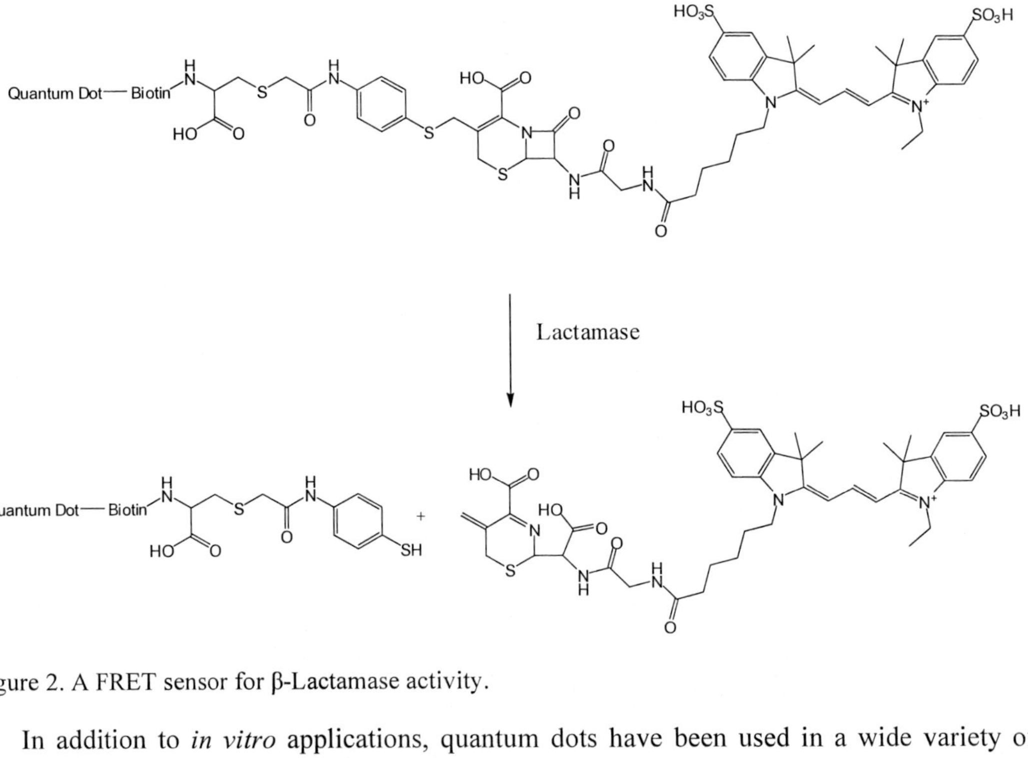

Figure 2. A FRET sensor for β-Lactamase activity.

In addition to *in vitro* applications, quantum dots have been used in a wide variety of deep tissue[85] and *in vivo* applications[86-91] including tracking metastatic tumor cell extravasation[92] and sentinel lymph node mapping during surgery.[93,94] Unlike fluorescent dyes, the multivalent nature of quantum dots enables multiple copies of the same ligand to be attached to the surface of the dot; alternately, several different ligands may be attached to the surface of the dot. For instance, specificity can be introduced through the conjugation of an appropriate ligand, while another surface modification may confer an additional characteristic to the quantum dot. An example of this may include the addition of paramagnetic coatings for use in MRI.[95,96] The multivalent nature of quantum dots may lead to the development of smart drug targeting nano conjugates with enhanced sensitivity and selectivity for desired biological targets, and nano particle based gene delivery systems.

ANTIBODY CONJUGATED QUANTUM DOTS

1. Quantum Dot Based Fluorescent Assays

One of the widest applications of quantum dots as biological labels to date has been to label biological targets with antibodies. Given the low signal to noise ratios of fluorescent dyes and their susceptibility to photobleaching, it is not surprising that a variety of sandwich based assay systems for many different cellular components and toxins have been developed to incorporate quantum dots. In these systems, the antibody may be attached to the surface of modified poly acrylamide (AMP) coated quantum dots via a biflunctional linker such as sulfo-SMCC (illustrated in Figure 3).[97] Alternately, the antibody may be attached to an adaptor protein[98] which has been conjugated to the dots surface.

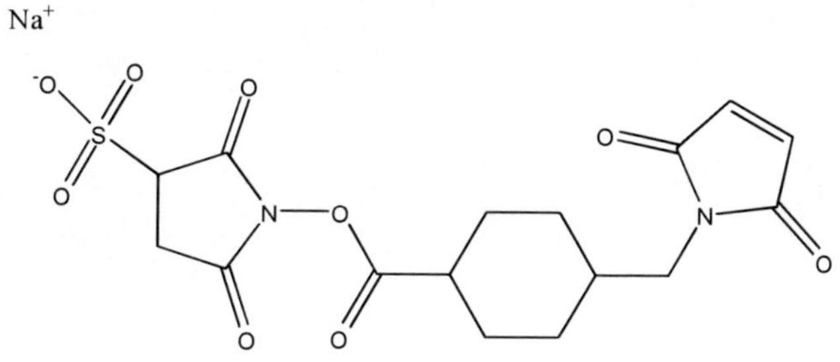

sulfo-SMCC

Figure 3. The structure of sulfo-SMCC.

These conjugates have been used for the detection of staphylococcal enterotoxin B (SEB) in plate based assays,[98] and in continuous flow immunoassays for the detection of the explosive 2,4,6-trinitrotoluene (TNT) in aqueous samples.[98] The narrow fluorescent emission spectra of quantum dots has been exploited by many groups and multiplex toxin assays have been developed with antibody conjugated quantum dots using four colors of quantum dots.[99] Another method of antibody conjugation uses biotinylated antibodies which are attached to quantum dots via an interaction with avidin or streptavidin. Avidin conjugated quantum dots have been used as an assay in conjunction with biotinylated antibodies to detect SEB and cholera toxin.[100] Pegilated streptavidin conjugates have been used in the development of tissue microarray sandwich assays for the detection of proteins expressed by malignant cells[101,102] as well as sufamethazine residue in chicken muscle tissue[103] and an immunosensor for the detection of prostate-specific antigen.[104]

Antibody conjugated quantum dots have also been employed in one recent application which provides an improved detection methodology for respiratory syncytial virus (RSV) infections.[105] This study utilized viral fusion (F) and attachment (G) proteins, which are incorporated into the host cell membrane following viral infection, as antigenic markers of RSV infection. The enhanced sensitivity and improved signal to noise of quantum dot probes permits the detection of viral infection as early as one hour following infection, even at an exceedingly low multiplicity of infection (Figure 4). Traditional western blot or PCR methodologies, currently utilized for clinical detection, can require as many as four days of cell culture to verify infection, but antiviral treatment is only effective if administered early in the course of infection. Consequently, quantum dots are being pursued in a clinical diagnostic application and may ultimately provide a method for early, rapid detection of RSV infection.

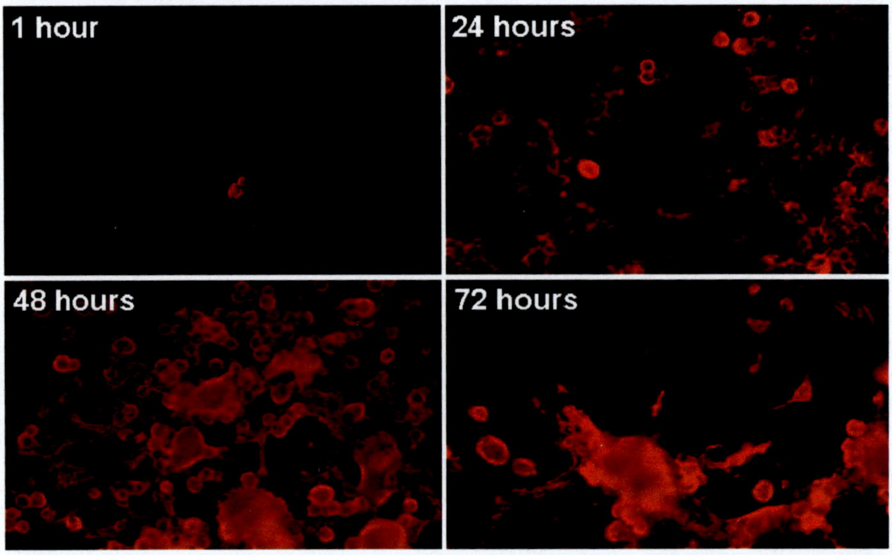

Figure 4. Progression of RSV infection as monitored by antibody conjugated quantum dots. A series of images showing the detection of RSV fusion proteins with antibody conjugated quantum dots at various time points after infection. Traditional methodologies for RSV detection typically require at least four days of cell culture before detection is possible. Quantum dots, however, show remarkable sensitivity and are able to identify an RSV infection after only one hour. Adapted from Bentzen et. al.[105]

Quantum dots conjugated to antibodies may also be used in western blot detection systems, facilitating ultra sensitive and quantitative detection for multiplexed proteomic analysis.[106,107] Given their size-tunable fluorescent emission, it is possible to develop assays where an antibody is conjugated to one color quantum dot while an analyte is bound to a quantum dot of another color. The resulting antigen/antibody immunocomplex has a color that is a combination of these two colors. This principle has been demonstrated with quantum dots conjugated to BSA and the corresponding anti-BSA antibody (IgG).[108] A flouroimmuno assay as also been developed which relies upon Förster resonance energy transfer (FRET) to detect human estrogen receptor-β (ER-β). In this assay, quantum dots with a fluorescent emission at 565 nm were used as a FRET donor, while a polyclonal anti-ER-β antibody labeled with either an Alexa Fluor® 568 or an Alexa Fluor® 633 dye served as the acceptor.[109] Another application targeted *Listeria monocytogenes*, an important food born pathogen, by detection of Internalin A (In1A) and Internalin B (In1B), proteins which promote host cell invasion. Fluorescent microscopy on fixed cell samples was performed using a primary anti-In1A antibody, followed by a biotinylated secondary antibody and subsequent streptavidin quantum dot detection. [110] Formalin fixed paraffin embedded tonsil and lymphoid tissue has also been imaged using streptavidin conjugated quantum dots in conjunction with an appropriate primary antibody and multiple biotinylated secondary antibodies.[111] Finally, formalin fixed Purkinje cells in the glia have been labeled in cerebellum tissue sections with an antibody specific for the glial fibrillaray acidic protein (GFAP).[112]

2. Quantum Dot-Antibody Based Live Cell Assays

In addition to sandwich assays, antibody conjugated quantum dots have been used to label live and fixed cells. Specific labeling of live cells with avidin conjugated quantum dots and biotinylated antibodies has been demonstrated for HeLa cells which were transfected to express the an extracellular loop of the multidrug transporter P-glycoprotein (pgp).[113] In an elegant experiment, Dahan showed that the glycine receptor in live cells could be labeled with primary antibody (mAb2b), a biotinylated anti-mouse Fab fragment and streptavidin conjugated quantum dots.[114] Cancer markers in live Her2 cells[115], TrkA receptor expression in live PC12 cells[116], and single bacterial pathogens such as *E.coli* DH5α[117] have been labeled using antibodies and streptavidin conjugated quantum dots. Additionally, SiHa cells have been labeled with quantum dots coated in streptaividin and conjugated to anti-EGFR. These conjugates have been utilized in a diagnostic test for use in early cervical cancer detection.[118] The narrow fluorescent emissions of quantum dots have also been used to expand the capabilities flow cytometry. In a recent article, Chattopadhyay et al. demonstrated the simultaneous detection of 17 different fluorescent emissions, each corresponding to a specific antigen expressed in T-cells, utilizing quantum dot fluorphores.[119] Streptavidin conjugated quantum dots and dots coated in an amphiphillic polymer have been used in conjunction with a biotinylated anti-CD33 antibody to label and track human leukemic, bone marrow and cord blood cells via flow cytometry.[120] Furthermore, a fluorescent assay using biotinylated antibodies and streptavidin conjugated quantum dots has been developed which permits the simultaneous detection of *Escherichia coli* 0157:H7 and *Salmonella* Typhimurium.[121] In addition to these cell based assays, antibodies have been conjugated to quantum dots and used to image tumors *in vivo*.[122]

PEPTIDE CONJUGATED QUANTUM DOTS

Peptides may potentially be better suited to serve as targeting ligands for biological applications due to their considerably smaller size compared to antibodies. This size difference allows tens or even hundreds of peptides to be attached to the surface of a single quantum dot. Consequently, peptide conjugated quantum dots may exhibit stronger binding affinity and better targeting efficacy as a result of this polyvalent effect. A variety of peptides have been conjugated to quantum dots and used to image live cells. Some notable peptide-based quantum dot applications include angiotensin II binding to the angiotensin I receptor,[123] biotinylated epidermal growth factor (EGF) detecting the erbB/HER receptor,[124] and neuronal growth factor conjugated quantum dots binding TrKA receptors in PC12 cells.[125]

Cai et al. have conjugated pegilated cadmium telluride quantum dots to the arginine-glycine-aspartic acid (RGD) peptide, illustrated in Figure 5, for *in vivo* imaging of integrin $\alpha_v\beta_3$-positive tumor vasculature in a murine xenograft model.[126,127] Integrin $\alpha_v\beta_3$ plays a key role in tumor angiogenesis and metastasis and it is significantly upregulated in invasive tumor cells of certain cancer types (glioblastoma, melanoma, breast, ovarian, and in prostate cancer and in almost all tumor vasculature) but not in quiescent endothelium and normal tissue. Integrins expressed on endothelial cells modulate cell migration during angiogenesis, and inhibition of integrin $\alpha_v\beta_3$ has been shown to prevent tumor growth and cause tumor regression. Several integrin $\alpha_v\beta_3$ inhibitors are currently in clinical trials as therapeutics, and the ability to image this receptor *in vivo* would have potential applications in both cancer imaging and image guided surgery. Near IR quantum dots were functionalized by attaching a thiolated derivative of the cyclic peptide RGD to the pegilated surface of the cadmium telluride quantum dots via a SMCC linker. A thiolated derivative of RGD was synthesized by reacting the lysine ε-amino residue with S-acetylthioglycolic acid N-hydroxysuccinimide ester (SATA), followed by a thiol deprotection using hydroxylamine under neutral conditions. The resulting conjugates were injected into U87MG tumor-bearing mice and a rapid uptake in tumor tissue was observed relative to background fluorescence. The quantum dots were also observed to accumulate in the liver, spleen, bone marrow and lymph nodes.

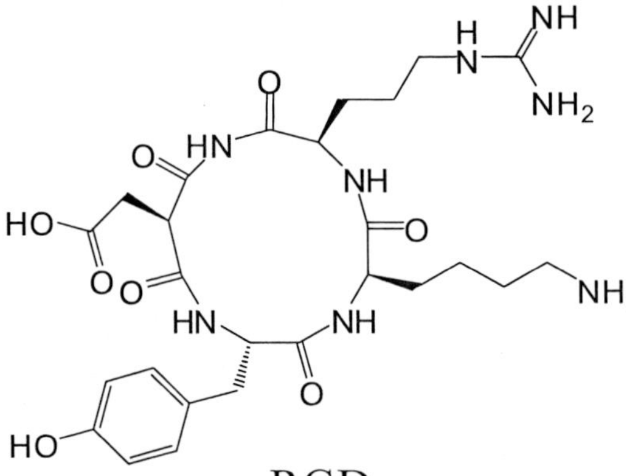

RGD

Figure 5. RGD a cyclic peptide that has high affinity for integrin $\alpha_v\beta_3$.

A recent cancer therapy application has utilized peptide conjugated quantum dots targeted to specific heat shock proteins as potential tumor imaging agents. Heat shock proteins are a set of chaperone proteins involved in many intra- and intercellular processes, including protein synthesis and folding, vesicular trafficking, and antigen presentation and processing. Glucose-regulated protein is a member of this family and is usually expressed and located in the edoplasmic reticulum. However stressful conditions, such as heat exposure, lead to an increase in expression of this protein and the presence of a surface membrane bound form of glucose-regulated protein. In addition, the over expression of this protein has been shown to occur in cancer. Consequently, this receptor has been identified an attractive target for drug delivery of chemotherapeutic agents for cancer. A number of different cyclic peptides have been identified that bind to this protein and are internalized by cancer cells. One of these, the peptide designated Pep42, was covalently attached to the surfaces of modified amphiphillic poly acrylamide (AMP) coated quantum dots using an EDC coupling. The binding and internalization of these conjugates in the highly metastatic human melanoma Me6652/4 or Me6652/56 cells was then studied by a FACS analysis.[128]

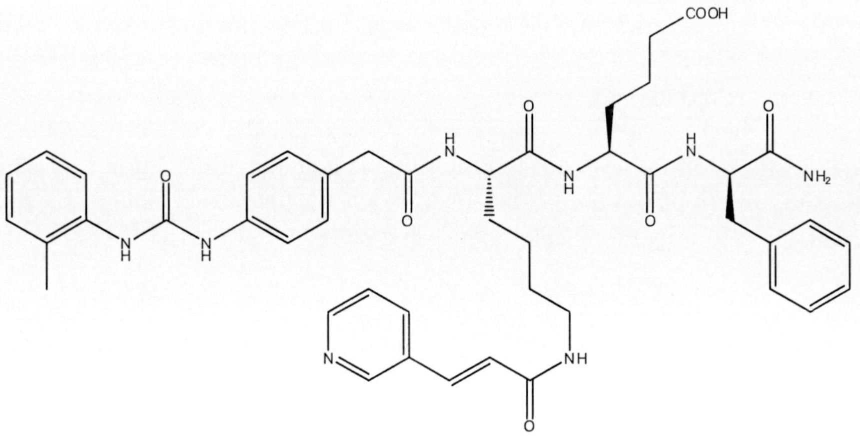

LLP1A (1)

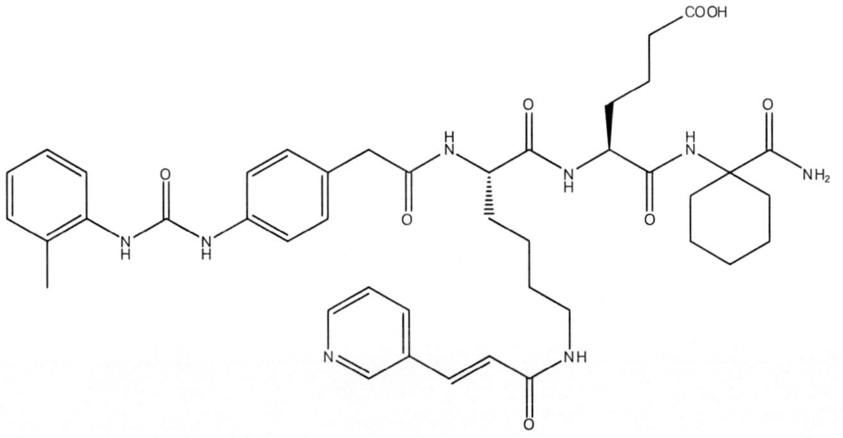

LLP2A (2)

Figure 6. The chemical structures of LLP1A and LLP2A.

Phage-display peptide libraries are an established approach for identifying specific peptide sequences with a high affinity for cellular targets. In one such application, the multiple cell binding tetrameric peptide TP H1299.2 was identified and used in conjunction with streptavidin quantum dots to image live H1299 cells.[129] High affinity and specific targeting peptides for the $\alpha_4\beta_1$ integrin receptor have also been identified using phage-display peptide libraries. The endogenous ligands for the $\alpha_4\beta_1$ integrin receptor are the QIDS and ILDV motifs of vascular cell adhesion molecule-1 (VCAM-1) and fibronectin, respectively.

This receptor is expressed in proliferating endothelial cells in angiogenesis during tumor development and therefore is also an attractive target for imaging and therapeutic agents for cancer. Using one compound on one bead combinatorial chemistry, several high affinity peptidomimetics have been identified; two of these compounds, LLP1A and LLP2A, are shown in figure 6. The binding affinities of LLP2A and LLP2B for the $\alpha_4\beta_1$ integrin receptor were measured using an $\alpha_4\beta_1$-mediated adhesion assay with Jurkat cells and an immobilized CS-1 peptide (a 25-amino acid linear peptide of fibronectin that interacts with the $\alpha_4\beta_1$ integrin receptor). The IC_{50} of LLP2A and LLP1A were found to be 2 ± 1.4 pM and 22 ± 18 pM respectively. A biotinylated version of LLP2A was subsequently used to image Jurkat cells expressing the $\alpha_4\beta_1$ intergrin receptor by incubating the cells in the presence of the biotinylated LLP2A followed by quantum dot 605 conjugates. In addition LLP2A biotin was also conjugated to streptavidin-quantum dots and used for in *vivo* imaging which showed strong binding to $\alpha_4\beta_1$ integrin receptor expressing cells Molt-4 cells.[130]

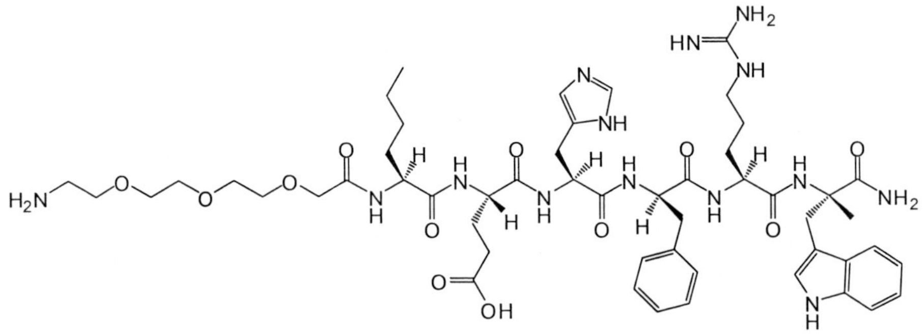

NDP

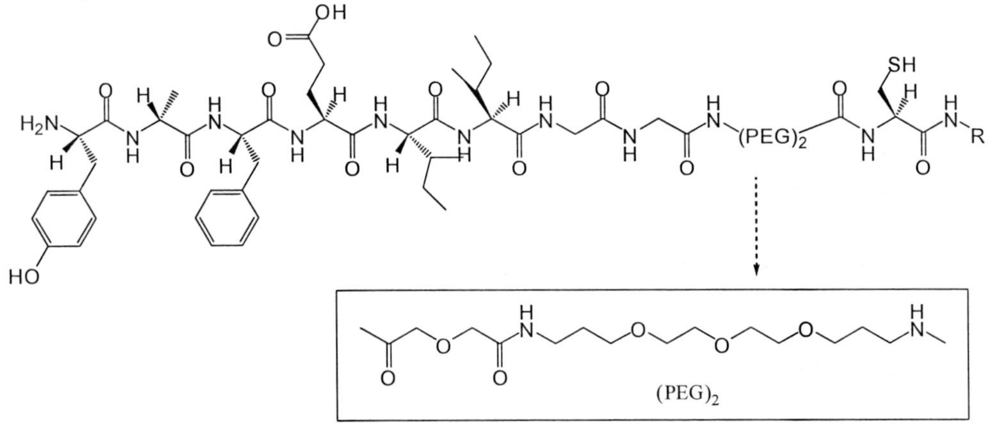

Deltorphin-II

Figure 7. NDP and Deltorphin-II ligands used to label G-protein coupled receptors in live cells.

Zhou et al. have recently reported the development of peptide labeled quantum dots for imaging G protein coupled receptors in live cells. In their system, they attached the peptides either via an EDC coupling to carboxylic acids on the quantum dot surface or via a SMCC cross linker. These dots were coated in a low molecular weight (~1200 Da) diblock copolymer which encompassed acrylic acids as hydrophilic segments and amino-octyl side chains as hydrophobic segments. Two different peptides were employed to introduce specificity (Figure 7). The NDP peptide is an α-MSH analog that has a high affinity for human melanocortin receptor, while the Deltrophin-II analog is specific for δ-opioid receptors. After conjugation, the quantum dots were used to label live either HEK cells expressing the NDP receptor or live CHO cells expressing the δ-opioid receptor.[131]

Small Molecule Conjugated Quantum Dots

The first report of small molecule quantum dot conjugates employed in a biological imaging application was published by Rosenthal et al.[72] In this work, quantum dots were used to image the serotonin transporter (SERT) following surface modification with pegilated serotonin ligands[132], illustrated in Figure 8. Attachment directly to the surfaces of quantum dots was carried out via the thiol terminus. These conjugates antagonized the serotonin transporter protein (SERT) with an IC_{50} of 115 μM and facilitated fluorescence imaging of SERT expression in transfected HEK-293 cells.

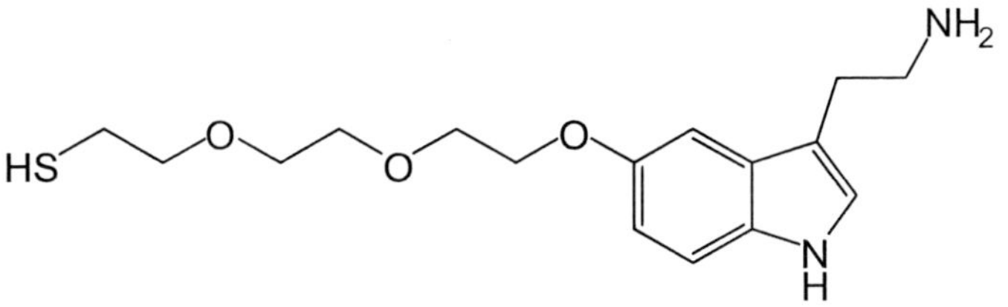

Figure 8. The serotonin ligand conjugated to mercapto acetic acid coated quantum dots and used to image SERT expressing cells.

Following the early success with these serotonin ligand conjugates, we have since synthesized a variety of ligands that are targeted to SERT, the dopamine transporter (DAT) and the $5HT_{2A}$ receptor. These were conjugated to quantum dots using a variety of methodologies including biotin-streptavidin interactions, EDC couplings to the surfaces of AMP dots and thiol exchange reactions to the surfaces of mercapto acetic acid conjugated quantum dots. A selection of these compounds is shown in Figure 9. All of these ligands have been subsequently tested against their biological targets and shown to be biologically active.[133-137]

Additionally, recent efforts have resulted in a pegilated derivative of muscimol designed to interact specifically with the $GABA_C$ receptor. Following conjugation to AMP quantum dots (Figure 10), Gussin et al. demonstrated that this pegilated muscimol derivative could be used to image $GABA_C$ expressed in *Xenopus Laevis* oocytes. Quantum dot conjugates specifically labeled $GABA_C$ expressing oocytes and AMP coated quantum dots gave very little nonspecific labeling with transfected and untransfected oocytes, as illustrated in Figure 11.[138]

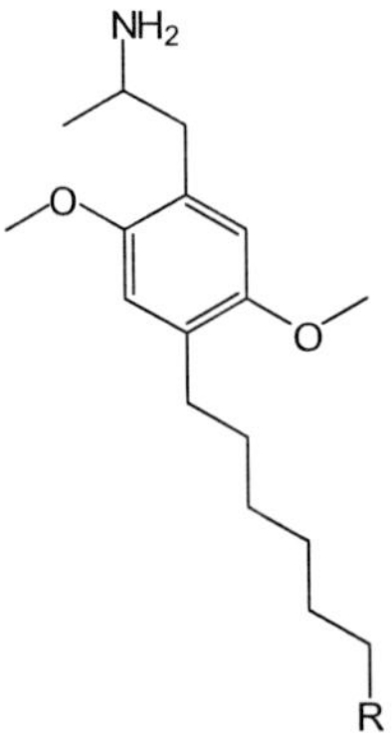

(I) R = SH
(II) R = PEG600NH$_2$

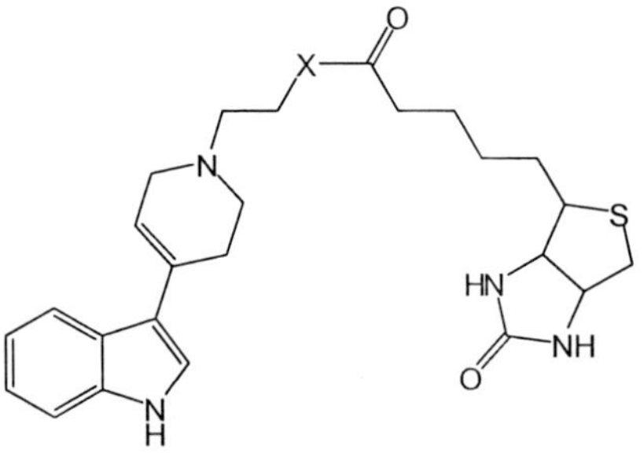

(III) X = NH
(IV) X = NHCOCH$_2$PEG3400NH

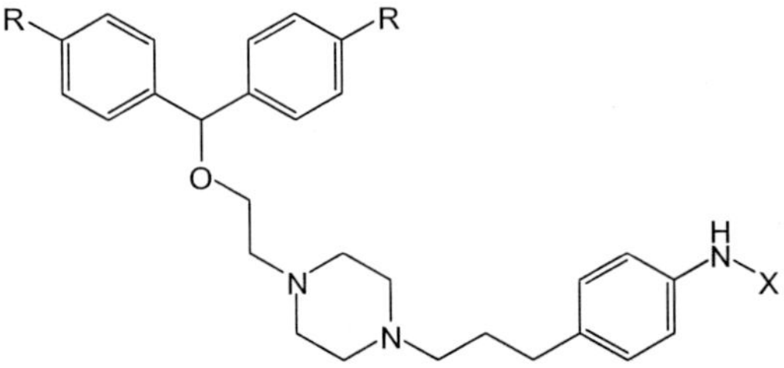

(V) R = H, X = $C_{11}H_{22}$SH
(VI) R = F, X = $C_{11}H_{22}$SH
(VIII) R = F, X = Biotin
(X) R = H, X = Biotin
(XI) R = F, X = $C_{11}H_{22}$NHBiotin

Figure 9. Ligands that have been conjugated to quantum dots.

In addition to our work in the small molecule conjugates field, Clarke et al. have conjugated a dopamine derivative to the surface of quantum dots, via acid base conjugation chemistry, and have subsequently used these conjugates to target cells expressing the D2 receptor. [139] These conjugates were readily internalized by D2 expressing HEK293 and 3T3 cells, and could be blocked by a 10 fold excess of free dopamine. This internalization was shown to be specific as these conjugates did not bind to cells lacking the D2 receptor. This report also detailed the formation of a dopamine quinone complex and described its observed cellular toxicity. Dopamine can undergo a light insensitive oxidization to form a quinone, as shown schematically in Figure 12. This compound is toxic, causing oxidative damage to > 90% of the cells. However, the addition of mercapto ethanol, a reducing agent which inhibits quinone formation, results in enhanced labeling and normal cell viability. Furthermore, they have proposed an energy transfer process by which a molecule in close proximity to the quantum dot surface may act as a photosensitizer and lead to the generation of free radicals such as $^1O_2$. Consequently, the ability of these dopamine conjugates to induce oxidative damage in cellular systems may ultimately permit quantum dot photodynamic therapy applications.[139]

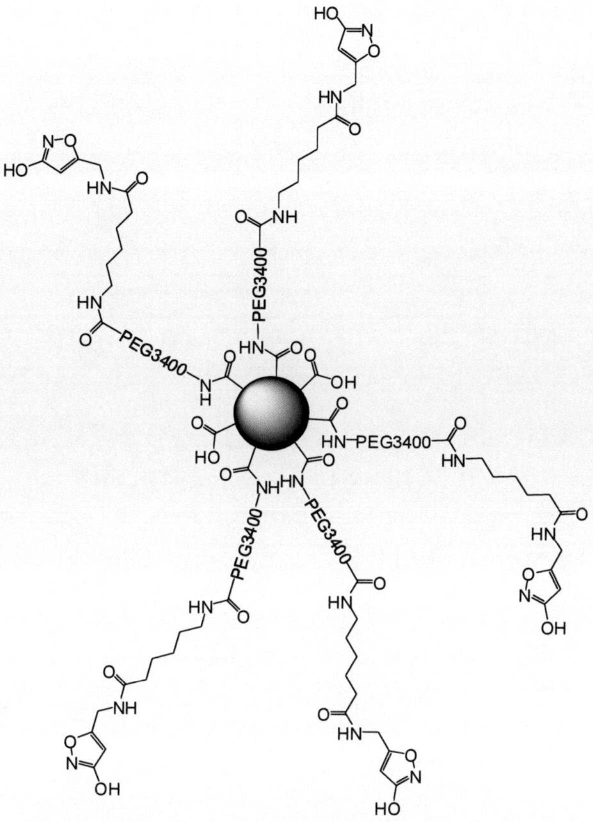

Figure 10. Schematic illustration of AMP dots conjugated with a muscimol ligand.

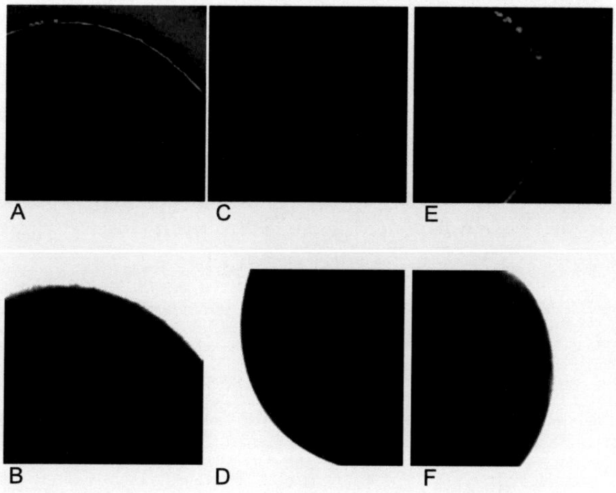

Figure 11. Bright field and fluorescent images of oocytes incubated with a 34 nM solution of either muscimol conjugated dots or a 34 nM solution of AMP dots. Panels A and B are the fluorescent and bright field images of a $GABA_C$ expressing Oocyte that was incubated with a 34 nM solution of muscimol conjugated dots. Panels C and D shows a fluorescent and bright field image of a $GABA_C$ expressing Oocyte incubated with a 34 nM solution of AMP dots. Panels E and F show the fluorescent and bright field images of an untransfected Oocyte incubated with muscimol conjugated AMP dots.

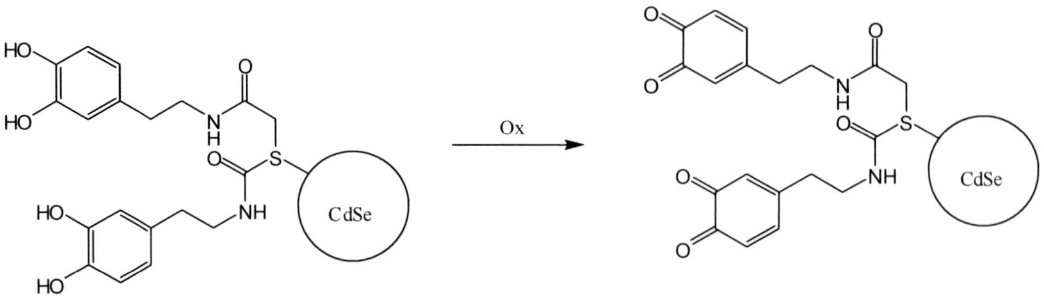

Figure 12. Oxidation of a dopamine ligand on the surfaces of quantum dots.

FUTURE APPLICATIONS OF QUANTUM DOTS IN DRUG DEVELOPMENT AND MEDICINAL CHEMISTRY

Cell based assays may be used as part of the drug discovery process, and it is likely that small molecule or peptide conjugated quantum dots will form the basis of such assays. The nature of the quantum dot surface is crucially important for the development of such probes, requiring excellent colloidal stability and photostability in a wide range of buffers with little or no nonspecific adsorption to a variety of cell types. In addition, the ligands must have high affinity, in the low nanomolar or sub nanomolar range, and selectivity for the desired receptor. The multiplexing afforded by quantum dots would enable adaptation to a high throughput format.

Many different drug development fluorescent assay based platforms can be envisaged. For example transfected cells may be plated out in multi well plates and incubated with quantum dots conjugated to a specific antagonist or agonist. The resultant fluorescently labeled cells could be incubated with a wide range of test compounds for an allotted time period and subsequently washed with buffer. Any displacement of the quantum dot conjugates would result in a reduction of fluorescent intensity and would indicate that the test compound is biologically active.

To be useful in the clinic for *in vivo* applications, it is apparent that quantum dots must first be capable of demonstrating little to no cellular toxicity. Currently, however, there is little information present in the literature regarding added toxicity as a result of quantum dot exposure. Both cadmium and selenium are known toxins and, additionally, cadmium is a suspected carcinogen. Cadmium has a half life of 15-20 years in humans and is systemically transported around the body, with the ability to cross the blood brain barrier, eventually accumulating primarily in the liver and kidneys. The possibility that cadmium may leak from quantum dots and have a deleterious effect on cellular physiology has been studied in the literature. These studies, however, were limited to cadmium selenide cores[140] lacking any zinc sulfide shell or the wide variety of capping ligands routinely used with quantum dots. Loric et al. found that CdTe quantum dots had variable toxicity in PC12 rat cytoma cells depending upon their surface modification, In this study, quantum dots coated with mecaptoacetic acid and cystine had an observed toxicity at concentrations of 10 μg/ml while uncoated quantum dots were cytotoxic at 1 μg/ml. Additionally, cytotoxicity was significantly greater for small quantum dots with a positive charge than larger quantum dots with a similar charge. The distribution inside the cell was also affected by the size of the quantum dot, as small quantum dots were capable of entering the nucleus while larger quantum dots remained in the cytosol.[141] Other groups have suggested that the quantum dot capping material may be responsible for added cytotoxicity. Notably, Hoshino et al. found that mercaptoundecanoic acid alone caused toxicity in murine T-cell Lymphoma EL-4 cells.[87] Several *in vitro* and *in vivo* studies have been cited in the literature as demonstrating a lack of evidence for quantum dot cytotoxicity including Ballou et al.[88], Dubertret et al.[44] and Jaiswal et al.[142]

Conclusion

Quantum dots are increasingly finding a diverse range of applications in enzyme assays as well as fluoro immuno assay based applications. This review has highlighted numerous applications where the unique photophysical properties of quantum dot fluorophores have allowed unprecedented insight into biological processes. These properties enable long periods of illumination and high quantum yields permit detection at the sub nanomolar range. The narrow emission spectra of quantum dots facilitate their used in several multiplexed assay systems, and they have been used as fluorescence donors in many FRET based assay systems. Additionally, the multivalent nature of their surfaces may be useful for the development of nano vectors for drugs and gene therapy. As the size of quantum dots is larger than 3.5 nM, generally agreed to be the maximum particle size for renal clearance,[89] their application in some *in vivo* imaging systems may be limited, especially those where a low background fluorescence or a rapid clearance is required. The continued development of alternate surface modifications should facilitate improved biologically inert probes with enhanced colloid stability and reduced nonspecific cellular interactions. Incorporating quantum dots in whole cell assays capable of simultaneously screening a wide variety of drug candidates will move conventional drug development approaches beyond the current single target approach. Consequently, quantum dots are likely to be of great benefit in future drug discovery applications.

References

[1] Bruchez, M. , Jr.; Moronne, M.; Gin, P.; Weiss, S.; Alivisatos, A. P. *Science*, 1998, 281, 2013-2016

[2] Chan, W. C. W.; Nie S. *Science*, 1998, 281, 2016-2018

[3] Tsay, J. M.; Pflughoefft, M.; Bentolila, L. A.; Weiss, S *J. Am. Chem. Soc.* 2004, 126, 1926-1927

[4] Zheng, J.; Petty, J. T; Dickson, R. M. *J. Am. Chem. Soc.* 2003, 125, 7780-7781

[5] Yang, P.; Lű, M.; Xű, D.; Yaun, D; Zhou, G. *Chemical Physics Letters* 2001, 336, 76-80

[6] Agostiano, A.; Catalano, M.; Curri, M. L.; Della Monica, M.; Manna, L.; Vasanelli, L. *Micron,* 2000, 31, 253-258

[7] Schroedter, A.; Weller, H.; Eritja, R.; Ford, W. E.; Wessels, J. M. *Nano Letters*, 2002, 2, 1363-1367

[8] McBride, J.; Treadway, J.; Feldman, L. C.; Pennycook, S. J.; Rosenthal, S. J. *Nano Lett.,* 2006, 6, 1496-1501
[9] Rosenthal S. J., McBride J., Pennycook S.J., Feldman L. C., *Surface Science Reports,* 2007, 62,111-157.
[10] Watson A.; Wu X.; Bruchez M. *Biotechniques*, 2003, 34, 2, 296
[11] Yu W. W., Qu L., Gua W., Peng X., *Chem. Mater.*, 2003, 15, 2854-2860
[12] Ward A. S. J., Prausnitz J. M., Parak W. J., Zanchet D., Gerion D., Milliron D., Alivisatos A. P., *J. Phys. Chem. B*, 2002, 106, 5500-5505
[13] Cai W.; Shin D. W.; Chen K.; Gheysens O.; Cao Q.; Wang S. X.; Gambhir S. S.; Chen X. *Nano Letters*, 2006, 669-676
[14] Alivisatos, A.P. *J. Pys. Chem.* 1996, 100, 13226-13239
[15] Murray, C. B.; Norris, D. J.; Bawendi, M. G. *J. Am. Chem. Soc.* 1993, 115, 8706-8715
[16] Hines, M. A.; Guyot-Sionnest, P. *J. Phys. Chem.* 1996, 100, 468-471
[17] Michalet X., Pinaud F., Lacoste T. D., Dahan M., Bruchez M. P., Alivisatos A. P., Weiss S., *Single Molecules*, 2001, 4, 261-276
[18] Li J.J, Wang A., Guo W., Keay J. C., Mishima T. D., Johonson M. B., Peng X., *Journal of the American Chemical Society*, 2003, 125, 12567-12575
[19] Gerion D., Pinaud F., Williams S. C., Parak W. J., Zanchet D., Weiss S., Alivisatos A. P., *J. Phys. Chem. B*, 2001, 105, 8861-8871
[20] Liu W. S., *Journal of Bioscience and Bioengineering*, 2006, 1, 1-7
[21] Mitchell G.P., Mirkin C. A., Letsinger R. L., *J. Am. Chem. Soc.*, 1999, 121, 8122-8123
[22] Mattoussi H., Mauro J. M., Golman E. R., Anderson G. P., Sundar V. C., Mikulec F. V., Bawendi M. G., *J. Am. Chem. Soc.*, 2000, 122, 12142-12150
[23] Clapp A. R., Golman E. R., Mattoussi, *Nature Protocols*, 2006, 1, 1258-1266
[24] Sukhanova A., Devy J., Venteo L., Kaplan H., Artemyev M., Oleinikov V., Klinov D., Pluot M., Cohen J. H. M., Nabiev I., *Anal. Biochem.*, 2004, 324, 60-67
[25] Kim S., Bawendi M. G., *J. Am. Chem. Soc.*, 2003, 125, 14652-14653
[26] Pinaud F., King D., Moore H. P., Weiss S., *J. Am. Chem. Soc.,* 2004, 126, 6115-6123
[27] Gao X., Cui Y., Levenson R. M., Chung L. W., Nie S., *Nat. Biotechnology*, 2004, 22, 969-976
[28] Wang X. S., Dykstra T. E., Salvador M. R., Manners I., Scholes G. D., Winnik M. A., *J. Am. Chem. Soc,* 2004, 126, 7784-7785
[29] Jiang W., Mardyani S., Fischer H., Chan W. C. W., *Chem. Mater.*, 2006, 18, 872-878
[30] Skaff H., Sill K., Emrick T., *J. Am. Chem. Soc.*, 2004, 126, 11322-11325
[31] Watson A., Wu X., Bruchez M., *Biotechniques*, 2003, 34, 296-303
[32] Wu X., Liu H., Liu J., Haley K. N., Treadway J. A., Larson J. P., Ge N., Peale F., Bruchez M. P., *Nat. Biotechnology*, 2003, 21, 41-46
[33] Petruska M. A., Bartko A. P., Klimov V. I., *J. Am. Chem. Soc.*, 2004, 126, 714-715
[34] Potapova I., Mruk R., Prehi S., Zentel R., Basché T., Mews A., *J. Am. Chem. Soc.*, 2003, 125, 320-321
[35] Lemon B. I, Crooks R. M., *J. Am. Chem. Soc.*, 2000, 122, 12886-12887
[36] Zhang C., O'Bien S., Balogh L., *J. Phys. Chem. B*, 2002, 106, 10316-10321
[37] Chen F., Gerion D., *Nano Lett.*, 2004, 4, 1827-1832
[38] Yu W. W., Chang E., Drezek R., Colvin V. L., *Biochemical and Biophysical Research Communications*, 2006, 348, 781-786

[39] Schroedter A., Weller H., Ertjia R., Ford W. E., Wessels J. M., *Nano Lett.*, 2002, 2, 1363-1367
[40] Schroedter A., Weller H., Eritja R., Ford W. E., Wessels, *Nano Lett.*, 2002, 2, 1363-1367
[41] Bentzen E. L.; Tomlinson I. D.; Mason, J. N.; Gresch, P.; Wright, D.; Sanders-Busch, E.; Blakely, R.; Rosenthal, S. J. *Bioconjugate Chemistry*, 2005, 16, 1488-1494
[42] Hanaki K., Momo A., Oku T., Komoto A., Maenosono S., Yamaguchi Y., Yamamoto K., *Biochem. Biophys. Res. Comm.*, 2003, 302, 496-501
[43] Maysinger D., Lovrić J., Eisenberg A.,, Savić R., *Eur. J. of Pharmaceutics and Biopharmaceuticals*, available online September 2006.
[44] Dubertret B., Skourides P., Norris D. J., Noireaux V., Brivanlou A. H., Libchaber A., *Science*, 2002, 298, 1759-1762
[45] Osaki F., Kanamori T., Sando S., Sera T., Aoyama Y., *J. Am. Chem. Soc.*, 2004, 126, 6520-6521
[46] Fan H., Leve E. W., Scullin C., Gabaldon J., Tallant D., Bunge S., Boyle T., Wilson M. C., Brinker C. J., *Nano Lett.*, 2005, 5, 645-648
[47] Boldt K., Bruns O. T., Gaponik N., Eychműller, *J. Phys. Chem. B*, 2006, 110, 1959-1963
[48] Hosho A., Fujioka K., Oku T., Suga M., Sasaki Y. F., Ohta T., Yasuhara M., Suzuki K., Yamamoto K., *Nano Lett.*, 2004, 4, 2163-2169
[49] Smith A. M., Duan H., Rhyner M. N., Ruan G., Nie S., *Phys. Chem. Chem. Phys.*, 2006, 8, 3895-3903
[50] Bäumle M., Stamou D., Segura J. M., Hovius R., Vogel H., *Langmuir*, 2004, 20, 3828-3831
[51] Mason J. N., Farmer H., Tomlinson I. D., Schwartz J. W., Savchenko V., DeFelice L. J., Rosenthal S. J., Blakely R. D., *Journal of Neuroscience Methods*, 2005, 143, 3-25
[52] Åkerman M. E., Chan W. C. W., Laakkonen P., Bhatia S. N., Ruoslahti E., *PNAS*, 2002, 99, 12617-12621
[53] Minet O., Dressler C. Beuthan J., *J. of Flourescence*, 2004, 14, 241-247
[54] Chunyang A., Hui M., Nie S., Yao D., Lei J., Dieyan C., *Analyst*, 2000, 125, 1029-1031
[55] Medintz., Deschamps J., R., *Current opinions in Biotechnology*, 2006, 17, 17-27
[56] Månsson A., Sundberg M., Balaz M., Bunk M., Nicholls I. A., Omling P., Tågerud S., Montelius L., *Biochem. Biophys. Res. Comm.*, 2004, 314, 529-534
[57] Gac S., Vermes I., Van Den Berg A., *Nano Lett.*, 2006, 6, 1863-1869
[58] Tan W. B., Jiang S. Zhang Y., *Biomaterials*, 2007, 28, 1565-1571
[59] Alivisatos P., *Nature Biotechnology*, 2004, 22, 47-52
[60] Patolsky F., Gill R., Weizmann Y., Mokari T., Banin U., Willner I., *J. Am. Chem. Soc.*, 2003, 125, 13918-13919
[61] Gerion D., Parak W. J., Williams S. C., Zanchet D., Micheel C. M., Alivisatos A. P., *J. Am. Chem. Soc.*, 2002, 124, 7070-7074
[62] Xiao Y., Barker P. E., *Nucleic acids Res.*, 2004, 32, No3e28
[63] Parak W. J., Gerion D., Zanchet D., Woerz A. S., Pellegrino T., Micheel C., Williams S. C., Seitz M., Bruehi R. E., Bryant Z., Bustamante C., Bertozzi C., Alivisatos A. P., *Chem. Mater.*, 2002, 14, 2113-2119

[64] Srinivasan C., Lee J., Papadimitrakopoulos F., Silbart L. K., Zhao M., Burgess D. J., *Molecular Therapy*, 2006, 14, 192-201

[65] Tholouli E., Hoyland J. A., Vizio D. D., O'Connell F., MacDermott S A., Twomey D., Levenson R., Liu Yin J. A., Golub T. R., Loda M., Byers R., *Biochem. Biophys. Res. Comm.,* 2006, 348, 628-636

[66] Zhang C. Y., Yeh H. S., Kuroki M. T., Wang T. H., *Nature Materials*, 2005, 4, 826-831

[67] Fu A. Micheel C. M., Cha J., Chang H., Yang H., Alivisatos A. P., *J. Am. Chem. Soc.,* 2004, 126, 10832-10833

[68] Dyadyusha L., Yin H., Jaiswal S., Brown T., Baumberg J. J., Booy F. P., Melvin T., *Chemcomm.,* 2005, 3201-3203

[69] Chakrabarti R., Klibanov A. M., *J. Am. Chem. Soc.*, 2003, 125, 12531-12540

[70] Bryde S. Grunwald I., Hammer A. Krippner-Heidenreich A., Schiestel T., Brunner H., Tovar G. E. M., Pfizenmaier K., Scheurich P., *Bioconjugate Chem.*, 2005, 16, 1459-1467

[71] Manchester M., Singh P., *Advanced Drug Delivery Reviews*, 2006, 58, 1505-1522

[72] Rosenthal S. J., Tomlinson I. D., Schroter S., Adkins E., Swafford L., Wang Y., DeFelice L. J., Blakely R. D., *J. Am. Chem. Soc.,* 2002, 124, 4586-4594

[73] Derfus A. M., Chan W. C. W., Bhayia S. N., *Adv. Mater.*, 2004, 16, 961-966

[74] Langerholm B., Christoffer W., Miaomiao E., Lauren A., Ly A., Daneith, Lui H., Hongjian, Bruchez M. P., Waggoner A. S., *Nano. Lett.*, 2004, 4, 2019-2022

[75] Delehanty J. B., Medintz I. L., Pons T., Brunel F. M., Dawson P. E., Mattoussi H., *Bioconjugate Chem.*, 2006, 17, 920-927

[76] Osaki F., Kanamori T., Sando S., Sera T., Aoyama Y., *J. Am. Chem. Soc.*, 2004, 126, 6520-6521

[77] Silver J., Ou W., *Nano Lett.*, 2005, 5, 1445-1449

[78] Hasegawa U., Nomura S., M., Kaul S. C., Hirano T., Akiyoshi K., *Biochemical and Biophysical Res. Comm.*, 2005, 331, 917-921

[79] Jaiwal J. K., Goldman E. R., Mattoussi H., Simon S. M., *Nature Methods*, 2004, 1, 73-78

[80] Chen F., Gerion D.,*Nano Lett.*, 2004, 4, 1827-1832

[81] Medintz I. L., Clapp A. R., Brunel F. M., Tiefenbrunn T., Uyeda H. T., Chang E. L., Deschamps J. R., Dawson P. E., Mattoussi H., *Nature Materials*, 2006, 5, 581-589

[82] Chang E., Miller J. S., Sun J., Yu W. W., Colvin V. L., Drezek R., West J. L., *Biochem. And Biophys. Res. Comm.*, 2005, 324, 1317-1321

[83] Zhang Y., So M. K., Rao J., *Nano Lett.*, 2006, 1988-1992

[84] Xu C., Xing B., Rao J., *Biochemical and Biophysical Research Communications*, 2006, 344, 931-935

[85] Medintz I. L., Uyeda H. T., Golman E. R., Mattoussi H., *Nature Materials*, 2005, 4, 435-446

[86] Gao X., Cui Y., Leveson R. M., Chung L. W. K., Nie S., *Nature Biotechnol.*, 2004, 22, 969-977

[87] Hoshino A., Hanaki K., Suzuki K., Yamamoto K., *Biochem. Biophys. Res. Commun.*, 2004, 314, 46-53

[88] Ballou B., Langerholm B. C., Ernst L. A., Bruchez M. P., Waggoner A. S., *Bioconjugate Chem.*, 2004, 15, 79-86

[89] Larson D R., Zipfel W. R., Williams R. M., Clark S. W., Bruchez M. P., Wise F. W., Webb W. W., *Science*, 2003, 300, 1434-1436

[90] Stroh M., Zimmer J. P., Duda D. G., Levchenko T. S., Cohen K. S., Brown E. B., Scadden D. T., Torchilin V. P., Bawendi M. G., Fukumura D., Jain R. K., *Nature Medicine,* 2005, 678-682

[91] Frangioni J. V., *Current Opinions in Chemical Biology*, 2003, 7, 626-634

[92] Voura E. B., Jaiswal J. K., Mattoussi H., Simon S. M., *Nature Medicine*, 2004, 10, 993-998

[93] Ballou B., Ernst L. A., Andreko S., Harper T., Fitzpatrick J. A. J., Waggoner A. S., Bruchez M. P., *Bioconjugate Chem.,* in press

[94] Kim S., Lim Y. T., Soltesz E. G., De Grand A. M., Lee J., Nakayama A., Parker J. A., Mihaljevic T., Laurence R. G., Dor D. M., Cohn L. H., Bawendi M. G., Frangioni J. V., *Nature Biotechnology*, 2004, 22, 93-97

[95] Van Tilborg G. A. F., Mulder W. J. M., Chin P. T. K., Storm G., Reutelingsperger C. P., Nicolay K., Strijkers G. J., *Bioconjugate Chem.,* 2006, 17, 865-868

[96] Mulder W. J. M. Koole R. K., Brandwijk R. J., Storm G., Chin P. T. K., Strijkers G. J., de Mello Donegá C., Nicolay K., Griffioen A. W., *Nano Lett.*, 2006, 6, 1-6

[97] Wolcott A., Gerion D., Visconte M., Sun J., Schwartzberg A., Chen S., Zhang J. Z., *J. Phys. Chem. B*, 2006, 110, 5779-5789

[98] Golman E. R., Anderson G. P., Tran P. T., Mattoussi H., Charles P. T., Mauro J. M., *Anal. Chem.*, 2002, 74, 841-847

[99] Goldman E. R., Clapp A. R., Anderson G. P., Uyeda H. T., Mauro J. M., Medintz I. L., Mattoussi H., *Anal. Chem.*, 2004, 76, 684-688

[100] Goldman E. R., Balighian E. D., Mattoussi H., Kuno M. K., Mauro J. M., Tran P. T., Anderson G. P., *J. Am. Chem. Soc.*, 2002, 124, 6378-6382

[101] Geho D., Lahar N., Gurnani P., Huebschman M., Herrmann P., Espina V., Shi A., Wulfkuhle J., Garner H., Petricoin E., Liotta L. A., Rosenblatt K. P., *Bioconjugate Chem.*, 2005, 16, 559-566

[102] Ghazani A. A., Lee J. A., Klostranec J., Xiang Q., Dacosta R. S., Wilson B. C., Tsao M. S., Chan W.C.W., *Nano Lett.*, 2006, 6, 2881-2886

[103] Ding S., Chen J., Jiang H., He J., Shi W., Zhao W., Shen J., *J. Agri. And Food Chem.*, 2006, 54, 6139-6142

[104] Kerman K., Endo T., Tsukamoto M., Chikae M., Takamura Y., Tamiya E., *Talanta*, 2006, doi:10.1016/j.talanta.2006.07.027

[105] Bentzen, E. L., House, F., Utley, T. J., Crowe, J. E., Jr., Wright, D. W., *Nano Lett.*, 2005, 5, 591-595

[106] Bakalova R., Zhelev Z., Ohba H., Baba Y., *J. Am. Chem. Soc.*, 2005, 127, 9328-9329

[107] Ornberg R. L., Harper T. F., Liu H., *Nature Methods*, 2005, 2, 79-81

[108] Wang S., Mamedova N., Kotov N. A., Chen W., Studer J., *Nano Lett.*, 2002, 2, 817-822

[109] Wei Q., Lee M., Yu X., Lee E. K., Seong G. H., Choo J., Cho Y. W., *Analytical Biochemistry*, 2006, 258, 31-37

[110] Tully E., Hearty S., Leonard P., O'Kennedy R., *International Journal of Biological Macromolecules,* 2006, 39, 127-134

[111] Fountaine T. J., Wincovitch S. M., Geho D. H., Garfield S. H., Pittaluga S., *Modern Pathology*, 2006, 19, 1181-1191

[112] Giepmans B. N. G., Deerinck T. J., Smarr B. J., Jones Y. Z., Ellisman M. H., *Nature Methods*, 2005, 2, 743-749

[113] Jaiswal J. K., Mattoussi H., Mauro J. M., Simon S. M., *Nature Biotechnology*, 2003, 21, 47-51

[114] Dahan M., Lévi S., Luccardini C., Rostaing P., Riveau B., Triller A., *Science*, 2003, 302, 442-445

[115] WuX., Liu H., Liu J., Haley K. N., Treadway J. A., Larson J. P., Ge N., Peale F., Bruchez M. P., *Nature Biotechnology*, 2003, 21, 41-46

[116] Rajan S. S., Vu T. Q., *Nano Lett.*, 2006, 6, 2049-2059

[117] Hahn M. A., Tabb J. S., Krauss T. D., *Anal. Chem.*, 2005, 77, 4861-4869

[118] Niba D. L., Rahman M. S., Carlson K. D., Richards-Kortum R., Follen M., *Gynecologic Oncology*, 2005, 99, 589-594

[119] Chattopadhyay P. K., Price D. A., Harper T. F., Betts M. R., Yu J., Gostick E., Perfetto S. P., Goepfert P., Koup R. A., De Rosa S. C., Bruchez M. P., Roederer M., *Nature Medicine*, 2006, 12, 972-977

[120] Garon E. B., Marcu L., Luong Q., Tcherniantchouk O., Crooks G. M., Koeffler H. P., *Leuk. Res.*, 2006, doi:10.1016/j.leukres.2006.08.006

[121] Yang L., Li Y., *Analyst*, 2006, 131, 394-401

[122] Sage L., *Anal. Chem.,* 2004, 453A

[123] Tomlinson I. D., Mason J. N., Blakely R. D., Rosenthal S.J., 2005, *Methods Mol. Biol.,* 303, 51-60

[124] Lidke D. S., Nagy P., Heintzmann R., Arndt-Jovin D. J., Post J. N., Grecco H. E., Jares-Erijman E. A., Jovin T. M., *Nat. Biotechnology*, 2004, 22, 198-203

[125] Vu T. Q., Maddipati R., Blute T. A., Nehilla B. J., Nusblat L., Desai T. A., *Nano Lett.*, 2005, 5, 603-607

[126] Cai W., Shin D. W., Chen K., Gheysens O., Cao Q., Wang S. X., Gambhir S. S., Chen X., *Nano Lett.*, 2006, 6, 669-676

[127] Cheng Z., Wu Y., Xiong Z., Gambhir S. S., Chen X., *Bioconjugate Chem.,* 2005, 16, 1433-1441

[128] Kim Y., Lillo A. M., Steiniger S. C. J., Liu Y., Ballatore C., Anichini A., Mortarini R., Kaufmann G. F., Zhou B., Felding-Habermann B., Janda K. D., *Biochemistry*, 2006, 45, 9434-9444

[129] Oyama T., Rombel I. T., Samli K. N., Zhou X., Brown K. C., *Biosensors and Bioelectronics,* 2006, 21, 1867-1875

[130] Peng L., Liu R., Marik J., Wang X., Takada Y., Lam K. S., *Nature Chemical Biology*, 2006, 2, 381-389

[131] Zhou M., Nakatani E., Gronenberg L. S., Tokimimoto T., Wirth M. J., Hruby V. J., Roberts A., Lynch R. M., Ghosh I., *Bioconjugate Chem.,* 2007, 18, 323-332

[132] Tomlinson I. D., Kippeny T., Siddiqui N., Rosenthal S. J., *J. Chem. Res.,* 2002, 527

[133] Tomlinson I. D., Gies A. P., Gresch P., Dillard J., Orndorff R. L., Sanders-Bush E., Rosenthal S.J., *Bioorganic and Medicinal chemistry Letters*, 2006, 16, 6262-6266

[134] Tomlinson I. D., Mason J. N., Blakely R. D., Rosenthal S. J., *Bioorganic and Medicinal Chem. Lett.,* 2006, 16, 4664-4667

[135] Tomlinson I. D., Mason J. N., Blakely R. D., Rosenthal S. J., *Bioorganic and Med. Chem. Lett.,* 2005, 15, 5307-5310

[136] Tomlinson I. D., Burton J. N., Mason J. N., Blakely R. D., Rosenthal S. J., *Tetrahedron*, 2003, 59, 8035-8047

[137] Tomlinson I. D., Grey J. L., Rosenthal S. J., *Molecules*, 2002, 7, 777-790

[138] Gussin H. A., Tomlinson I. D., Little D. M., Warnement M. R., Qian H., Rosenthal S. J., Pepperberg D. R., *J. Am. Chem. Soc.,* 2006, 128, 15701-15713

[139] Clarke S. J., Hollmann C. A., Zhang Z., Suffern D., Bradforth S. E., Dimitrijevic N. M., Minarik W. G., Nadeau J. L., *Nature Materials*, 2006, 5, 409-417

[140] Derfus A. M., Chan W. C. W., Bhatia S. N., *Nano Lett.,* 2004, 4, 11-18

[141] Lovric J., Bazzi H. S., Cuie Y., Fortin G. R. A., Winnik F. M., Maysinger D., *J. Mol. Med.,* 2005, 83, 377-385

[142] Jaiswal J. K., Mattoussi H., Mauro J. M., Simon S. M., *Nat. Biotechnol.,* 2003, 21, 47-51

In: New Developments in Medicinal Chemistry
Editors: Marta P. Ortega and Irene C. Gil ISBN 978-1-60456-810-3

Chapter 5

MEDICINAL CHEMISTRY OF COPPER AND VANADIUM BIOACTIVE COMPOUNDS

Susana B. Etcheverry[1,2] and Patricia A.M. Williams[2,*]
[1]Cátedra de Bioquímica Patológica, Facultad de Ciencias Exactas, Universidad Nacional de La Plata, La Plata, Argentina
[2]Centro de Química Inorgánica (CEQUINOR) (UNLP-CONICET), Facultad de Ciencias Exactas, Universidad Nacional de La Plata, La Plata, Argentina

ABSTRACT

Transition metals play a fundamental role in different living systems. In particular, in aerobic organisms, the presence of copper is essential for the function of many enzymes related to cellular respiration, iron homeostasis, neurotransmitter production, peptide biogenesis, connective tissue biosynthesis, and antioxidant defense. Copper compounds are reported to act as antioxidant, anti-inflammatory, antimicrobial, antiparasitic, anticonvulsant and antitumoral agents. In addition, in vertebrates, copper deficiency causes skeletal alterations. Vanadium, another transition metal, is present in trace amounts in higher animals. Even though its essentiality has not yet been clearly established, experimental results both in vivo and in vitro suggest that vanadium compounds may participate in important biological functions acting as insulinmimetic, osteogenic and antitumoral compounds. Once absorbed, vanadium and copper are distributed among tissues and stored mainly in bone. In this chapter, the behavior of these metal derivatives on bone-related cells in cultures is discussed in detail. Two cellular lines, MC3T3E1 derived from mouse calvaria, and UMR106 from rat osteosarcoma, have been used as a model for normal and tumoral bone processes. To expand the studies on antineoplastic metal drug activity, experiments with copper and vanadium compounds have been undertaken on two tumoral cell lines of human colon adenocarcinoma (Caco-2 and TC7). Different copper complexes with pharmacologically active ligands such as the antihypertensive drug losartan and a derivative of the antiparasitic santonin, santonic acid, were synthesized and tested in vitro in the mentioned cell lines. Both complexes

* Correspondence request: Patricia A.M. Williams, Facultad de Ciencias Exactas, Universidad Nacional de La Plata, 47 y 115 (1900) La Plata, Argentina. E.mail: williams@quimica.unlp.edu.ar.

improve the antitumoral effect of free copper ions. This behavior agreed with the morphological cellular alterations. On the other hand, the biological effects of vanadyl(IV) complexes with the flavonoids quercetin and hesperidin were discussed and compared with a vanadium(IV) derivative of the structural related ligand, the disaccharide trehalose. In the tumoral cell lines these compounds were deleterious. The effects on cellular differentiation (specific alkaline phosphatase activity and collagen type I production) were also described for the osteosarcoma cells. Moreover, for the complexes with quercetin and trehalose, the effect on the activation of ERK (extracellular regulated kinase) cascade was investigated using specific antibodies in order to identify one of the possible mechanisms of action. Altogether, these promising results of a first stage in medicinal chemistry on metal-based drugs merit further investigation in animal models.

INTRODUCTION

Biological and Pharmacological Relevance of Copper and Vanadium

1.a. Copper

Copper is an essential oligoelement that participates in numerous biological processes, being a cofactor or prosthetic group of several enzymes. Copper concentrations are higher in the liver, brain, heart, muscle and kidney (Mason, K.E., 1979). The physiological role of copper in humans is well recognised, as well as its toxicology side effects, and there are several studies on the mechanisms of Cu homeostasis by the identification and characterisation of novel Cu transporters and Cu binding proteins. Copper is involved in a variety of physiological processes and pathological conditions (Camakaris, J. et al., 1999). Free copper is bound to transport proteins (ceruloplasmin and albumin), storage proteins (metallothioneins) or copper-containing enzymes. Copper deficiency produces different diseases such as Menkes' kinky hair syndrome with degeneration of the central nervous system and hair abnormalities. It is a disease due to a defect in intestinal copper absorption. Copper excess produces Wilson's disease associated with cirrhosis of the liver and neurological manifestations, occurring predominantly during the first few decades of life (Baran EJ, 1994; Sorenson, J.R.J., 1985; Harris, E.D.,1983; Gubler, C.J., 1953; Solomons, N.W., 1985).

The metal is a constituent of many enzymes, such as ceruloplasmin (in which copper plays an important role in the iron oxidation), lysil-oxydase (essential for collagen and elastin formation), superoxide dismutase (antioxidant protection), cytochrome C oxidase (energy production), and tyrosinase (pigmentation) (Solomons, N.W., 1985).

Copper deficiency is characterized by anemia, neutropenia (Nagano, T. et al., 2005; Harless, W. et al., 2006) and skeletal abnormalities (Fong, T. et al., 2007). Copper exerts a vital role in the biosynthesis of bone and connective tissues. The reduction of copper levels affects the activity of lysyl oxidase, which produces the necessary cross-linking of connective tissue, reducing collagen and elastin formation. Therefore, a reduction in the enzyme activity affects numerous tissues such as skin, bones and blood vessels (Gacheru S.N. et al., 1990; Romero-Chapman, N. et al., 1991). It has been empirically established that broken bones heal faster when copper supplements are supplied (Dollwet H.H.A.; Sorenson J.R.J, 1985). Insufficient copper intake has also been shown to lower bone calcium levels during long-term deficiency (Danks, D.M., 1980).

Cardiovascular diseases are related with the level of copper in the human diet. Copper deficiency can increase the plasma cholesterol concentration (with an increase in LDL-cholesterol and a decrease in HDL-cholesterol) giving rise to cardiovascular risk (Klevay L.M., 1975; Klevay L.M. et al., 1987; L.M. Klevay, 1987). In addition, a deficiency of copper depresses Cu-Zn SOD activity with an observed increase in the level of lipid hydroperoxides in liver mitochondria (Baleuska P.S. et al., 1981). The decrease in antioxidant protection caused by copper deficiency produces a decrease in the activity of copper-dependent antioxidant enzymes, inducing a wide range of disturbances in other antioxidant enzyme systems, such as catalase and glutathione peroxidase (Allen K.G.D. et al., 1988). Moreover, copper(II) ions added to hydrogen peroxide generate hydroxyl radicals capable of degrading deoxyribose with the formation of thiobarbituric acid-reactive products (TBARS), and this damage can be inhibited by catalase, OH· radical scavengers and specific metal ion chelators. In this case, copper ions appear to bind to the proteins that prevent formation of OH· radicals in free solution (Gutteridge, J.M.C.; Wilkins, S. 1983). Other results of superoxide-promoted Fenton chemistry substantiate the previously suggested models and site specific OH· radical production. The combined action of the copper ions and the superoxide radical requires the complexation of the metal ions with the amino acid residues of the enzyme. They might act as catalysts in a model reaction, giving rise to secondary OH· radicals that are formed in the immediate vicinity of the biological target (Samuni, M. et al., 1981).

The role of copper in controlling inflammation processes has also been established (Milanino, R. et al., 1979, 1985, 2006; Sorenson, R.J.R., 1982, Lewis, A.J. 1984). The increase in serum copper is a physiological response to inflammation (Sorenson, R.J.R., 1977) and the main copper-containing enzyme, ceruloplasmin, is significantly elevated in inflammatory conditions and has anti-inflammatory activity as part of the so-called acute phase response during stress such as inflammation (Frieden, E., 1986). Additionally, it has been shown that copper deficiency increases the severity of experimentally-induced inflammation (Sorenson, R.J.R., 1984). The effect of stress on cellular copper enzyme activity levels is largely unknown. Possibly, certain enzyme activities could increase because ceruloplasmin is thought to transport copper to nonhepatic copper enzymes. In contrast, certain copper enzyme activities may decrease because of the depletion of some copper pools during high rates of copper incorporation into ceruloplasmin (Di Silvestro, R.A.; Marten, J.T., 1990).

Rheumatoid arthritis (RA) is a chronic, destructive inflammatory polyarticular joint disease. It is characterized by massive synovial proliferation and subintimal infiltration of inflammatory cells which, along with angiogenesis, leads to the formation of a very aggressive tissue called pannus. Expansion of the pannus induces bone erosion and cartilage thinning, leading to the loss of joint function. The rheumatoid pannus can thus be considered a local tumor. In human synovial fluid, noncaeruloplasmin-bound copper is thought to be loosely bound to albumin or to amino acids, such as histidine, and able to catalyse the free-radical-mediated reactions that have been implicated as an important part of the destructive process that occurs within the rheumatoid joint. However, subsequent studies suggest that it appears unlikely that copper plays a significant role in those transition-metal-catalysed radical reactions (Winyard, P.G. et al., 1987). Copper bracelets have been used in the past for the treatment of arthritis. Recent studies have shown that some people with arthritis seem to have difficulty metabolizing copper from the food they eat, leading to increased pain. The dissolved copper from the bracelet bypasses the oral route by entering the body through the skin and it has been demonstrated that copper supplementation reduces rheumatoid arthritis (Jackson, G.E. et al., 1978). In contrast, it has been suggested that copper deficiency induced by complexation with Tetrathiomolybdate is a potent approach both to inhibit the progression of rheumatoid arthritis with minimal adverse effects (Omoto, A et al., 2005). These two aspects of the physiological activity of copper have been considered to be paradoxical.

The serum level of copper is also often elevated in animals and humans with cancer (S. Inutsuka et al., 1978; Willingham, W.M.; Sorenson J.R.J., 1986) and it has been suggested that it is a physiological response designed to activate copper enzymes in cancer cells to inhibit their growth. Indeed, numerous copper complexes that demonstrate SOD-mimetic properties, including copper salicylate, have been shown to possess anticancer, anticarcinogenic, and antimutagenic effects both in vitro and in vivo (Sorenson, R.J.R., 1987).

In summary, reduced copper contents produces a lose in main anti-oxidant defense, less activity in CuZnSOD, a drop in energy production, defects in the immune system and generate more degenerative diseases.

1.b. Copper Compounds: Antiinflammatory, Antioxidant, Antimicrobian, Antitumoral

Copper complexes have also shown potential to the treatment of numerous chronic diseases such as gastrointestinal ulcers, cancers, and diabetes. Many copper complexes demonstrate superoxide dismutase (SOD) activity (Sorenson, R.J.R., 1989) and possess anti-inflammatory activity, with almost always significantly stronger activity and lower toxicity than their parent compounds (Sorenson, R.J.R., 1976). Sorenson reported that Cu-complexes of anti-inflammatory drugs were more active in animal models than either their parent inorganic Cu(II) salt or the parent non steroidal anti-inflammatory drugs (NSAIDs) (Sorenson, J.R.J., 1982; Korolkiewicz, Z. et al., 1989; Miche, H. et al., 1997; Kovala-Demertzi et al., 2004; Halova-Lajoie et al., 2006). The enhanced potency of Cu-NSAID complexes, compared with their individual components, was corroborated in 1989, when a Cu-salicylate complex was reported to be significantly more active than copper nitrate mixed with Na-salicylate. Among inactive substances activated by copper(II), 3,5-diisopropylsalicylic acid (Dips) has attracted great interest because of its radioprotectant, anti-inflammatory, anticancer, antimutagenic, antidiabetic, analgesic, antineoplastic and anticonvulsant activities (Baquial, J.G.L. et al., 1995; Brumas, V. et al., 2007).

The pharmacological activity was proposed to be due to the inherent physico-chemical properties of the complex itself rather than just that of its constituents, since the amount of Cu in such complexes does not correlate with anti-inflammatory activity. Also, many complexing agents, which are pharmacologically inactive by themselves, display anti-inflammatory properties in association with copper(II) (Sorenson, J.R.J. 1976; 1982; 1983; 1989; Berthon, G., 1993; Weder, J.E., 2002). This may be for two reasons: either the ligand simply behaves as a carrier that brings copper to the therapeutic target (regardless of its subsequent interactions with endogenous ligands) or the complex *per se* interferes with the inflammatory process, where copper acts as a specific catalyst in that case.

Sorenson and other researches reported extensively on the anti-epileptic, anticancer, anti-diabetic, radiation injury protective, antibacterial, and, most significantly for NSAIDs, the gastric sparing activities of Cu(II) complexes (Sorenson, J.R.J. 1980; 1989). There are several possible mechanisms for the anti-inflammatory activity of copper complexes. The copper may induce lysyl oxidase activity, modulate prostaglandin synthesis, induce or mimic superoxide dismutase activity, decrease the permeability of human synovial lysosomes and modulate the physiological effects of histamine (Auer, D., 1989; Jackson, G.E. et al., 2000).

Inflammation and inflammation-associated diseases, such as reumathoidal arthritis and oesteoarthritis are treated using SOD as a pharmaceutical (Henke, S.L., 1999). SOD provides complete protection against degradation of Hyaluronic acid by superoxide and prevents arthritis (McCord J.M., 1974; J.R.J. Sorenson, 1976, 1978; Petrone, W.F. et al, 1980; May P.J. et al., 1981). A number of Cu-containing SOD mimics have been developed to try to improve the efficiency of the natural enzyme, facilitating oral administration. (De Alvare, L.R., 1976; Younes, M. et al., 1977; Weser, U. et al., 1978; Harrison, J.R., 1986; Steinkühler, C. et al., 1988; Sorenson, R.J.R., 1989; Henke, S.L.,1999; Ferrer, E.G. et al., 2007; Etcheverry, S.B. et al., 2007).

Some pyridine carboxylic acids have antimicrobial activity. In some cases the suppression of the mutagenicity of an active molecule can be obtained not only by the introduction of different chemical groups able to change the structure and the chemical properties of the molecule but also by the introduction of a metal species which produces a coordination complex, as it was previously detected for different complexes. Some copper complexes with pyridine carboxylic acids have the ability to inhibit the growth and metabolism of microorganisms or to kill them. On the other hand, some Copper complexes showed a similar or reduced activity as compared to the ligand itself. In several cases, the genotoxic properties of the ligands disappeared in the complexes (Carcelli, M. et al, 1995). Moreover, complexes of Schiff bases (Valent, A. et al., 2002) and those containing 2,2'-bipyridyl ligand (Senthil Kumar, R. et al, 2008), have these properties.

It has been suggested that an increase in superoxide due to the inhibition of SOD would affect both the growth and survival of parasited cells. Some compounds with complexing ability towards copper ions, that behave as SOD inhibitors and in which the mode of complexation could be related to the enzyme inhibition patterns showed antiparasitical activity (Rodríguez-Ciria, M. et al., 2007). Another antiparasitic reported Copper complex is the sinefungin compound that interacts with nucleic acids and cleaves phosphodiester bonds. The resulting complex activates H_2O_2, generating radical species, that cause oxidative damage to nucleic acids and their components (Cappannelli, M. et al., 2007).

In epileptic patients serum Cu concentration is elevated while Cu content is markedly reduced in brain tissue. Many classes of Cu chelates including amino acids, salicylates, and Schiff bases as well as Cu chelates of well-known antiepileptic drugs are more effective anticonvulsant drugs than the parent ligand. It has been suggested that Cu chelates formed in vivo are the active forms of antiepileptic drugs (Viossat, B. et al., 2003). For instance, Valproic acid (2-propylpentanoic acid) in the form of its sodium salt has a wide spectrum of activity as an anticonvulsant drug. Copper(II) complexes of anticonvulsant and antiinflammatory ligands have been found to be more active and desirable drugs than the parent ligands themselves (Sorenson, J.R.J., 1989).

In the search of less toxic metal-based antineoplastic drugs, essential metal complexes such as copper-based drugs have been developed, and many new complexes have demonstrated great antitumor potential. Copper complexes may have relatively lower side effects than platinum-based drugs, and are suggested to be able to overcome inherited or acquired resistance of cisplatin (Wang, T. et al., 2006). The cytotoxicity of these complexes can be explained by different mechanisms of action. Cancer cells have a less than normal superoxide dismutase (SOD) activity and the treatment with bovine native Cu-SOD decreased the growth of several solid tumors. As a consequence a variety of copper complexes presenting SOD-like activity were examined for anticancer treatments (Oberley, L.W. et al., 1979; Sorenson, J.R.J., 1987; Urquiola, C., 2008).

2. *Vanadium: Functions—Insulin Mimetic, Osteogenic and Antitumoral Agent*

Vanadium is a micro trace element considered to play important roles in living systems. It is essential for lower animal forms of life and plants. Its essentiality in human beings is still under discussion (Nielsen, F.H.; Uthus, E.O., 1990). Nevertheless, vanadium also produced toxic effects that depend on its concentration and its oxidation states, both cationic and anionic form (Sigel, H.; Sigel A., 1995). Vanadium is accumulated and retained in bones, and in this tissue, the halflife of vanadyl(V) amounts to about a month (Melchior, M. et al., 1999). The anionic form, Vanadate (VO_3^-, oxidation state +5) has more serious side effects because it is faster absorbed and can replace the bioequivalent anion phosphate in cellular metabolism since it can enter the cell via phosphate and sulfate channels (Scior, T. et al., 2005). Vanadyl sulfate is considerably less toxic but, only sparingly absorbed in the intestines (Rehder, D., 2003). For these reasons, scientists try to develop vanadium complexes with organic ligands in order to obtain better results from an increase of assimilation and pahrmacological actions under physiological conditions. On the other hand, Vanadium compounds are characterized by a broad spectrum of action in vivo and in vitro. They have antidiabetic properties, and influence processes related to mitogenic cell responses (proliferation, differentiation, apoptosis, neoplastic transformation) (Sakurai, H., 2002). However, both anti- (Djorkjevic, C., 1995. Liasko, R., 1998) and pro-neoplastic (Wang, H., 1995) properties of vanadium have been reported.

Their insulin-mimetic activity is manifested in their ability to normalize changes observed in both clinical and experimental diabetes (i.e. hyperglycaemia, hyperlipidaemia, reduced cell sensitivity to insulin) through the regulation of carbohydrate and lipid metabolism and the removal of secondary symptoms of this disease (as e.g. retinopathy, cardiomyopathy, nephropathy). Type 1 diabetes mellitus (DM) is treated only by daily insulin injections while type 2 DM can be treated by synthetic drugs. Clinical studies on vanadium administration showed that in type 1 diabetic patients the daily insulin requirement decreased significantly and in type 2 diabetic patients, the sensitivity of insulin increased by the enhancement of the inhibiting effect of insulin on hepatic glucose production, and the stimulation of peripheral glucose uptake (Brichard, S.M.; Henquin, J.C., 1995). Vanadium compounds are potent phosphotyrosine phosphatase (PTP) inhibitors, which are associated with the insulin receptor kinase (IRK). Vanadium inhibition allows IRK auto-phosphorylation and activation in the absence of insulin, followed by the signal transduction cascade and insulin responses (Cam, M.C. et al., 1999). Vanadate also mimics the late effects of insulin related to mitogenesis, by activating the MAP (mitogen activated protein)-kinase cascade (Etcheverry, S.B. et al., 1997; Cheta, A. et al., 2003).

The first orally active vanadyl(IV) complexes were proposed in 1990 (Sakurai, H., 2005). Vanadium complexes with cysteine-methylester and oxalate reduced the high levels of blood glucose in strptozotocin-induced hyperglycemic type 1 diabetic rats (STZ-rats). Numerous vanadium compounds have been proved to have great potential for the pharmacotherapy of diabetes (Heyliger, C.E. et al., 1985; Sakurai, H.A. et al., 1990; McNeill, J.H. et al., 1991; Orvig, C. et al., 1995; Poucheret, P. et al., 1998; Thompson, K.H., 1999). The ligands maltol (3-hydroxy-2-methyl-4-pyrone) and its derivatives (Mc. Neill et al. 1992; Thompson, K.H. et al.,

2002), picolinic acid (2-pyridinecarboxylic acid) (Yasui, H. et al., 2002), dipicolinic acid, acetylacetonate (Crans, D.C., 2000; Reul, B.A. et al., 1999) have increased efficacy and decreased toxicity of vanadium compounds (Crans, D.C., 2004). Orally active insulin-mimetic vanadyl complexes with the $VO(L_4)$ coordination mode (L= 4O; 2O,2N; 2N,2S; 4S; 2O,2S) have been synthesized (Sakurai, H. et al., 2002) for pharmaceutical use to treat or improve both types of DM.

Once absorbed, Vanadium mainly accumulates in kidneys, bones, liver and to a lesser extent in lungs, with bones being probably the main tissue of vanadium accumulation (Sakurai, H., 1994). Bone is a dynamic, highly vascularised tissue with a unique capacity to heal and remodel throughout life (Salgado, A.J. et al., 2004). Bone-forming cells (osteoblasts) and bone resorbing cells (osteoclasts) are involved in the turnover of bone matrix components and the maintenance of a balance between formation and resorption. Disbalance in skeletal cell activity can result in severe bone alterations. Vanadium compounds with specific properties to induce osteogenesis have been described. Their participation on the growth and differentiation of osteoblasts in culture has been studied. They stimulate type-I collagen production and also induced the formation of nodules of mineralization in normal osteoblastic cells (Barrio, D.A. et al., 2006; Ferrer, E.G. et al., 2006; Cortizo, A.M. et al., 2006).

The anticarcinogenic effects of vanadium, in combination to its low toxicity, established also by its administration in humans, suggest vanadium as a candidate antineoplastic agent against human cancer (Evangelou, A.M., 2002). Vanadium plays a potential role in limiting the initiation event of hepatocarcinogenesis in rats (Chakraborty, T. et al., 2007). Results of both the in vivo and in vitro studies demonstrate that vanadium has the potential to be developed into an anti-breast cancer drug in the near future. Histological finding showed substantial repair of hyperplastic lesions (Ray, S. et al., 2006).

In general, metallocenes inhibit DNA synthesis and are anti-mitotic. Organometallic complexes of vanadium(IV) with cyclopentadienyl ($C_5H_5-{=}Cp-$) moieties, vanadocenes (VCp_2), exhibit potent cytotoxic activity against human cancer cells primarily via oxidative damage (Ghosh, P. et al., 2000). We have determined the ability of vanadyl(IV) compounds to exhibit more deleterious behavior on tumor osteoblastic cell line proliferation than on the normal one (Etcheverry, S.B. et al., 2002; Barrio, D.A. et al., 2003; Molinuevo, M.S. et al., 2004). On the other hand, it has also been demonstrated that vanadium compounds in oxidation states V and III, such as peroxovanadate(V) complexes (Djordkevic, C. et al., 1985; Posner, B.I. et al., 1994) as well as V(III) compounds with aminoacids (Evangelou, A. et al., 1997; Osińska-Królicka, I. et al., 2004) also displayed anti-tumoral activity.

3. In Vitro Studies: Cellular Models

Cells in culture provide a simple model to study a variety of biological and pharmacological processes under control conditions. This system is useful to study the effects of different substances (growth factors, hormones, natural products and synthetic drugs) on the growth and development of different cell types. In particular, this in vitro system is useful to investigate the possible mechanisms of action used by different compunds to exert their biological or pharmacological effects.

MC3T3E1 cells: MC3T3E1 is a clonal osteogenic cell line derived from newborn mouse calvaria and selected based on high alkaline phosphatase (ALP) activity in the confluent state. Cells in the proliferative state display a fibroblastic morphology and grow in multiple layers. On day 21, when they are incubated with ascorbic acid (AA) and beta-glycerol-phosphate

(Franceschi R.T. et al., 1994), clusters of cells showing typical osteoblastic morphology can be observed in the cultures. Then these clusters begin to mineralize and nodules of the mineral phase of hard tissues can be seen in them. Such nodules increase in number and size with incubation time and become easily identifiable also with the naked eye by day 40-50. In the central part of the grown nodules there are osteocytes embedded in the mineralized bone matrix. Osteoblasts are placed at the periphery of the bone spicules and are surrounded by lysosome-rich cells and fibroblasts. Numerous matrix vesicles scattered around the osteoblasts and young osteocytes are also observed. Matrix vesicles and plasma membranes of osteoblasts, newly formed osteocytes, and lysosome-rich cells show strong alkaline phosphatase (ALP) activity. Inside these vesicles high level of calcium ions are determined which contribute to the solid phase formation together with the phosphate anion, the product of ALP reaction. Minerals are initially localized in the matrix vesicles and then deposited on collagen fibrils. Minerals consist of calcium and phosphorus, and also some hydroxyapatite crystals. These results indicate that MC3T3E1 cells have the capacity to differentiate into osteoblasts and osteocytes and to form calcified bone tissue in vitro (Sudo, H. et al., 1983). For these reasons, this cell line is very interesting since it reproduces along its evolution in culture systems all the stages of bone development in vivo: proliferation, differentiation and mineralization.

Control of osteoblast growth and development is a very complex process characterized for the participation of different hormones, cytokines, growth factors and also different ions and substances of diverse nature. Specially, this control can be characterized from receptor-mediated events to nuclear messengers controlling gene transcription. From this analysis, it is possible to formulate a model to explain the reciprocal relationship between growth and differentiation. Central to this model are putative tissue specific transcriptional switches that may repress proliferation and permit the regulation of mature osteoblast phenotypic characteristics. Tissue specific transcription factors determine the capacity to express osteoblastic characteristic, whereas receptor activated signaling cascades, namely, cAMP/protein kinase A (PKA), receptor serine/threonine kinase, and vitamin D receptor-dependent pathways, regulate mature osteoblast-specific gene expression. Activated differentiation switches also may feedback to transcriptionally repress proliferation. Conversely, in preosteoblasts, in which differentiation switches are turned off, distinct signaling cascades involving tyrosine kinases, protein kinase C (PKC), and calcium/calmodulin regulate proliferation. Proliferating preosteoblasts also exhibit negative modulation of maturation either through inactivation of putative tissue-specific transcription factors and/or through suppression of genes expressed in mature osteoblast. Thus, the final outcome of transcriptional regulation of osteoblast function results from complex interactions between signalling pathways and permissive differentiating transcription factors. This model serves as a useful conceptual framework to further investigate the differential control of osteoblast proliferation and differentiation that may lead to improved pharmacologic ways to manipulate bone formation in vivo (Siddhanti, S.R.; Quarles, L.D., 1994; Barrio, D.A.; Etcheverry, S.B., 2006).

MC3T3E1 osteoblast-like cells in culture develop features of the osteoblastic phenotype and form many cell layers embedded in extracellular matrix which can mineralize. Osteoblasts secrete a complex extracellular matrix (ECM) containing collagenous and noncollagenous proteins, bone morphogenetic proteins (BMPs), and growth factors. Osteoblast-specific gene expression requires ascorbic acid (AA and beta-glycero phosphate to produce the assembly of a collagenous ECM. Matrix responsiveness requires an alpha2beta1 integrin-collagen interaction and mitogen-activated protein kinase (MAPK) activity, which phosphorylates and activates the osteoblast-specific transcription factor Cbfa1 (Barrio, D.A.; Etcheverry, S.B., 2006). MC3T3E1 cells constitutively express BMP-2, BMP-4, and BMP-7. Noggin, a specific BMP inhibitor, reversibly blocked AA-induced gene expression, indicating that BMP production by MC3T3E1 cells is necessary for differentiation. Autocrine BMP production as well as integrin-mediated cell-collagen interactions is both required for osteoblast differentiation, and both these pathways require MAP kinase activity (Xiao, G. et al., 2002)

The relationship between collagen matrix formation and osteoblast-specific gene expression has been shown through kinetic studies which revealed that ascorbic acid increases proline hydroxylation in the intracellular procollagen pool and stimulates the cleavage of type I collagen propeptides in the process of collagen synthesis. Ascorbic acid also increases the rate of procollagen secretion from the osteoblast cell line to culture medium. Different expression times have been observed for other proteins constituents of the ECM in the presence of AA. These results indicate that the induction of osteoblast markers by ascorbic acid does not require the continuous hydroxylation and processing of procollagens and suggest that a stable, possibly matrix-associated signal is generated at early times after ascorbic acid addition that allows subsequent induction of osteoblast-related genes (Franceschi, R.T. et al., 1994).

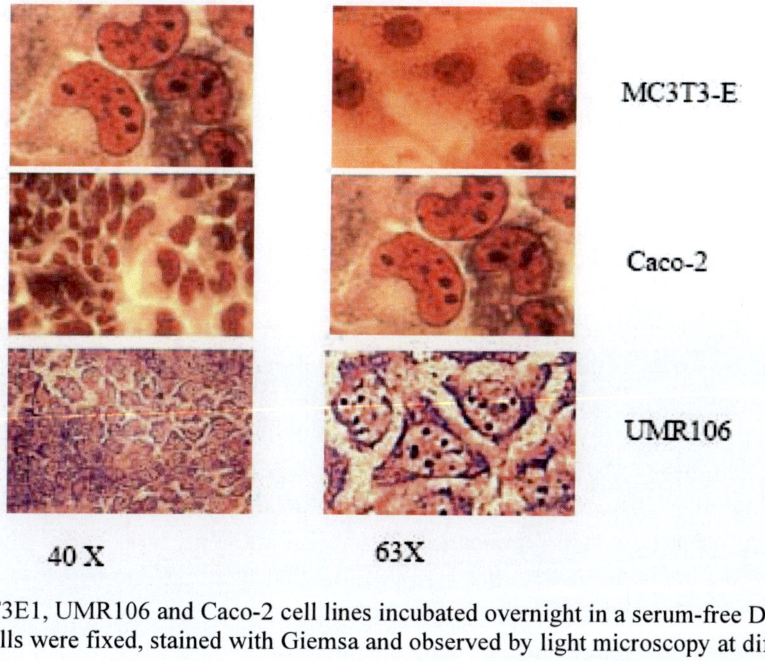

Figure 1. MC3T3E1, UMR106 and Caco-2 cell lines incubated overnight in a serum-free DMEM medium. The cells were fixed, stained with Giemsa and observed by light microscopy at different magnifying power.

The complexity and versatility of the differentiation process of preosteoblasts into mature osteoblasts depends on several factors such as hormones, growth factors, metal ions, etc. but also the effects caused by different factors and substances may be different when they are added to different stages of development of this cell line. For instance, the expression of the parathyroid hormone and its receptor (PTH/PTHrP), increases with osteoblastic differentiation. Some experiments previously reported suggest that the transforming growth factor TGF beta 1 inhibits osteoblastic maturation with more persistent effects found in less differentiated osteoblastic cells (McCauley, L.K. et al., 1995; 1996).

The influence of different metal ions in the proliferative and differentiative process of osteoblasts may be explained, at least partially through the experiments made by Quarles, L.D. et al., 1997). From this report, the presence of a cation-sensing mechanism in osteoblasts is suggested by the ability of specific cations to stimulate osteoblastic proliferation in culture and to induce de novo bone formation in some experimental models. Thus, study examines whether extracellular cations stimulate osteoblasts through the G protein-coupled calcium receptor (CaR). They showed that CaR agonists, calcium (Ca^{2+}), gadolinium (Gd^{3+}), aluminum (Al^{3+}), and neomycin, stimulated DNA synthesis in murine-derived MC3T3E1 preosteoblasts, whereas magnesium (Mg^{2+}), nickel (Ni^{2+}), cadmium (Cd^{2+}), and zinc (Zn^{2+}) had no effect. With the exception of Mg^{2+}, the cation specificities and apparent affinities were similar to that reported for CaR. Similar results were obtained by this group in other cell lines also in malignant cell lines such as UMR106 osteoblast-like cells derived from a rat osteosarcoma. These findings suggest that an extra-cellular cation-sensing mechanism is present in murine-derived osteoblasts that are functionally similar to CaR.

Cell death through apoptosis is a well-known mechanism for maintaining homoeostasis in many developmental and pathological processes. There is some presented evidence for the occurrence of apoptosis during the formation of bone-like tissue in vitro. Because apoptosis is a fundamental regulatory event during bone tissue differentiation, our findings emphasize the importance of thyroid hormones in bone maintenance and development. (Varga, F. et al., 1999).

The monolayer of MC3T3E1 in culture and stained with Giemsa shows the aspect of typical fibroblasts. As it has been established, in the proliferative stage these cells are preosteoblasts. Cells are stellate in shape and display slender lamellar expansions that connect each cell with its neighborings. The cytoplasms have numerous vacuoles and inclusions and the nuclei are round with thick chromatin granules inside (Figure 1).

UMR106 cells: Another cellular model used in in vitro studies with osteoblast-like cells from murine origin is the UMR106 line, derived from a rat osteosarcoma induced by ^{32}P. Grafts in the same rat colony preserve this cell line. When used together with a non-transformed osteoblast cell line like MC3T3E1 as control, this line is very interesting to investigate the effects of different drugs on a malignant osteosarcoma (Partidge, N.C. et al., 1983). This immortalized cell line, has the phenotype of differentiated osteoblasts but is unable to mineralize. It expresses high levels of ALP activity and produces type I collagen, but these osteoblast-like cells cannot produce bone mineral in culture. They have been characterized biochemically and it has been shown that this cell line expresses receptors for insulin, GF and PTH (Ituarte, E.A. et al., 1989; Thomas D.M. et al., 1996 a,b). These cells also have some of the features of primary osteoblasts such as ALP activity and adenylate cyclase (AC) and activation by PTH. Both parameters were preserved through successive passages. UMR106 cell line is a tumoral line with a high degree of differentiation and preserved mature osteoblastic properties. In control cultures (without addition of drugs to the culture medium) and with Giemsa staining, these cells show a polygonal morphology with well-stained nuclei and cytoplasms showing numerous processes among neighboring cells.

Many of the nuclei display a kidney-like shape and also some mitotic figures can be seen due to the active proliferating process of this cell line (Figure 1).

Caco-2 and TC-7 cells: Caco-2 cells, which express spontaneous enterocytic differentiation at confluency, are one of the most relevant in vitro models to study the differentiation and regulation of cell intestinal functions (Jumarie, C; Malo, C., 1991). Intestinal epithelium lies on the extracellular matrix (ECM) and is a very dynamic tissue that is continuously renewed (Teller, I.C.; Beaulieu, J.F., 2001). In the differentiation stage, they exhibit several of the morphological and biochemical characteristics of small intestinal enterocytes (Sambuy, Y. et al., 2005). This cell line displays the establishment of cell polarity and the expression of brush-border membrane enzyme markers (sucrase, maltase, lactase, alkaline phosphatase, gamma-glutamyltransferase, aminopeptidase N, and dipeptidyl-dipeptidase IV).

In vitro, Caco-2 cells undergo spontaneous differentiation in a period of ca 13-15 day culture. These cells can be maintained in serum-free medium and this system allows the study of the factors involved in the regulation of the differentiation of enterocyte in vitro. It is also a very interesting model to study the effects of different substances that are absorbed through an epithelial barrier. In the proliferative step they display the following morphological features: cells displayed polyhedral form with big nuclei and numerous nucleoles. Like UMR106, several intercellular connections and some mitotic figures are observed (Figure 1). TC-7, a subclone of Caco-2 is undistinguishable from the parental line on morphological base. They are characterized for a shorter time of differentiation (ca 7 days).

Copper and Vanadium Compounds

Potential Pharmacological Effects: In Vitro Studies

1. Copper(II) Compounds: Losartan and Santonic Acid—Antitumoral Actions

High blood pressure is the most important modifiable risk factor for stroke and other vascular diseases. Antihypertensive drugs were then used to lower blood pressure for stroke prevention. Losartan potassium, 2-butyl-4-chloro-1-[p-(O-1H-tetrazol-5-ylphenyl) benzyl] imidazole-5-methanol monopotassium salt (Figure 2) is a potent and orally effective pharmaceutical product used for the treatment of arterial hypertension (Wong, P.C. et al., 1991). It has been designed as a peptidomimetic of the hormone angiotensin II, and acts by selectively binding to and blocking the angiotensin II type 1 receptor, thus interfering with the rennin-angiotensin system, an important regulator of normal blood pressure (Moen, M.D.; Wagstaff, A.J., 2005). It has a good oral bioavailabitliy, rapidly absorbed reaching maximum concentrations 1–2 hours post-administration (Sica, D.A. et al., 2005). Taking into account that Losartan possess a number of properties, independent of its antihypertensive effects, that may be associated with decreased vulnerability of the plaque, myocardium, and blood (Díez, J., 2006), it has been interesting to undertake in our laboratory the study of the anti-neoplastic properties of the ligand and its copper(II) complex (Etcheverry S.B., 2007).

Sesquiterpene lactones are the active constituents of a variety of medicinal plants used in traditional medicine for the treatment of inflammatory diseases. In recent years, their anti-cancer property has attracted a great deal of interest and extensive research work has been carried out to characterize the anti-cancer activity, the molecular mechanisms, and the potential chemopreventive and chemotherapeutic application (Siyuan, Z. et al., 2005). Santonic acid ((-)-2,3,3a,4,5,6,7,7a-octahydro-α-3a.5trimethyl-6,8-dioxo-1,4-methano-1H-indene-1-acetic acid) (Figure 2) is derived from the sesquiterpene santonin, isolated from Artemisia, and acts as an antihelmintic agent. It is destructive to worms and used for removing internal parasitic worms in animals and humans (Brunskill, A.P.J. et al., 1999). It has then been very attractive to carry out the study of its anti-tumoral activity and the improvement of this potential property when complexing with Cu(II).

Losartan copper complex, $[Cu(Los)_2(H_2O)_3]_2$ was prepared upon slowly mixing aqueous solutions of copper(II) acetate and losartan potassium salt (molar ratio 1:2) (Etcheverry, S.B. et al., 2007). The light blue solid compound was filtrated, washed with water and dried in an oven at 60°C. In order to synthesize the complex of copper with Santonic acid, $[Cu_2(sant)_4(H_2O)_2].2½ H_2O$, a methanolic solution of the ligand has been added to a heated and stirred suspension of $CuCO_3.Cu(OH)_2$ (Williams et al., 2008). Green monocrystals where obtained upon recrystallization of the obtained green powder from ethyl alcohol.

Both compounds have been characterized by diffuse refectance spectra and FTIR spectroscopy. It can be concluded that Cu(II) binds to the N atoms of tetrazol moiety of Losartan and that the carboxylate group of santonic acid bridge two different copper ions in the complex. By electron paramagnetic spectroscopy, it has been demonstrated that both complexes are binuclear and the EPR spectra of the "mononuclear" impurities allowed the determination of the coordination spheres in each case (N and O_{water} for losartan and O for santonic acid). The latter structure was confirmed by RX diffraction measurements of the monocrystal.

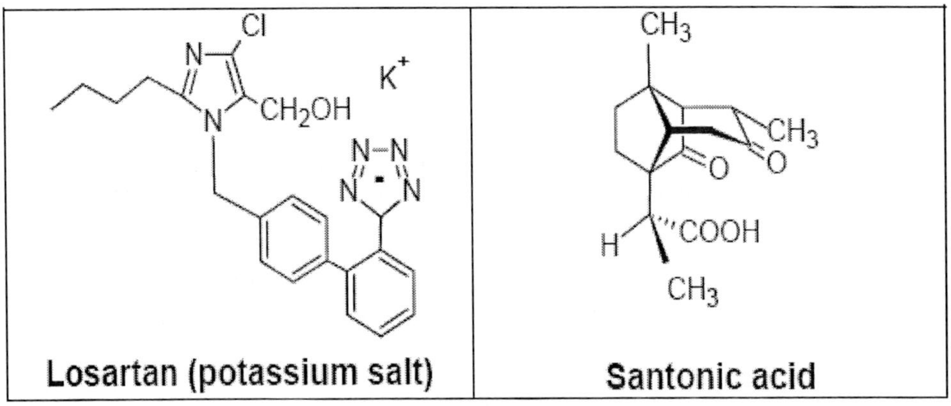

Figure 2. Schemes of losartan (potassium salt) and santonic acid.

The effect of both complexes on normal (MC3T3E1) and tumoral osteoblasts (UMR106) has been tested (Figure 3). Losartan and santonic acid do not exert any effects on cell proliferation of both osteoblastic lines (only at high doses the inhibition produced is of ca. 10%). Interestingly, copper(II) ions exerted a more deleterious activity on normal osteoblasts. The inhibition of proliferation has been improved by copper complexation of losartan in both cellular lines in a dose response manner. On the other hand, only the santonic acid complex behaves as an inhibitory agent on the tumoral osteoblasts proliferation, being more toxic (ca. 60% inhibition) than in the case of the normal bone cells (ca. 20% inhibition).

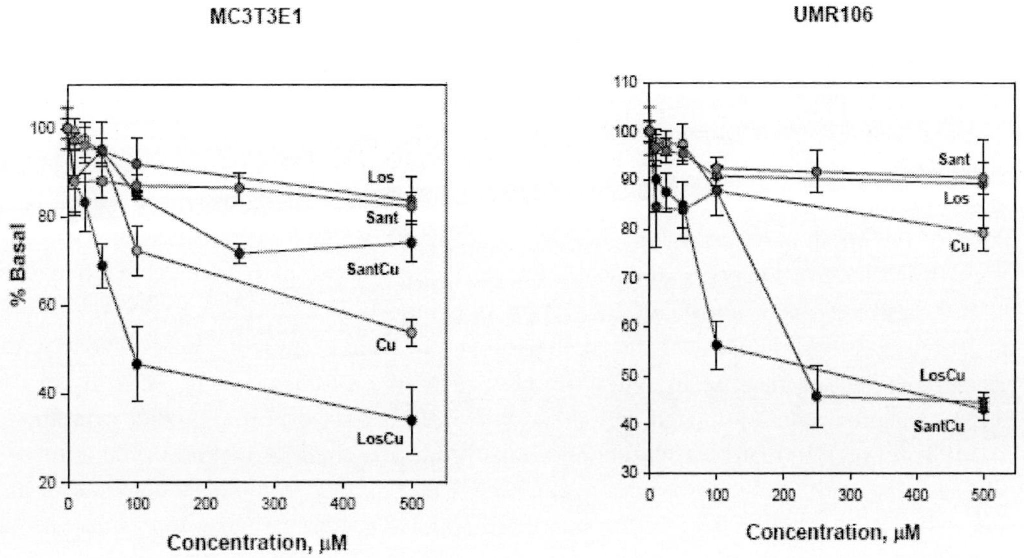

Figure 3. Effect of losartan, santonic acid, copper(II) and its complexes on normal and tumoral ostoblastic cell proliferation. Cells were incubated in serum-free DMEM alone (basal) or with different concentrations of the compounds at 37°C for 24 hours. The proliferation has been tested by the crystal violet bioassay. Results are expressed as % basal and represent the mean ± SEM, n = 9.

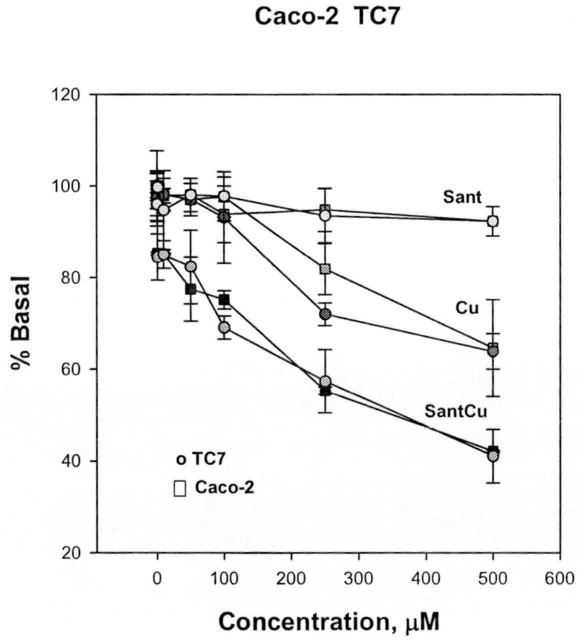

Figure 4. Effect of santonic acid, copper(II) and its complex on human adenocarcinoma Caco-2 and TC7 cell proliferation. Cells were incubated in serum-free DMEM alone (basal) or with different concentrations of the compounds at 37°C for 24 hours. The proliferation has been tested by the crystal violet bioassay. Results are expressed as % basal and represent the mean ± SEM, n = 9.

Taking into account the toxicity of the santonic copper complex on tumoral cells, its effect upon human colon adenocarcinoma cell lines, Caco-2 and TC-7 (a subclone of Caco-2 cells), has been determined. The improvement of the antitumoral action of copper cation by complexation with santonic acid is shown in the Figure 4 where the concentration of the complex is considered per copper atom. Therefore, in the three tested tumoral cell lines, SantCu exert a potential antitumoral effect.

Cellular morphologies are altered in concordance to the proliferative behavior. The study has been perfomed after overnight incubation in a serum-free medium with or without the addition of the compounds, and then the cells were fixed, stained with Giemsa and observed by light microscopy. Losartan and santonic acid did not produce noticeable changes in cellular morphology. A 500 μM $CuCl_2$ solution added to MC3T3E1 cell lines and the losartan complex, produced a decrease in the cell number per field, a gradual loss of connections between cells and an increase in cytoplasm condensation. A 500 μM santonic acid complex solution generates more elongated cell shape but slight decrease in the number of cells. For the UMR106 line the treatment with copper ions produce a small reduction in the number of surviving cells, but the higher toxicity of the LosCu and SantCu complexes generates a great reduction in the number of living cells and a strong condensation of the cytoplasms with loss of connections between cells compared to basal or control conditions (without addition of compound). The cells display a neuron like shape when a 500 μM solution of SantCu was added.

The morphologies of Caco-2 cell lines and their subclone, TC7, where similar. Santonic acid did not induce morphological changes. The effect of SantCu was more marked than the effect of the metal ions alone. Both compounds produce rather bad defined cytoplasmic borders, irregular vacuoles immersed in the cytoplasm and the nuclei were more condensed and strongly stained. The number of cells per field diminishes from copper ions to the SantCu complex, indicating a more severe cytotoxic action for the complex than for the copper ions.

2. Vanadium-Related Compounds: Trehalose, Quercetin and Hesperidin—Biological Actions

Flavonoids occur in most plant species, e.g., dried green tea leaves contain approximately 30% flavonoids by weight. Flavonoids have been shown to have antibacterial, anti-inflammatory, antiallergic, antimutagenic, antiviral, antineoplastic, anti-thrombotic, and vasodilatory activity. The potent antioxidant activity of flavonoids, their ability to scavenge hydroxyl radicals, superoxide anions, and lipid peroxy radicals, may be the most important function of flavonoids, and underlies many of the above actions in the body (Tripoli, E. et al., 2007) Quercetin (3,3',4',5,7-pentahydroxyflavone) (Figure 5) exhibits anti-tumor and antioxidant activities (Petta, P.J., 2000; Harwood, M. et al., 2007). We have synthesized the complex of Quercetin with the vanadyl(IV) cation and found very interesting biological properties (Ferrer, E.G. et al., 2006).The complex behaves as antitumoral and osteogenic. The complex has also been tested for insulin-enhancing activity, (Shukla, R. et al., 2004), the reduction of oxidative stress induced by vanadium toxicity (Shukla, R. et al., 2006) and the improvement of carbohydrate metabolism and reduction of oxidative stress (Shukla, R. et al., 2007).

Another interesting vanadium(IV) compound is the complex formed between vanadyl(IV) cation and the disaccharide trehalose. Trehalose (α-D-glucopyranosyl-α-D-glucopyranoside) (Figure 5) is present in plants, algae, fungi, yeasts, bacteria, insects and other invertebrates and functions in many organisms as an energy source or a protectant compound against the effects of freezing or dehydration. It also possesses physical and/or chemical properties that are different than other sugars, and is safe for use as a food ingredient, in health, beauty and for pharmaceutical products (Richards, A.B. et al, 2002). We have prepared the VO(IV) complex with the trehalose dihydrate and it presents insulin mimetic, anti neoplastic and osteogenic properties (Barrio, D.A. et al., 2003).

Taking into account the biological activities of vanadyl(IV) complexes with flavonoids and disaccharides, we have prepared the complex of a related ligand, a flavonoid linked to a disaccharide, hesperidin (Etcheverry, S.B. et al., 2007). The flavonoid hesperidin is a flavanone glycoside (glucoside) comprised of the flavanone (a class of flavonoids) hesperitin and the disaccharide rutinose (Figure 5). Hesperidin is the predominant flavonoid in lemons and oranges. The peel and membranous parts of these fruits have the highest hesperidin concentrations. Hesperidin is named hesperetin 7-rhamnoglucoside, hesperetin-7-rutinoside and (S)-7-[[6-0-(6-deoxy-alpha-L-mannop-yranosyl)-beta-D-glucopyranosyl]oxy]-2,3-dihydro-5-hydroxy-2-(3-hydroxy-4-methoxyphenyl)-4*H*-1-benzopyran-4-one (Miller, A.L., 1996). The complex of hesperidin with vanadyl(IV) cation displays antitumoral and antioxidant activities.

The vanadyl complex of threhalose, TreVO, $Na_6[VO(Tre)_2].4H_2O$ was prepared mixing $VOCl_2$ and trehalose (in water, 1:2 ratio) at pH 13. The green powder obtained after the addition of absolute ethanol was washed and stored in an oven at 60°C. The quercetin complex, QuerVO, $[VO(Quer)_2EtOH]_n$ has been synthesized by refluxing $VOCl_2$ with an ethanolic solution of quercetin (3 h, pH 4). The green solid was washed and air dried. VOHesp, $[VO(Hesp)(OH)_3]Na_4.3H_2O$ was prepared in a similar way than trehalose. The three compounds were characterized by physicochemical techniques. The V=O stretching bands are indicative of the coordination sphere especially in the case of carbohydrate bonds. For TreVO this band appeared at 941 cm^{-1} and in the case of HespVO, at 931 cm^{-1}, indicating a direct interaction of the metal center with the sugar moiety of hesperidin. On the other hand, the position of ν(V=O) for QuerVO (977cm^{-1}) is typical for an oxygenated sphere around the vanadium center. The polymeric nature of QuerVO was determined by EPR powder spectroscopy (broad quasi isotropical signal, characteristic of magnetically extended VO(IV) centers. Small amount of monomeric impurities allow establishing the equatorial coordination sphere of the complex that includes four different oxygenated moieties. On the contrary, the EPR data confirmed the monomeric nature of HespVO. The Spin Hamiltonian parameters are in accordance with other vanadyl-saccharide complexes, confirming that the metal binds the rutinose moiety of hesperidin.

When the vanadyl(IV) compounds were tested on UMR106 cell proliferation they exhibited an antiproliferative effect in dose response manner (Figure 6). HespVO produced the more deleterious effect in comparison with the other complexes. On the contrary, free VO(IV) caused a slight stimulatory effect and the free ligands did not cause any change respect to basal conditions.

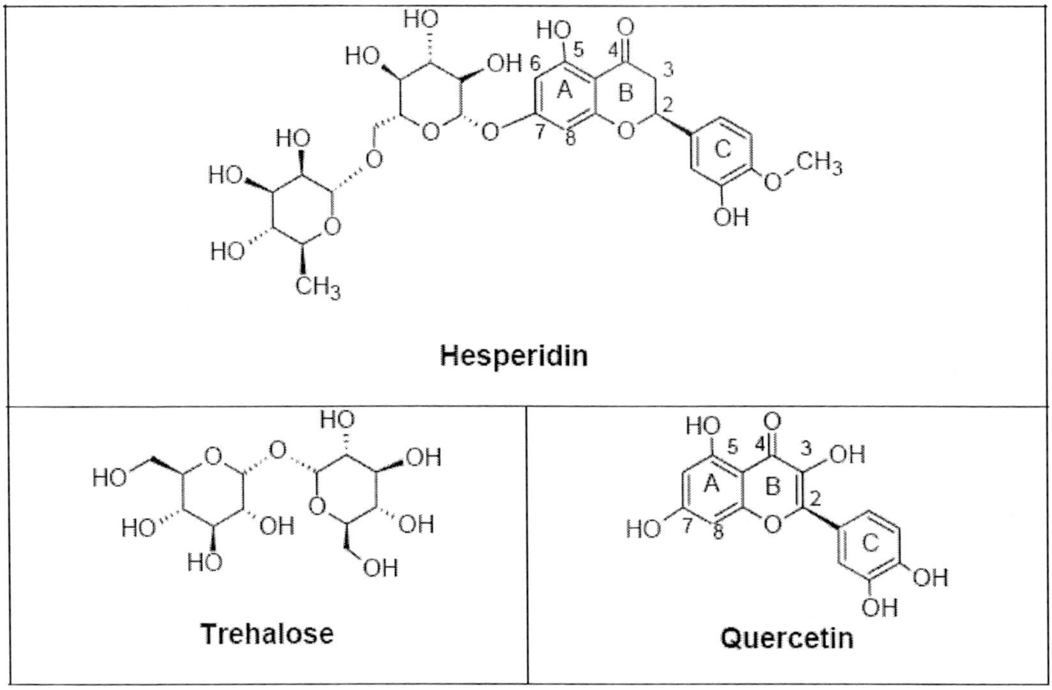

Figure 5. Schemes of hesperidin, trehalose and quercetin.

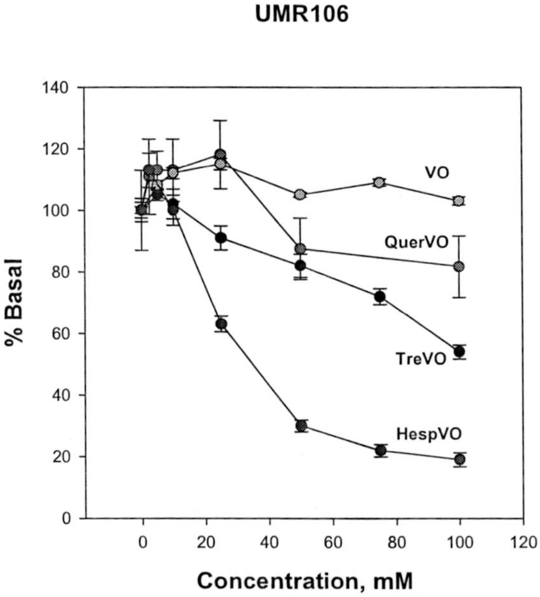

Figure 6. Effect of VO(IV) and its complexes with quercetin, trehalose and hesperidin on tumoral ostoblastic cell proliferation. Cells were incubated in serum-free DMEM alone (basal) or with different concentrations of the compounds at 37 °C for 24 hours. The proliferation has been tested by the crystal violet bioassay. Results are expressed as % basal and represent the mean ± SEM, n = 9.

Extracellular matrix (ECM) proteins play a critical role in many cellular functions, from spreading, migration, proliferation, differentiation and apoptosis. ECM is composed of a rich variety of different proteins such as laminin-1, fibronectin, collagens, etc. Laminin-1 is as an effective attachment protein for a large variety of cultured cell types and enhances cell survival, proliferation and differentiation (Basson, M.D. et al., 1996; Basson, M.D. et al., 2000; Ekblom, P. et al., 2003). Fibronectin increases adhesion-dependent cell growth (Sottile, J.; Hocking, D.C., 2002) and proliferation (Schwarzbauer, J.E.; Sechler, J.L., 1999; Danen, E.H.J.; Yamada, K.M., 2001). Besides, adhesion, spreading and proliferation are positively regulated by collagen-1, the most abundant component of ECM (Basson, M.D. et al., 1996). From the above, the effects of ECM components on cell growth and differentiation has been well established on the basis of several in culture studies. Besides, at hard tissue level, the specific activity of alkaline phosphatase (ALP) is another crucial marker of osteoblast differentiation. ALP is a widely distributed transmembrane enzyme that is expressed in most cells and tissues of vertebrates. Various isoenzymes and isoforms of ALP are known at present but only the specific action of the bone isoform is understood (Whyte, M.P., 2002).

Osteoblasts during the course of their differentiation into mature phenotype show a very high activity of ALP in their cellular membranes. This enzyme hydrolyzes phosphate esters providing in that way the phosphate concentration needed for the precipitation of the mineral phase of bones. The effect of several vanadium complexes with organic ligands on osteoblast differentiation have been previously reported for our group using as markers of osteoblast differentiation the levels of ALP and the production of collagenous-1. For one of these compounds, a vanadyl(IV) complex with the disaccharide trehalose (TreVO), the study was also extended to the mineralization stage. TreVO was a good promoter of collagen production. Besides, it caused only a weak inhibition of ALP activity (Barrio, D.A. et al., 2003). As a whole, these two actions of TreVO in UMR106 osteosarcoma cells encouraged the study of this complex on the mineralization process. Taking into account that the immortalized cell line UMR106 is unable to reach the mineralization stages, long-term studies were carried out using the MC3T3E1 osteoblast-like cells to test the effect of TreVO on the formation of the mineral phase of bone. In the non-transformed osteoblasts MC3T3E1, TreVO also produced a slight decrease in ALP activity, an increase in collagen type I production and a promotion of the mineralization of ECM (Etcheverry, S.B.; Barrio, D.A., 2007). QuerVO was a weaker ALP inhibitor than vanadyl cation. In UMR106 cell line it inhibited this specific enzymatic activity in a similar way to TreVO and as it caused a significant stimulation of collagen type I production, it has been also considered as a potential osteogenic compound (Ferrer, E.G. et al., 2006). In UMR106 cells, the complex of Hesperidin with vanadyl(IV) cation, HespVO, produced a standard behavior on ALP specific activity. It behaved as a stronger inhibitory agent than TreVO and QuerVO. Opposite to the actions of trehalose and quercetin that did not cause any effect on ALP activity, hesperidin produced a slight decrease of the enzyme activity by itself. Considering collagen type I production as other important characteristic for osteoblast differentiation, it has been shown that HespVO and VO(IV) slightly promoted the synthesis of collagen up to 10.0 μM. The comparison of HespVO with TreVO and QuerVO indicate, in agreement with the cell culture results that TreVO and QuerVO may be considered as potential osteogenic agents but this is not the case for HespVO. The latter could be a potential antitumoral compound for further investigations in cancer treatments due to is cytotoxicity in two tumoral cell lines: UMR106 osteosarcoma cells and Caco-2 human adenocarcinoma cells (Etcheverry, S.B. et al., 2008).

Studies in different cell types have shown that vanadium compounds may use different signal transduction pathways to exert their biological and pharmacological actions (Swarup, G. et al., 1982; Tracey, A.S.; Gresser, M.J., 1986; Shisheva, A.; Shechter, Y., 1992; Pandey, S.K. et al., 1999). One of the major signaling pathways apparently involved in vanadium biological effects is the mitogen activated protein kinase (MAPK) cascade. In mammalian cells, the major subfamilies of structurally related MAPKs that have been identified are the extracellular signal-regulated kinases (ERKs), the c-jum N-terminal kinases (JNKs) and the p38 MAPkinases (p38s) (Cano, E., 1995; Marshall, C.J., 1995; Robinson, M.J.; Cobb, M.H., 1997). Besides, it has also been reported that the stimulation of the ras-ERK cascade by vanadyl sulfate is dependent on the activation of the phosphatidyl inositol-3 kinase (PI3-K) (Pandey, S.K. et al., 1999). Even though ERK pathway is classically recognized as a key transducer in the signal cascade that mediates cell proliferation and differentiation as well as protection from apoptosis (Xia, Z. et al., 1995), it has been shown that this signal cascade is also related with cell cycle arrest, antiproliferation and apoptotic and non apoptotic cell death (Blázquez, G. et al., 2000). On the other hand, vanadium compounds also show some toxic effects, which depend on several factors, as stated above in this chapter, such as the oxidation state, the coordination sphere and the cell type or tissue investigated (Domingo, J.L., 1996; Cortizo, A.M. et al., 2000).

To investigate the signal transduction pathways involved in the mechanism of action of the TreVO complex, its effect on the activation of ERK (through a phosphorylation cascade) was analyzed by Western blot using specific antibodies for ERK1/2 and PERK1/2. TreVO stimulated ERK phosphorylation in a dose-dependent manner from 50 to 1000 μM. The stimulation at a fixed concentration is also time-dependent. Indeed, the maximum effect of 500 μM was reached at 1 hour of incubation and began to decline thereafter. To try to elucidate the mechanism by which TreVO induced ERK phosphorylation, the potential role of both oxidative and MAPK pathways were investigated using specific inhibitors such as PD98059, wortmannin and a mixture of vitamins E and C (scavenger of reactive oxygen species (ROS)). Under these experimental conditions, ERK phosphorylation was induced by low doses (25μM) of TreVO in the MC3T3E1 cell line. The stimulatory effect (175% over basal, $P<0.001$) was totally blunted by either PD98059 or wortmannin. However, the mixture of vitamins E and C did not affect the TreVO-induced ERK activation. These results suggest that the effect of low doses of TreVO, which have shown to enhance proliferation in osteoblasts in culture (Barrio, D.A. et al., 2003), could be mediated by the PI3K-MEK-ERK pathway and may be independent of the formation of free radicals in these cells. On the other hand, the ERK phosphorylation induced by high concentrations of TreVO (1000 μM), which was deleterious for osteoblasts in culture, could be partially inhibited by wortmannin, PD98059 or a mixture of vitamins. An additive effect could be observed when the cells were incubated with the mixture of all the used inhibitors that totally blocked the effect of TreVO on ERK phosphorylation. The complex of vanadyl(IV) with the disaccharide trehalose is a very interesting vanadium compound from a pharmacological point of view, since at low doses it exerts beneficial effects on osteoblasts in culture, and at high concentrations it may act potentially as an antitumoral agent since its deleterious effect at these doses is higher in tumoral osteosarcoma cell line UMR106 than in the non-transformed osteoblasts MC3T3E1 (Barrio, D.A. et al., 2003). Altogether, these results suggest that the activation of ERK at higher concentrations of TreVO could occur through at least two mechanisms, one dependent on and the other independent of the oxidative stress. On the other hand, the complex of vanadyl(IV) cation and the flavonoid quercetin, QuerVO, also stimulates osteoblast proliferation and activates ERK cascade (Ferrer, E.G. et al., 2006), although in this case specific inhibitors have not been employed to finally prove the involvement of ERK cascade activation in the proliferative effect of the complex in osteoblast in culture.

Finally, the complex HespVO, which has shown potentially useful in activity in two tumoral cell lines (UMR106 and Caco-2), possibly exerts its antitumoral effect by enhancing the phosphorylation of some signaling transduction pathways involved in cell cycle arrest and apoptosis.

CONCLUSION

On the basis of the presented results, the synthesis of new organic derivatives of copper(II) and vanadyl(IV) cations are of great interest in medicinal chemistry. In particular, it is very useful to obtain compounds with organic ligands that display therapeutic actions by themselves. Their complexation with metal ions modulates the pharmacological and/or biological actions, improving them in most cases. Moreover, the cellular in vitro system of appropriate cell types is a good laboratory model for investigating the biochemistry and the possible mechanisms of action employed by the metal complexes to exert their actions. Specifically, the osteoblast cells in culture allow the investigation of the effect of copper and vanadium compounds on osteoblast growth, differentiation and mineralization processes. Moreover, the tests in two different tumoral cell lines allow selection of candidate compounds for further investigation in antineoplastic therapies. Collectively taken, the results indicate that the complexation of therapeutical ligands with biometals renders complexes that, on the basis of these experiments, seem to be good candidates for antitumoral, antidiabetic, antioxidant and osteogenic treatments.

ACKNOWLEDGMENTS

The authors acknowledge the support from UNLP, CICPBA, CONICET (PIP 6366) and ANPCyT (PICT 10968). PAMW is a member of the Carrera del Investigador, CICPBA and SBE is a member of the Carrera del Investigador CONICET, Argentina.

REFERENCES

Allen, K.G.D.; Arthur, J.R.; Morrice, P.C.; Nicol, F.; Mills C.F. *Proc. Soc. Exp. Biol. Med.* 1988, 187, 38-43.

Auer, D. In: Free Radicals, Inflammation and Copper: Conference Proceedings, 11th Bain-Fallon Memorial Lecture, Australian Equine Veterinary Association, Coffs Harbour, Australia, 1989.

Baleuska, P.S.; Russanov, E.M.; Kassabova, T.A. *Int. J. Biochem.* 1981, 13, 489-493.

Baquial, J.G.L.; Sorenson, J.R.J. *J. Inorg. Biochem.* 1995, 60, 133-148.

Baran, E.J. In: Química Bioinorgánica; Mc.Graw-Hill; Madrid; 1994.

Barrio, D.A.; Cattáneo, E.R.; Apezteguía, M.C.; Etcheverry, S.B. *Can. J. Physiol. Pharmacol.* 2006, 84, 765-775.

Barrio, D.A.; Williams, P.A.M.; Cortizo, A.M.; Etcheverry, S.B. *J. Biol. Inorg. Chem.* 2003, 8, 459-468.

Barrio, D.A.; Etcheverry, S.B. *Can. J. Physiol. Pharmacol.* 2006, 84, 677-686.

Basson, M.D.; Emenaker, N.J.; Sanders, M.A. *Cell. Physiol. Biochem.* 2000, 10, 27-36.

Basson, M.D.; Turowski, G.; Emenaker, N.J. *Exp. Cell Res.* 1996, 255, 301-305.

Berthon, G. Agents Actions 1993, 39, 210-217.

Blázquez, G.; Galve-Roperh, I.; Guzman, M. *FASEB J.* 2000, 14, 2315-2322.

Brichard, S.M.; Henquin J.C. *Trends Pharmacol. Sci.* 1995, 16, 265-270.

Brumas, V.; Miche, H.; Fiallo, M. *J. Inorg. Biochem.* 2007, 101, 565–577.

Brunskill, A.P.J.; Thompson, H.W.; Lalancette, A. *Acta Cryst.* 1999, C55, 566-568.

Cam, M.C.; Rodrigues, B.; McNeill, J.H. *Eur. J. Endocrinol.* 1999, 141, 546-554.

Camakaris, J.; Voskoboinik, I.; Mercer, J.F. *Biochem. Biophys. Res. Commun.* 1999, 261, 225-232.

Cano, E.; Mahadevan, L.C. *Trends Biochem. Sci.* 1995, 20, 117-122.

Cappannelli, M.; Gaggelli, E.; Jezowska-Bojczuk, M.; Molteni, E.; Mucha, A.; Porciatti, E.; Valensin, D.; Valensin, G. *J. Inorg. Biochem.* 2007, 101, 1005-1012.

Carcelli, M.; Mazza, P.; Pelizzi, C.; Pelizzi, G.; Zani, F. *J. lnorg. Biochem.* 1995, 57, 43-62.

Chakraborty, T.; Chatterjee, A.; Rana, A.; Dhachinamoorthi, D.; Kumar A.; Chatterjee, M. *Biochim. Biophys. Acta* 2007, 1772, 48-59.

Cheta, D.; Orasanu, G.; Nicolaie, T.; Iordachescu, D.; Buligescu, S.; Constantin, C.; Hassanain, M.; Coman, A.; Enache, M.; Negru, R.; Tica, V.; Timofte, D.; Gutu, D.; Panaite, C. *J. Cell. Mol. Med.* 2003, 7, 447-454.

Cortizo, A.M.; Bruzzone, L.; Molinuevo, S.; Etcheverry, S.B. *Toxicology* 2000, 147, 89-99.

Cortizo, A.M.; Molinuevo, M.S.; Barrio, D.A.; Bruzzone, L. *Int. J. Biochem. Cell Biol.* 2006, 38, 1171-1180.

Crans, D.C. *J. Inorg. Biochem.* 2000, 80, 123-131.

Crans, D.C.; Smee, J.J.; Gaidamauskas, E.; Yang, L. *Chem. Rev.* 2004, 104, 849-902

Danen, E.H.J.; Yamada, K.M. *J. Cell Phys.* 2001, 189, 1-13.

Danks, D.M. Copper Deficiency in Humans. In: "Biological Roles of Copper." CIBA Foundation Symposium-79. Exerpta Medica, Amsterdam, 1980. p. 209.

De Alvare, L.R.; Goda, K.; Kimura, T. *Biochem. Biophys. Res. Commun.* 1976, 69, 687-694.

Díez, J. *Clin. Therapeut.* 2006, 28, 832-848.

DiSilvestro, R.A.; Marten J.T. *J. Nutrition* 1990, 120, 1223-1227.

Djordjevic, C. *Met. Íons. Biol. Syst.* 1995, 31, 595-616.

Djordjevic, C.; Craig, S.A.; Sinn, E. *Inorg. Chem.* 1985, 24, 1281-1283.

Dollwet, H.H.A.; Sorenson, J.R.J. *Tr. Elem. in Med.* 1985, 2, 80-87.

Domingo, J.L. *Reprod. Toxicol.* 1996, 10,175-182.

Ekblom, P.; Lonai, P.; Talts, J.F. *Matrix Biol.* 2003, 22, 35-47.

Etcheverry, S.B.; Crans, D.C.; Keramidas, A.D.; Cortizo. A.M. *Arch. Biochem. Biophys.* 1997, 338, 7-14.

Etcheverry, S.B.; Barrio, D.A.; Cortizo, A.M.; Williams. P.A.M. *J. Inorg. Biochem.* 2002, 88, 94-100.

Etcheverry S.B.; Ferrer, E.G.; Naso, L.; Barrio, D.A.; Lezama, L.; Rojo, T.; Williams, P.A.M. *Bioorg. Med. Chem.* 2007, 15, 6418-6424.

Etcheverry, S.B.; Barrio, D.A. Vanadium and bone. Relevance of vanadium compounds in bone cells In: Vanadium: The versatile element. Kustin, K.; Costa Pesoa, J. Crans, D.C. (Eds) American Chemical Society Series 974. Vol 15: 204-216, 2007

Etcheverry, S.B., Ferrer, E.G., Naso, L.; Rivadeneira, J.; Salinas, V.; Williams, P.A.M. *J. Biol. Inorg. Chem.*, 2008, 13, 435-447.

Evangelou, A.; Karkabounas, S.; Kalpouzos, G.; Malamas, M.; Llasko, R.; Stefanou, D.; Vlahos, A.T.; Kabanos, T.A. *Cancer Lett.* 1997, 119, 221-225.

Evangelou A.M. *Crit. Rev. Onc. Hemat.* 2002, 42, 249-265.

Ferrer, E.G.; Salinas, M.V.; Correa, M.J.; Naso, L.; Barrio, D.A.; Etcheverry, S.B.; Lezama, L.; Rojo, T.; Williams, P.A.M. *J. Biol. Inorg. Chem.* 2006, 11, 791-801.

Ferrer, E.G.; López Tévez, L.L.; Baeza, N.; Correa, M.J.; Okulik, N.; Lezama, L.; Rojo, T.; Castellano, E.E.; Piro, O.E.; Williams. P.A.M. *J. Inorg. Biochem.* 2007, 101, 741-749.

Fong, T.; Vij, R.; Vijayan, A.; DiPersio, J.; Blinder, M. *Haematologica* 2007, 92,1429-1430.

Franceschi, R.T.; Iyer, B.S.; Cui Y. *J. Bone Miner. Res.* 1994, 9, 843-854.

Frieden, E. *Clin. Physiol. Biochem.* 1986, 4, 11-19.

Gacheru, S.N.; Trackman, P.C.; Shah, M.A.; O'Gara C.Y.; Spacciapoli, P.; Greenaway, F.T.; Kagan, H.M. *J. Biol. Chem.* 1990, 265, 19022-19027.

Ghosh, P.; D'Cruz, O.J.; Narla, R.K.; Uckun, F.M. *Clin. Cancer Res.* 2000, 6, 1536-1545.

Gubler, C.J.; Lahey, M.E.; Cartwright, G.E., Wintrobe, M.M. *J. Clin. Invest.* 1953, 32, 405-414.

Gutteridge, J.M.C.; Wilkins, S. Biochim. Biophys. Acta 1983, 759, 38-41.

Halova-Lajoie, B.; Brumas, V.; Fiallo M.M.L.; Berthon, G. *J. Inorg. Biochem.* 2006, 100, 362-373

Harless, W.; Crowell, E.; Abraham *J. Am. J. Hematol.* 2006, 81, 546-549.

Harris, E.D. In: Trace Elements in Health; Ed. J. Rose, Butterworths. London, 1983.

Harrison, J.R.; Rillema, D.P.; Ham, J.H.; Sando, J.*J. Cancer Res.* 1986, 46, 5571-5575.

Harwood M.; Danielewska-Nikiel, B; Borzelleca, J.F.; Flamm; G.W.; Williams, G.M.; Lines, T.C. *Food Chem. Toxicol.* 2007, 45, 2179-2205

Henke, S.L. Expert Opin. Ther. Pat. 1999, 9, 169-180.

Heyliger, C.E.; Tahiliani, A.G.; McNeill, J.H. *Science*, 1985, 227, 1474-1477.

Inutsuka, S.; Araki, S. *Cancer* 1978, 42, 626-631.

Ituarte, E.A.; Halstead, L.R.; Iida-Klein, A.; Ituarte, H.G.; Hahn, T.J. *Calcif. Tissue Int.* 1989, 45, 27-33.

Jackson, G.E.; May, P.M.; Williams, D.R. *J. Inorg. Nucl. Chem.* 1978, 40, 1189-1194.

Jackson, G.E.; Mkhonta-Gama, L.; Voyé, A.; Kelly, M. *J. Inorg. Biochem.* 2000, 79, 147-152.

Jumarie, C.; Malo, C. *J. Cell. Physiol.* 1991, 149, 24-33.

Klevay, L.M. *Med. Hypothesis* 1987, 24, 111-119.

Klevay, L.M.; Inman, L.; Johnson, L.K.; Lawler, M.; Mahalko, J.R.; Milne, D.B.; Lukaski, H.C.; Bolonchuk, W.; Sandstead, H.H. *Metabolism* 1984, 33, 1112-1118.

Korolkiewicz, Z.; Hac, E.; Gagalo, I.; Gorczyca, P.; Lodzinska, A. *Agents Actions* 1989, 26, 355–359.

Kovala-Demertzi, D.; Galani, A.; Demertzis, M.A.; Skoulika, S. *J. Inorg. Biochem.* 2004, 98, 358-364

Lewis, A. J. *Agents Actions* 1984, 15, 513–519.

Liasko, R.; Kabanos, T.A.; Karkabounas, S.; Malamas, M.; Tasiopoulos, A.J.; Stefanou, D.; Collery, P.; Evangelou, A. *Anticancer Res* 1998, 18, 3609-3613.

Marshall, C.J. *Cell* 1995, 80,179-185.

Mason, K.E., *J. Nutrit.* 1979, 109, 1979-2066.

May P.M.; Williams, D.R. Role of low molecular weight complexes in the control of rheumatoid arthritis. InMetal Ions in Biological Systems. (Ed. H. Sigel) Marcel Dekker, New York 1981, Vol. 12, 283-317.

McCauley, L.K.; Koh, A.J.; Beecher, C.A.; Cui, Y.; Decker, J.D.; Franceschi, R.T. *J. Bone Miner. Res.* 1995, 10, 1243-1255.

McCauley L.K.; Koh, A.J.; Beecher C.A.; Cui Y.; Rosol, T.J.; Franceschi, R.T. *J. Cell Biochem.* 1996, 61, 638-647.

McCord, J.M. *Science (*Washington, DC) 1974,185, 529-531.

McNeill, J.H.; Delgatty, H.; Batell, M.L *Diabetes* 1991, 40, 1448-1452.

McNeill, J.H.; Yuen, V.G.; Hoveyda, H.R.; Orvig, C. *J. Med. Chem.* 1992, 35,1489-1491.

Melchior, M.; Thompson, K.H.; Jong, J.M.; Rettig, S.J.; Shuter, E.; Yuen, V.G.; Zhou, Y.; McNeill, J.H.; Orvig, C. *Inorg. Chem.* 1999, 38, 2288-2293.

Miche, H.; Brumas, V.; Berthon, G. *J. Inorg. Biochem.* 1997, 68, 27-38.

Milanino, R.; Passarella, E.; Velo, G. P. Copper in the inflammatory process. In Advances in Inflammation Research. (Eds. G. Weissmann, B. Samuelsson, and R. Paoletti) Raven Press, New York 1979, Vol. 1, pp. 281–291.

Milanino, R.; Conforti, A.; Franco, L.; Marrella, M.; Velo, G.P. *Agents Actions* 1985, 16, 504-513.

Milanino, R.; Buchner, V. *Rev. Environ. Health* 2006, 21, 153-215.

Miller, A.L. *Alt. Med Rev.* 1996, 1, 103-111.

Moen, M.D.; Wagstaff, A.J. *Drugs* 2005, 65, 2657-2674.

Molinuevo, M.S.; Barrio, D.A.; Cortizo, A.M.; Etcheverry. S.B. *Cancer Chemother Pharmacol* 2004, 53, 163-172.

Nagano, T.; Toyoda, T.; Tanabe, H.; Nagato, T.; Tsuchida, T.; Kitamura, Kasai A.G. *Int. Med.,* 2005, 44, 554-559.

Néve, J.; Fontaine, J.; Peretz, A.; Famayer, J.P. *Agents Actions* 1988, 25, 146-155.

Nielsen, F.H.; Uthus, E.O. The essentiality and metabolism of vanadium. In: Chasteen, N.D. (ed) Vanadium in Biological Systems. Kluwer, Dordrecht, The Netherlands, 1990, pp. 51-62.

Oberley, L.W.; Buettner, G.R. *Cancer Res.* 1979, 39, 1141-1149.

Omoto, A.; Kawahito, Y.; Prudovsky, I.; Tubouchi, Y.; Kimura, M.; Ishino, H.; Wada, M.; Yoshida, M.; Cono, M.; Yoshimura, R.; Yoshikawa, T.; Sano, H. *Arthritis Res. Ther.* 2005, 7, R1174-R1182.

Orvig, C.; Thompson, K.H.; Battell, M.; McNeill, J.H. *Metal ions biol. Syst.*1995, 31, 575-594.

Osińska-Królicka, I.; Podsiadły, H.; Bukietyńska, K.; Zemanek-Zboch, M.; Nowak, D.; Suchoszek-Łukaniuk K.; Malicka-Błaszkiewicz, M. *J. Inorg. Biochem.* 98, 2004, 2087-2098.

Pandey, S.K.; The′ berge, J.F.; Bernier, M.; Srivastava, A.K. *Biochemistry* 1999,
38, 14667-14675.

Partidge, N.C.; Alcom, D.; Michelangeli, V.V.P.; Ryan, G.; Martin, T.J. *Cancer Res.* 1983, 45, 4308-4314.

Petrone, W.F.; English, D.K.; Wong, K.; McCord J.M. *Proc. Natl. Acad..Sci. USA.* 1980, 77, 1159-1163.

Pietta, P.J. *J. Nat. Prod.* 2000, 63, 1035-1042

Posner, B.I.; Faure, R.; Burgess, J.W.; Bevan, A.P.; Lachance, D.; Zhang-Sun, G.; Fantus, I.G.; Ng, J.B.; Hall, D.A.; Lum, B.S. *J. Biol. Chem.* 1994, 269, 4596-4604.

Poucheret, P.; Verma, S.; Grynpas, M.D.; McNeill, J.H. *Mol. Cell. Biochem.* 1998, 188, 73-80.

Quarles, L.D.; Hartle, J.E. 2nd; Siddhanti, S.R.; Guo, R.; Hinson, T.K. *J. Bone Miner. Res.* 1997, 12, 393-402.

Ray, S.; Ghosh, B.; Rana, A.; Chatterjee, M. *Int. J. Cancer* 2006, 120, 13-23.

Rehder, D. *Inorg. Chem. Comun.* 2003, 6, 604-617.

Reul, B.A.; Amin, S.S.; Buchet, J.P.; Ongemba, L.N.; Crans, D.C.; Brichard, S.M.; *Br. J. Pharmacol.* 1999, 126, 467-477.

Richards, A.B.; Krakowka, S.; Dexter, L.B.; Schmid, H.; Wolterbeek, A.P.M.; Waalkens-Berendsen, D.H.; Shigoyuki, A.; Kurimoto, M. *Food Chem. Toxicol.* 2002, 40, 871-898.

Robinson, M.J.; Cobb, M.H. *Curr. Opin. Cell. Biol.* 1997, 9, 180-186.

Rodríguez-Ciria, M.; Sanz, A.M.; Yunta, M.J.R.; Gómez-Contreras, F.; Navarro, P.; Sánchez-Moreno, M.; Boutaleb-Charki, S.; Osuna, A.; Castiñeiras, A.; Pardo, M.; Cano, C.; Campayo, L. *Bioorg. Med. Chem.* 2007, 15, 2081-2091.

Romero-Chapman, N.; Lee, J.; Tinker, D.; Uriu-Hare, J.Y.; Keen, C.L.; Rucker, R.R. *Biochem. J.* 1991, 275, 657-662.

Sakurai, H.A.; Tsuchiya, K.; Nukatsuka, M.; Sofue, M.; Kawada, J.; *J. Endocrinol.* 1990, 126, 451-459.

Sakurai H. *Envir. Health Perspect.* 1994, 3, 35-36.

Sakurai, H. *Chem. Rec.* 2002, 2, 237-248.

Sakurai, H.; Kojima, Y.; Yoshikawa, Y.; Kawabe, K.; Yasui, H. *Coord. Chem. Rev.* 2002, 226,187-198.

Sakurai, H. *Clin. calcium* 2005, 15, 49-57.

Salgado, A.J.; Coutinho, O.P.; Reis, R.L. *Bioscience* 2004, 4, 643-765.

Sambuy, Y.; De Angelis, I.; Ranaldi, G.; Scarino, M.L.; Stammati, A.; Zucco, F. *Cell. Biol. Toxicol.* 2005, 21, 1-26.

Samuni, A.; Chevion, M.; Czapski, G. J. Biol. Chem 1981, 256, 12632-12635.

Schwarzbauer, J.E.; Sechler J.L. *Curr. Opin. Cell Biol.* 1999, 11, 622-627.

Scior, T.; Guevara-Garcia, A.; Bernard, P.; Do, Q.; Domeyer, D.; Laufer, S. Mini *Rev. Med. Chem.* 2005, 5, 995-1008.

Senthil Kumar, R.; Sasikala, K.; Arunachalam S. *J. Inorg. Biochem.,* 2008, 102, 234-241.

Shisheva, A.; Shechter, Y. *FEBS Lett.* 1992, 23, 93

Shukla, R.; Barve, V.; Pakhye, S.; Bhonde, R. *Bioorg. Med. Chem. Lett.* 2004, 14, 4961-4965.

Shukla, R.; Barve, V.; Pakhye, S.; Bhonde, R. *Biometals* 2006, 19, 685-693.

Shukla, R.; Pakhye, S.; Modak, M.; Ghaskadbi, S.S.; Bhonde, R. *Rev. Diabet. Stud.* 2007, 4, 33-43.

Sica, D.A.; Gehr, T.W.; Ghosh, S. *Clin. Pharmacokinet.* 2005, 44, 797-814.

Siddhanti, S.R.; Quarles, L.D. *J. Cell. Biochem.* 1994, 55, 310-320.

Sigel, H.; Sigel, A. (Eds.) Vanadium and its Role in Life. In: Metal Ions in Biological Systems, Marcel Dekker, New York, vol. 31, 1995.

Siyuan, Z.; Yen-Kim, W.; Choon-Nam, O.; Han-Ming, S. *Curr. Med. Chem.* 2005, 3, 239-249.

Solomons, N.W. *J. Am. Coll. Nutr.* 1985, 4, 83-105.
Sorenson, J.R.J. In: H. Sigel (Ed.), *Met. Ions Biol. Syst.* 1976, 14, 77-124.
Sorenson, J.R.J. *J. Med. Chem.* 1976, 19, 135-148.
Sorenson, J.R.J. *J. Pharm. Pharmac.* 1977, 2, 450-452
Sorenson, J.R.J. *Prog. Med. Chem.* 1978, 15, 211-260.
Sorenson, J.R.J. Patent Application: US 75-563778 19750331, CAN 94:25225: Anti-inflammatory and Anti-ulcer Compounds and Processes, 1980.
Sorenson, J.R.J. The anti-inflammatory activities of copper complexes. In Metal Ions in Biological Systems. (Ed. H. Sigel) Marcel Dekker, New York 1982. Vol. 14, pp. 77-124.
Sorenson, J.R.J. (Ed.) Inflammatory Diseases and Copper, The Humana Press, Clifton, NJ 1982, 289-301.
Sorenson, J.R.J.; Oberley, L.W.; Crouch, R.K.; Kensler, T.W.; Kishore, V.; Leuthauser, S.W.C.; Oberley T.D.; Pezeshk, A. *Biol. Tr. Elem. Res.* 1983, 5, 257-273.
Sorenson, J.R.J.; Kishore, V.; Pezeshk, A.; Oberley, J.W.; Leuthauser, S.W.C.; T. Oberley, D. *Inorg. Chim. Acta* 1984, 91, 285-294.
Sorenson, J.R.J.; Kishore, V. *Tr. Elem. Med.* 1984, 1, 93-102.
Sorenson, J.R.J. *Med. Bio.* 1985, 63, 40-43.
Sorenson, J.R.J. (ed.), Biology of Copper Complexes. Humana Press, Clifton, NJ. 1987.
Sorenson, J.R.J. In: Milanino, R.; Rainsford, K.D.; Velo, G.P. (Eds.), Copper and Zinc in Inflammation, Kluwer Academic Publishers, Dordrecht, 1989, p. 69.
Sorenson, J.R.J. *Prog. Med. Chem.* 1989, 26, 437-568.
Sottile. J.; Hocking, D.C. *Mol. Biol. Cell* 2002, 13, 3546-3559.
Steinkühler, C.; Mavelli, I.; Melino, G.; Rossi, L.; Weser, U.; Rotilio, G. Ann. *N.Y. Acad. Sci.* 1988, 551, 133-136.
Sudo, H.; Kodama, H.A.; Amagai, Y.; Yamamoto, S.; Kasai, S. *J. Cell. Biol.* 1983, 96, 191-198.
Swarup, G.; Cohen, S.; Garbers, D.I. *Biochem. Biophys. Res. Commun.* 1982, 107, 1104-1109.
Teller, I.C.; Beaulieu, J.F. *Expert. Rev. Mol. Med.* 2001, 28, 1-18.
Thomas, D.M.; Hards, D.K.; Rogers, S.D.; Ng, K.W.; Best, J.D. *J. Bone Miner.* Res. 1996a, 11, 1312-1320.
Thomas, D.M.; Maher, F.; Rogers, S.D.; Best J.D. *Biochem Biophys. Res. Commun.*1996b, 218, 789-793.
Thompson, K.H. *BioFactors* 1999, 10, 43-51.
Thompson, K.H.; Tsukada, Y.; Xu, Z.; Battell, M.; McNeill, J.H.; Orvig, C. *Biol. Trace Elem. Res.* 2002, 86, 31-44.
Tracey, A.S., Gresser, M.J. *Proc. Natl. Acad. Sci. USA* 1986, 83, 609-613.
Tripoli, E. , La Guardia, M.; Giammanco, S.; Di Majo, D.; Giammanco, M *Food Chem.* 2007,104, 466-479.
Urquiola, C.; Gambino, D.; Cabrera, M.; Lavaggi, M.L.; Cerecetto, H.; González, M.; López de Cerain, A.; Monge, A.; Costa-Filho, A.J.; Torre, M.H. *J. Inorg. Biochem.* 2008, 102, 119-126.
Valent, A.; Melník, M.; Hudecova, D.; Dudova, B.; Kivekäs, R.; Sundberg, M.R. *Inorg. Chim.Acta* 2002, 340, 15-/20.
Varga, F.; Luegmayr, E.; Fratzl-Zelman, N.; Glantschnig, H.; Ellinger, A.; Prinz, D.; Rumpler, M.; Klaushofer, K. *J. Endocrinol.* 1999, 160, 57-65.

Viossat, B.; Daran, J.C.; Savouret, G.; Morgant, G.; Greenaway, F.T.; Dung , N.H.; Pham-Tran, V.A.; Sorenson, J.R.J. *J. Inorg. Biochem.* 2003, 96, 375-385 and references therein.

Wang, H.; Scott, R.E. *Mol. Cell. Biochem.* 1995,153, 59-67.

Wang, T.; Guo, *Z. Curr. Med. Chem.* 2006,13, 525-537.

Weder, J.E.; Dillon, C.T.; Hambley, T.V.; Kennedy, B.J.; Lay, P.A.; Biffin, J.R.; Regtop, H.L.; Davies, N.M. *Coord. Chem. Rev.* 2002, 232, 95-126.

Weser, U.; Richter, C.; Wendel, A.; Younes, M. *Bioinorg. Chem.* 1978, 8, 201-213.

Whyte, M.P. "Hypophosphatasia". In: Scriver CR et al. (Eds). The metabolic and molecular basis of inherited disease. 7 ed. New York. McGraw Hill, Chapter 207, 2002, CDRom

Williams, P.A.M.; Zinczuk, J.; Barrio, D.A.; Piro, O.E.; Nascimento, O.R.; Etcheverry, S.B. *Bioorg. Med. Chem.* 2008, 16, 4313-4322.

Willingham W.M.; Sorenson, J.R.J. *Tr. Elem. Med.* 1986, 3, 139-140.

Winyard, P.G.; Pall, H.; Lunec, J.; Blake, D. R. *Biochem. J.* 1987, 247, 245-246.

Wong, P.C.; Barnes, T.B.; Chiu, A.T.; Christ, D.D.; Duncia, J.V.; Herblin, W.F.; Timmermans, P.B.M.W.M. *Cardiovasc. Drug Rev.* 1991, 9, 317-339.

Xia, Z.; Dickens, M.; Raingeaud, J.; Davis, R.J.; Greenberg, M.E. *Science* 1995, 270, 1326-1331.

Xiao, G.; Gopalakrishnan, R.; Jiang, D.; Reith, E.; Benson, M.D.; Franceschi, R.T. *J. Bone Miner. Res.* 2002, 17, 101-110.

Yasui, H.; Tamura, A.; Takino, T.; Sakurai, H. *J. Inorg. Biochem.* 2002, 91, 327-338.

Younes, M.; Weser, U. Biochem. *Biophys. Res. Commun.* 1977, 78, 1247-1253.

In: New Developments in Medicinal Chemistry
Editors: Marta P. Ortega and Irene C. Gil
ISBN 978-1-60456-810-3

Chapter 6

ANTIMALARIAL PEROXIDES: FROM ARTEMISININ TO SYNTHETIC PEROXIDES

Jernej Iskra *

Laboratory of Organic and Bioorganic Chemistry,
Department of Physical and Organic Chemistry,
"Jozef Stefan" Institute, Jamova 39, Ljubljana, Slovenia

ABSTRACT

Malaria remains on of the most prevalent and deadly of the insect-borne communicable diseases. Since the parasite *Plasmodium* developed resistance to classical drugs of quinine type, artemisinin-based peroxides have become the most important antimalarial drugs. Various artemisinin derivatives have been developed to improve the antimalarial performance, while insights into its mechanism of action reveal that the pharmacophore unit in artemisinin is a peroxide group bound to the 1,2,4-trioxane skeleton. However, it is not economical to synthesize artemisinin directly and currently, the global demand for artemisinin cannot be met by the low yield obtained from the cultivation of herb *Artemisia annua*. Therefore, current research focuses on the development of easily accessible, cheap and effective antimalarial drugs based on synthetic peroxides. The main synthetic analogues of artemisinin belong to a 1,2,4-trioxane type of organic peroxides, that together with structurally related 1,2,4,5-tetraoxanes and 1,2,4-trioxolanes constitutes the core of antimalarial synthetic peroxides, while various endoperoxides, cyclic and acyclic perketals are also active antimalarial agents. Another interesting concept in antimalarial chemotherapy are chimeric compounds or hybrids; these are composed of two different antimalarial agents joined in one molecule. The following chapter discusses synthetic antimalarial peroxides with an emphasis on the development of synthetic compounds from 2000 onwards.

---

* e-mail: jernej.iskra@ijs.si

1. INTRODUCTION

Protozoan parasites of the genus *Plasmodium* are pathogens of malaria, the most widespread and the deadliest of the communicable diseases transferred to humans by a mosquito vector. It may well be that humans and malaria parasites have co-existed during the whole era of *Homo sapiens*, in Africa the latest genetic evidence suggests that the parasite population rapidly increased 10,000 years ago.[1] Despite this lengthy coexistence, our immune system has no durable defense against malaria. Currently, 40% of world's population is at threat from malaria and it is estimated that there are over 500 million episodes of clinical *Plasmodium falciparum* malaria each year.[2] Malaria also kills one child every 30 seconds resulting in a cumulative human death toll of over 3 million annually.[3] Today, malaria is endemic in underdeveloped regions and is often linked with poverty, although it is also the cause of poverty. The poor economic strength of those countries where malaria is endemic means they are unable to invest in effective control measures.

Intensive development of antimalarial drugs began when malaria became a player in major wars including the American Civil War, the World Wars and Vietnam.[4] Today, globalization has increased the awareness that poorer regions, like Africa, need help in fighting malaria. The situation is further complicated by climate change, and its potential for allowing the spread of malaria to the central or even northern parts of Europe and large parts of North America.[5] More detailed calculations show that while some areas will become malarious others will become unsuitable for vectors and as a result less dramatic change in malaria distribution is predicted.[6] Nevertheless, there is a growing awareness for the need to act against tropical diseases and various programmes like the *Special Programme for Research and Training in Tropical Diseases* run by the World Health Organization (TDR), together with international, national agencies and philanthropic foundations are creating an improved environment for a stable pipeline of drug discovery and the development of candidate drugs through public-private partnerships.[7-9]

Despite malaria's long history it was not until the beginning of 20th century that science began to understand this disease. The milestones of malaria research are marked by two Nobel prizes in physiology or medicine; one awarded in 1902 to the Scottish physician Sir Ronald Ross who demonstrated that malaria is transmitted by mosquitoes and the other to a French army doctor - Charles Louis Alphonse Laveran in 1907, who identified that a protozoan parasite later named *Plasmodium* was the cause of the disease. Only four types of *Plasmodiums* are identified as a source of human malaria, *P. falciparum*, *P. vivax*, *P. ovale* and *P. malariae*, with the first two being the most important ones. The life cycle of the parasite is complex and this is one reason why malaria is difficult to eradicate. The female anopheles mosquito is the primary host transmitting sporozoites when biting to a human who then becomes the secondary host (Figure 1). Once in the body, it takes approximately one hour for the sporozoites to reach the cells of the liver. Once present, the pre-erythrocytic and exo-erythrocytic stages result in the development of schizonts. After 5-7 days these schizonts release thousands of merozoites which invade the host's red blood cells and begin the erythrocytic phase. Asexual reproduction produces schizonts through trophozoites and discharges merozoites into the blood within 48h. Alternatively, sexual reproduction occurs when some merozoites develop into male and female gametocytes that are ingested by the mosquito when it feeds on an infected human host's blood. These gametocytes fuse to produce sporozoites and the cycle starts over.[10]

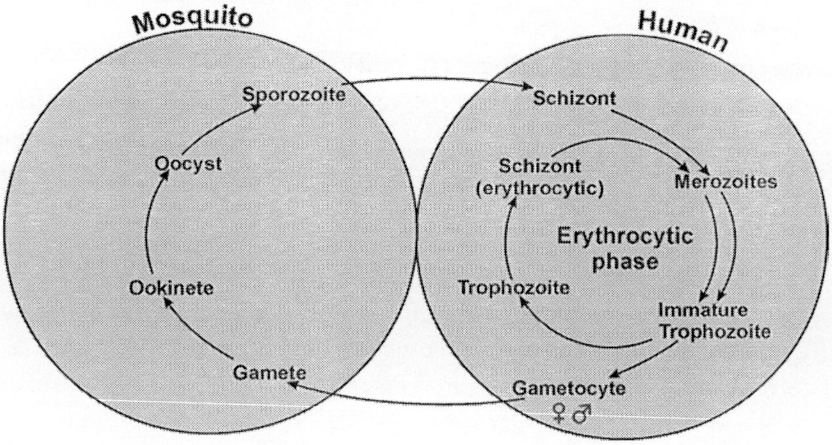

Figure 1. Life cycle of malaria parasite.

Despite this complex life cycle providing the parasite with a good defense system it does offer several possible targets for disease control. These include: vector control, immunotherapy, vaccination and drug therapy. Classical drugs of the quinine-type act on the asexual stage of parasite's life cycle. Unfortunately, the parasite has developed a resistance to these agents. This resistance requires a new approach to control malaria and various agents are in development that target parasite proteases, the mitochondrial system, parasite membrane synthesis, and parasite transporters.[10-15]

Among the agents that target the erythrocytic stage of *Plasmodium,* artemisinin and its derivatives and synthetic peroxides occupy an important position. These compounds are active against chloroquine resistant strains of *Plasmodium falciparum* and are the first line-of-defense for this type of malaria. In following sections the development of artemisinin and its derivatives will be discussed with an emphasis on recent advances in the field of synthetic peroxides from 2000 onwards. The synthetic peroxides are grouped according to structural similarities and key synthetic strategies are presented. Wherever possible the antimalarial activities are presented to show how the structure of peroxide effects the activity of peroxides of similar structural type. The *in vitro* activities against chloroquine-resistant strain of *P. falciparum* are given by the IC_{50} values while the *in vivo* activities of compounds administered by orally (po), subcutaneously (sc) or intraperitoneally (ip) to *P. berghei* infected mice are represented by their ED_{50} or ED_{90} values obtained using the Peters protocol.

2. ARTEMISININ

Traditional medicine contains a wealth of knowledge concerning the effect that herbs have on human health, and nowhere is this truer than in traditional Chinese medicine, which has developed over several thousand years. During the period of Cultural Revolution (1966-1976) practitioners of traditional Chinese medicine were brought together with scientists to subject traditional knowledge to modern scientific securitization. Within the malaria programme, *qinghao* was among the first ten herbs to be screened because of Ge Hong (284-363), a renowned physician, who suggested that a drug derived from *qinghao* or *Artemisia annua* be used for "intermittent fever".[16] However, the initial screening within the programme gave poor results and it was not until 1971, that a group led by Professor Tu Youyou, by carefully following Ge Hong's original procedure of wringing out a cold water infusion of *qinghao*, were able to prepare an effective drug. This was followed in 1972 by the first report of the successful treatment of malaria patients with *qinghao*. The finding had a major effect on the development of antimalarial drug therapy; *qinghao* (*Artemisia annua)* or sweet wormwood as it is known in the west became an important source of antimalarial agents. In 1973, the active compound was isolated and named as qinghaosu or artemisinin **1**.[4,16] Artemisinin is insoluble in water which explains why it was so difficult to prepare the herbal drug, but by wringing out a cold infusion of the herb an emulsion of flavonoids and aromatic oils present in the plant help to solubilise the artemisinin. A possible synergism among them could also not be excluded. This also explains why it is difficult to use traditional tea preparations of *Artemisia annua*.[17-20] Such encouraging initial results led to further research on artemisinin and its derivatives with pharmaceutical production of the drug beginning in 1986. By 2004, the World Health Organization WHO had changed its policy on malaria treatment and recommended that an artemisinin-based combination therapy – ACT (main combinations: artemether/ lumefantrine, artesunate/ amodiaquine, artesunate/ mefloquine, artesunate/ sulfadoxine/ pyrimethamine) should represent the first-line treatment for malaria. Then in 2006, WHO basing its argument on published research reinforced the statement[21] that monotherapy with artemisinin derivatives should be abandoned and it requested that pharmaceutical companies stop marketing "single-drug" artemisinin malaria pills in order to prevent the mosquito developing a resistance to it.[22] The importance of artemisinin to antimalarial chemotherapy and the scientific community's continued interest in its development is reflected in the many review papers dedicated to it and its derivatives.[23-28]

The structure of artemisinin 1 is impressive with a 1,2,4-trioxane ring incorporated into a sesquiterpene skeleton. The pharmacophore unit is a 1,2,4-trioxane ring or more precisely a peroxide bond (

Figure 2). Unfortunately, although artemisinin 1 is a very active compound its insolubility makes it unsuitable for therapeutic use. One solution is to transform artemisinin into dihydroartemisinin (DHA) 2; this was first prepared in 1973 by the group led by Tu Youyou while characterizing the structure of 1 without realising they had discovered an even more potent agent.[16] Etherification or esterification of the 10-hydroxy group produced the first generation of artemisinin derivatives, of which the most important are artemether 3 and artesunate 5 (

Figure 2).[24] These first generation artemisinin derivatives are fast-acting agents and have a high potency, but they have a short half-life in plasma as result of their rapid hydrolysis to DHA 2 leading to elimination via glucoronidation by Phase II metabolism.[27] Hence the need for their combination with a longer-acting drug (ACT) to prevent recrudescence of malaria.

To improve artemisinin derivatives further, development has focused on derivatives with a prolonged half-life in plasma (Figure 3).[11,24] Another important factor is the cost of preparation, which must be as low as possible as the drug is mostly destined for developing countries. Toxicity, in particular neurotoxicity is a key consideration in drug development and it is important that any newly developed compounds have a sufficiently low lipophilicity to prevent them crossing the blood-brain barrier.

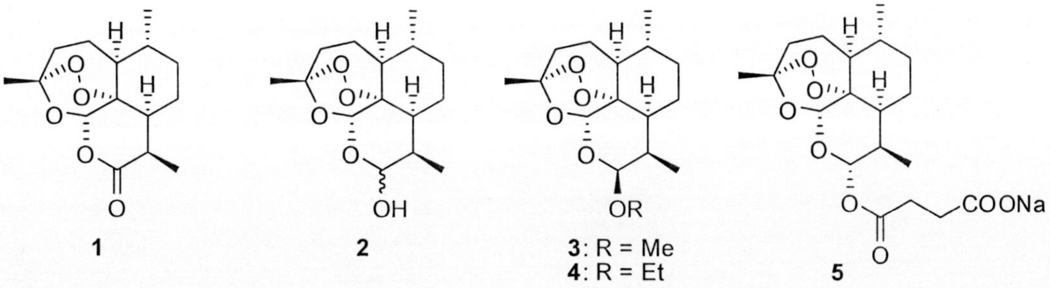

Figure 2.

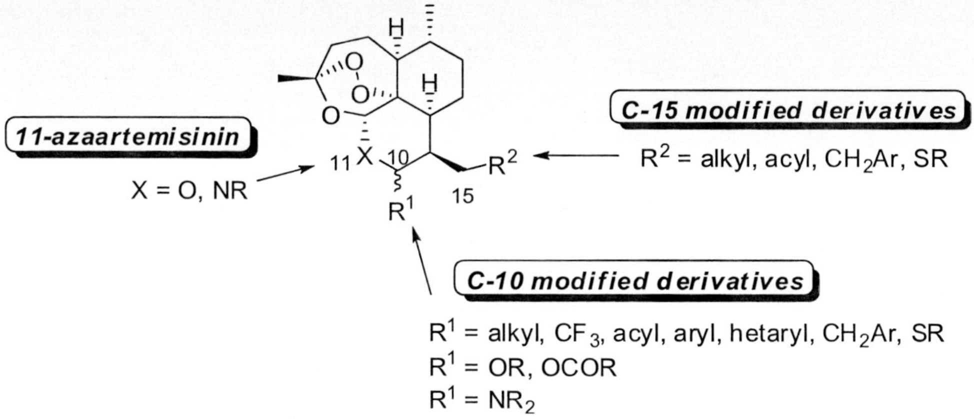

Figure 3.

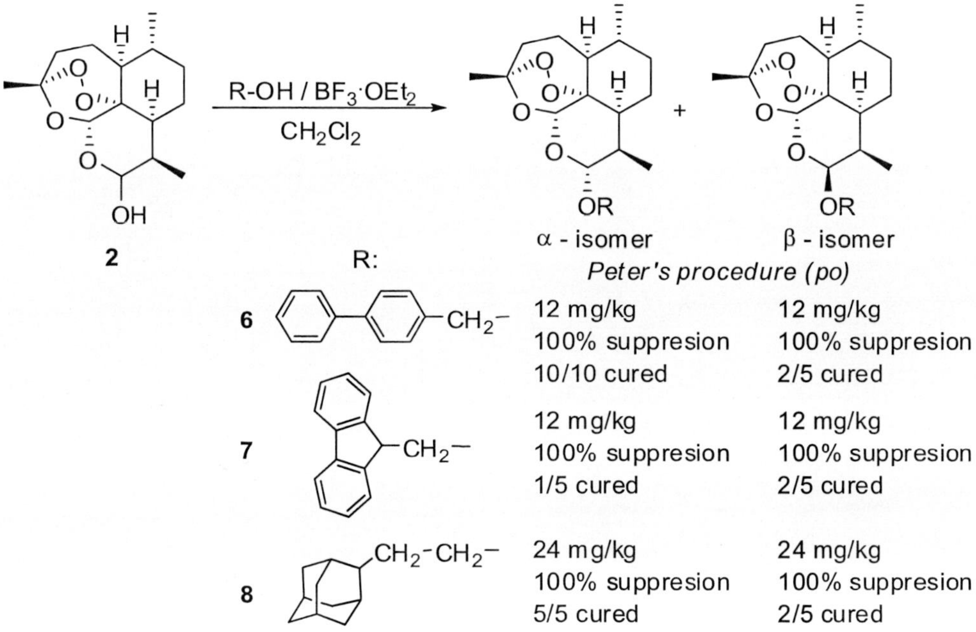

Scheme 1.

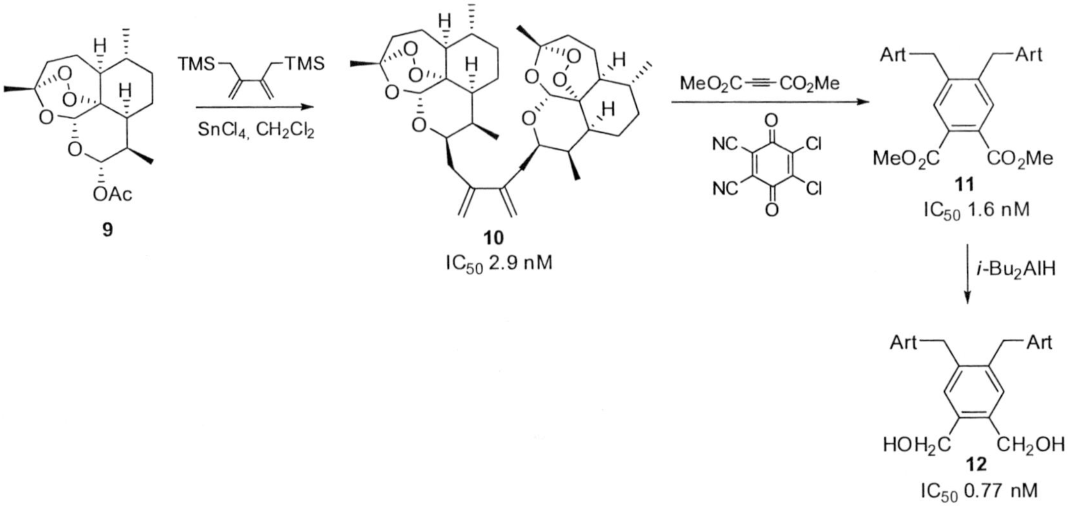

Scheme 2.

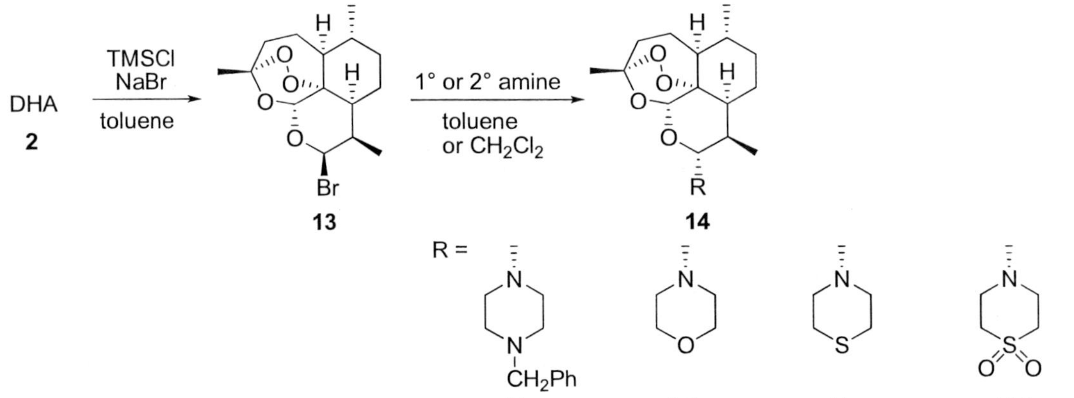

	14a	14b	14c	14d
logP	5.62	3.05	4.97	2.49
neurotoxicity	low	high	high	very low
ED90 (sc)	0.8 mg/kg	0.18 mg/kg	0.51 mg/kg	1.5 mg/kg
(po)	3.5 mg/kg	1.3 mg/kg	1.9 mg/kg	3.1 mg/kg

Scheme 3.

The majority of derivatives belong to a group of artemisinins functionalized at the C-10 position. These are prepared from DHA 2 where the C-10 alcohol group is altered to give ether, ester or carbonate functionality. A recent example shows that the incorporation of biphenyl 6, fluorenyl 7, and adamantyl 8 scaffold through an ether linkage gives an orally active antimalarial agent that is more active against *P. yoelii* than β-arteether (Scheme 1).[29] Interestingly, stereochemistry plays an important role in these compounds and it is the α-isomers of the ethers that have the highest activity contrary to arteether 4, where the β-isomer is more active. These derivatives are quite lipophilic with a logP > 6, which according to the authors results in increased oral bioavailability. No toxicity data for these C-10 derivatized DHAs 6-8 are reported.

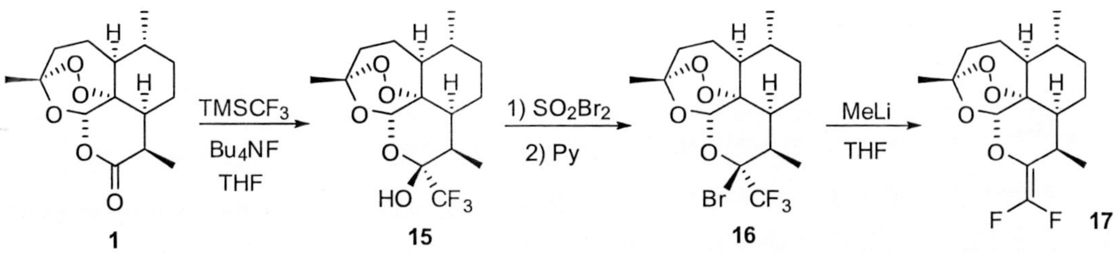

Scheme 4.

To prevent the hydrolysis of artemisinin to DHA 2 and consequently its elimination from the organism various C-10 alkyl and aryl derivatives have been synthesized. A recent example includes the use of artemisinin dimers 10-12 as longer-lived antimalarials that show curative activity after a single low parenteral dose (Scheme 2).[30,31]

Another possibility to prevent hydrolysis while at the same time incorporating a polar group to reduce lipophilicity and thus possible neurotoxicity is to introduce an alkylamino group at position 10 in artemisinin. 10-Alkylamino artemisinin derivatives 14 are very active, while neurotoxicity depends on the nature of the heteroatom or substituent in the alkylamino structure (Scheme 3).[32] Furthermore, lipophilicity is easily tuned by the correct choice of amine. The best candidate in this group - artemisone 14d, was prepared at a comparable price but with improved characteristics that allows a lower dosage. Preclinical screening showed that 14d has good properties with negligible cyto- and neurotoxicities making it a good candidate for clinical trials.

Another approach to turn artemisinin into a metabolically more stable compound is that described by the group of Bonnet-Delpon.[33] They substituted the 10-carbonyl group of artemisinin into the 10-*gem*-difluoroethylene group that sterically and electronically mimics the carbonyl group. In doing so were able to develop an artemisinin analogue 17 with improved stability, longer lasting antimalarial activity and a higher activity against *P. falciparum* FCB1 strain (IC_{50} 4.6nM) than its parent artemisinin 1 (IC_{50} 8.9nM). Despite these changes the structure of the derivative remains the same as does the recognition of the agent by the organism (Scheme 4).

3. Synthetic Peroxides

The main drawbacks of artemisinin and its first generation derivatives were successfully addressed by the creation of semisynthetic artemisinin derivatives, however their principle disadvantage *i.e.*, the necessity of artemisinin for their synthesis still remains. This is significant because the current demand for artemisinin globally exceeds supply, a fact made evident by the shortage in supply in 2004. Estimation for global demand in 2005 was 120 million of adult treatment courses.[34] The content of artemisinin in the leaves of *Artemisia annua* is low and is strongly dependent on growth conditions, region and strain.[35] After WHO recommended artemisinin-based combination therapy, the cultivation of *Artemisia annua* has expanded and hopefully the quantity and the yield will increase through improved agricultural practice and novel biotechnologies. An example of such approach is a collaboration between UC Berkeley, Amyris Biotechnologies and the Institute for One World

Health under the financial aid of the Bill and Melinda Gates Foundation. This collaboration was successful in developing a low cost way of producing artemisinin using genetically engineered yeast cells (*Saccharomyces cerevisiae)* to produce artemisic acid in a three-step oxidation of amorpha-4,11-diene.[36]

An more obvious solution to produce large amounts of artemisinin is organic synthesis, however the 10 steps required to synthesise artemisinin makes it uneconomical.[37-39] An alternative approach includes the preparation of synthetic organic peroxides that can mimic the action and/or structure of artemisinin.[23,26,40-42] This methodology will be more fully addressed later.

Having established that the peroxide bond is essential to the antimalarial activity of artemisinin, its mode of action could then be linked with the parasite's digestion of haemoglobin. It was the elimination of the dangerous heme that led to the logical conclusion that it is the interaction with Fe(II) within the erythrocytes that generates poisonous radicals resulting in artemisinin's antimalarial action. This mechanism is supported by experimental data, but the actual mode of action of the artemisinin derivatives remains a hot topic of debate within the scientific community as new data shows that artemisinin inhibits enzyme PfATP6, a key metabolic enzyme of SERCA-type Ca^{2+} ATPase.[22,43,44] However, it does pave the way for the development of synthetic peroxides because it was believed that the most important thing is the interaction with heme. Of course, almost any peroxide, being an oxidant, can react with the iron(II) species. Even simple peroxides, like hydrogen peroxide and *tert*-butyl hydroperoxide, are antimalarial agents.[45,46] Therefore, the organic scaffold around the peroxide bond gives the agent its correct activity in the organism. A variety of structures of synthetic antimalarial peroxides show that there are many ways of wrapping a peroxide bond with an organic scaffold to produce an antimalarial effect. In the following sections, peroxidic antimalarials are organized according to the structural element of the basic peroxidic bond.

3.1. Trioxanes

Artemisinin has a very interesting tetracyclic structure with a 1,2,4-trioxane ring incorporated into a sesquiterpene core. Having in mind the synthesis of artemisinin analogues a question immediately arises, i.e., whether or not its complicated structure is essential for antimalarial activity. This is important because a simpler structure usually means a less complicated synthesis. The most evident structure in artemisinin is the peroxidic 1,2,4-trioxane ring and various synthetic trioxanes evolved directly from artemisinin and were studied as potential antimalarial agents.

3.1.1. Tricyclic 1,2,4- trioxanes

To simplify the structure of artemisinin it would be necessary to know which ring could be left out without affecting its antimalarial activity. Research has shown that the lactone ring in 1 is superfluous and that the tricyclic analogue 18 is active (Figure 4).[47] Furthermore, by changing the 7-membered ring in 18 into a 6-membered one 19 the antimalarial activity is lost.[48] Whereas, analogue 20 with its ABC ring combination and without the 7-cyclic ring was active.[47]

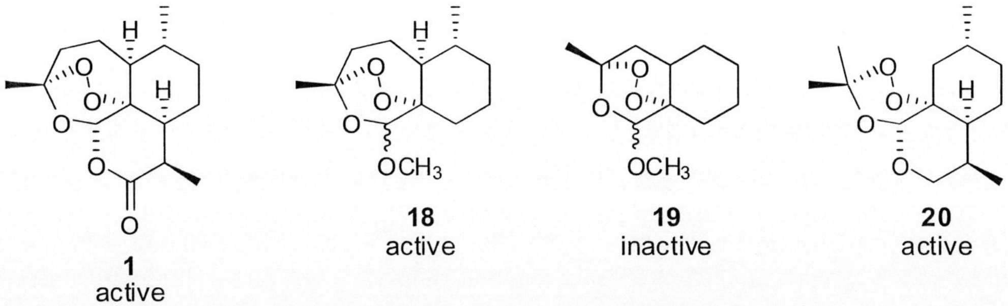

Figure 4.

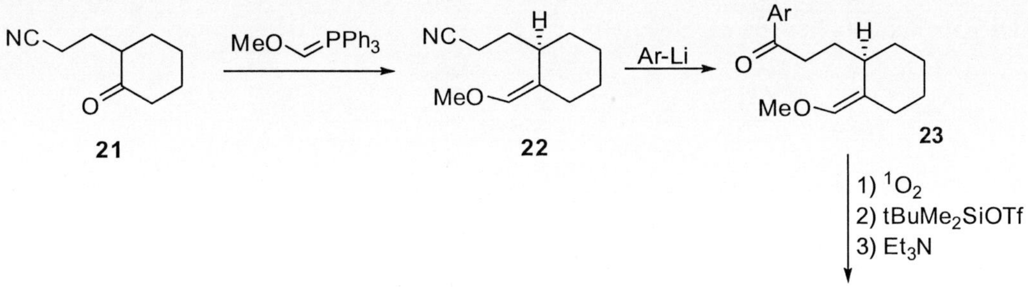

	IC_{50} (nM)		ED_{50} (mg/kg)	
	α	β	sc	po
24a: *p*-$HOCH_2Ph$	78	15	3.4	5.5
24b: *p*-$MeCOOCH_2Ph$	44	20	2.8	14
24c: *p*-$FPhCH_2OCH_2Ph$	42	23	3.5	3.8
24d: *p*-FPh	65	30	6.8	10

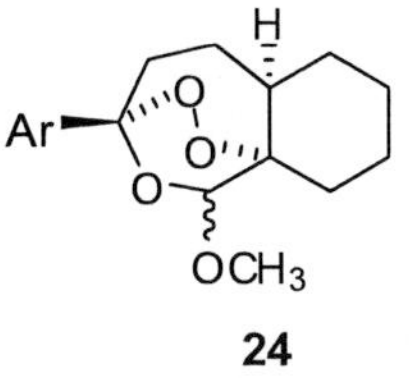

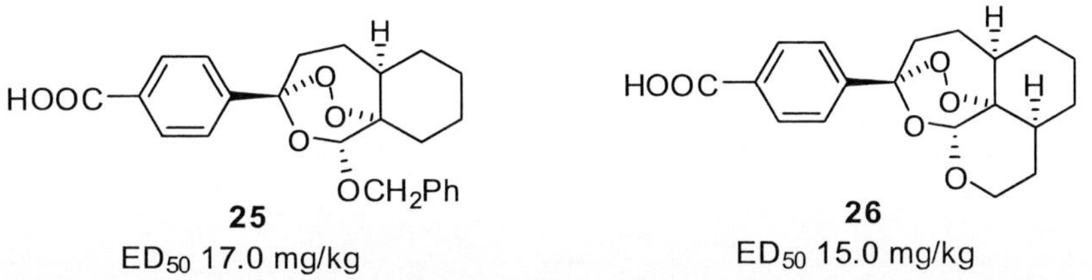

Scheme 5.

Posner's group presented a series of tricyclic trioxanes with retained rings A, C and D of the parent artemisinin scaffold. Their synthesis started with cyclohexanone that was converted to the nitrile 21. Wittig methoxymethylenation and arylation by aryl lithium produced the precoursor 23. A peroxide group was introduced using a singlet oxygen oxygenation followed by an acid catalyzed rearrangement to give 3-aryltrioxane 24 (Scheme 5).[49-51] Studies on the antimalarial activity of 3-aryltrioxanes 24 show that they are promising drug candidates and have been found to show good activity *in vitro* and especially *in vivo* studies. Furthermore they are non-toxic in therapeutic doses. Development of water-soluble trioxanes 25 and 26 following a similar methodology produce even more efficient derivatives with enhanced bioavailability.[52-54] The multi step synthesis of these 3-aryltrioxanes 24 limits their development but research into these substrates does provide valuable information on the mechanism of action of artemisinin-like compounds. It has also been observed that chirality does not influence antimalarial activity as both enantiomers of 24d have a similar *in vitro* activity against the *P. falciparum*.[55,56]

3.1.2. Bicyclic 1,2,4-trioxanes

It is not necessary to preserve the tetracyclic artemisinin skeleton for a compound to show good antimalarial activity. It appears that apart from the tricyclic trioxanes the bicyclic 1,2,4-trioxanes are also potent building blocks for antimalarial active compounds as indicated by the activity of 20 (Figure 4). *cis*-Fused cyclopenteno-1,2,4-trioxanes 29 are an example of bicyclic peroxides with potent antimalarial activity. They are synthesized from 1,4-diaryl-1,3-cyclopentadienes 27 by photooxygenation to introduce a peroxy function in the form of a 1,4-endoperoxide 28 that is transformed by acid-catalyzed ring-enlargement with either a ketone or aldehyde into 29 (Scheme 6).

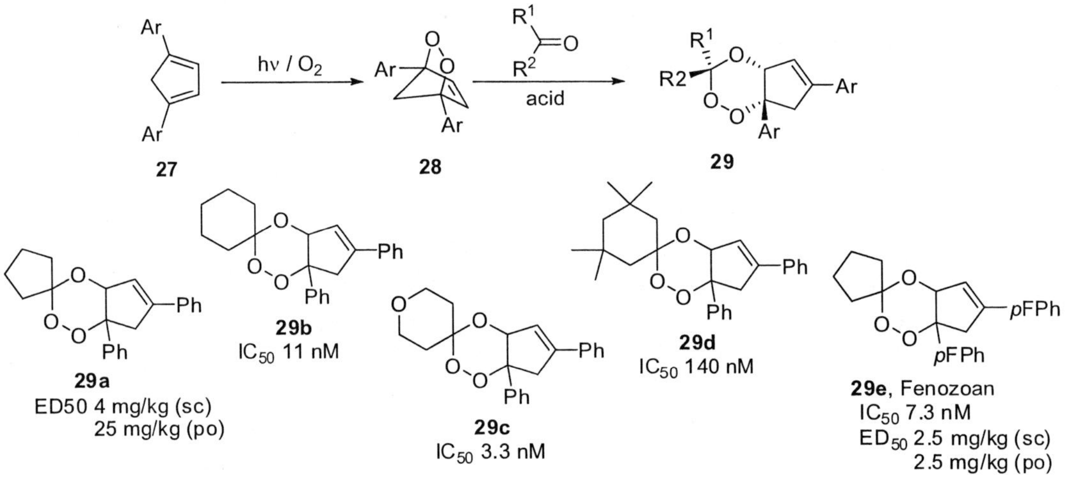

Scheme 6.

32

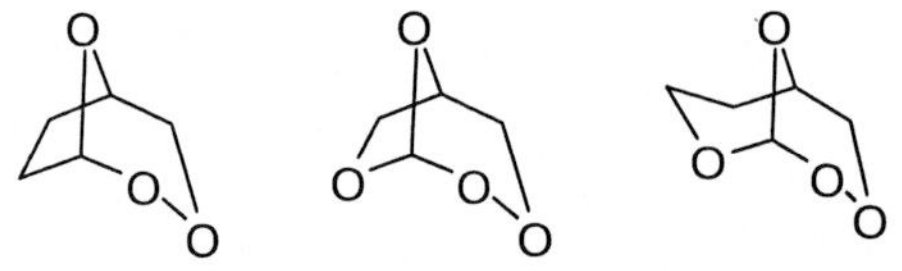

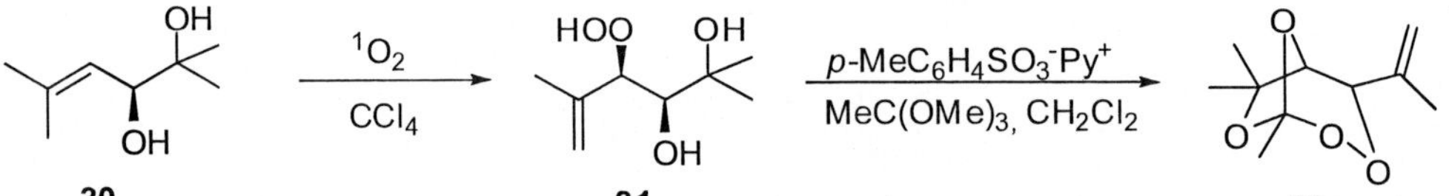

30 31 32a

Scheme 7.

Intensive research on these structures reveals an interesting effect that their structure has on their activity. A 3,3-spirocycloalkane substituent is important for the antimalarial activity, while the highest activity is observed for the cyclopentane ring 29a. On the other hand, the heteroatoms in the 3,3-cyclohexane ring does not influence to any great extent its activity, with the spiropyran analogue 29c having an even higher activity, while steric effects are more important – this can be judged from the lower activity of 29d. The most promising candidate among this family is Fenozoan 29e, which has all the important elements for good activity: 3,3-spirocyclopentane substituent, *cis*-fused 1,2,4-trioxane, cyclopentene ring and fluorine substituents on the *para* position in the aryl ring with the emphasized effect on *in vivo* activity. [57-59]

Singlet oxygen is the major source of peroxide functionality in the preparation of bicyclic peroxides containing the 1,2,4-trioxane structural unit. In a recently published synthetic strategy, ene-reaction was used to transform allylic alcohols 30 into unsaturated vicinal hydroperoxy alcohols 31 that were then transformed into various cyclic peroxides with a 1,2,4-trioxane ring in the bicyclic structure, as exemplified in 32a (Scheme 7).[60] Preliminary studies on the *in vitro* antimalarial activity against *P. falciparum* show that these bicyclic 1,2,4-trioxanes 32 posses moderate antimalarial activity and could serve as a skeleton for further derivatization.

3.1.3. Spiro 1,2,4-trioxanes

Alternatively, O'Neill's group reports constructing a trioxane ring using Mukaiyama hydroperoxysilylation of allyl alcohols by molecular oxygen as a source of the peroxide bond with a subsequent cyclization by another carbonyl compound (Scheme 8).[61]

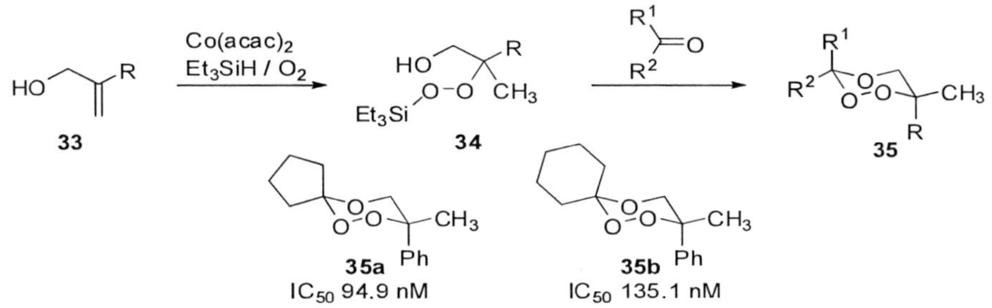

Scheme 8.

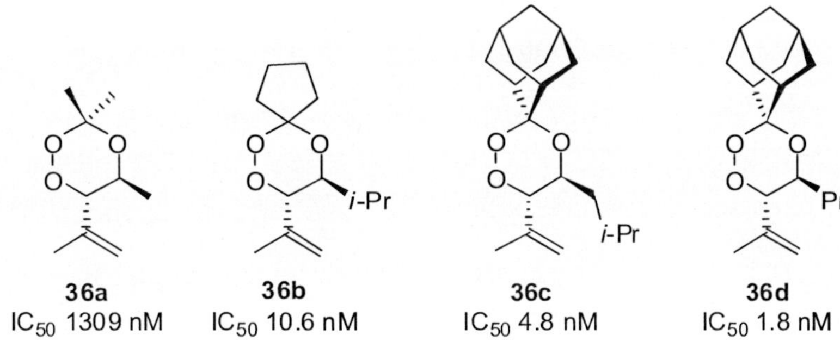

Scheme 9.

The majority of synthetic methods include singlet oxygen for the construction of trioxanes. In a similar strategy to that outlined in Scheme 7, spiro-1,2,4-trioxanes 36 were synthesized from an ene-reacton of allylic alcohols with the subsequent cyclization of the previously formed *vic*-hydroperoxy alcohols with ketones (cycloalkanones, adamantanone) to prepare compounds 36. These compounds display antimalarial activity on the nanomolar scale with low toxicity (Scheme 9).[62-64] Results show the necessity for constructing spiro bicyclic trioxanes over the *gem*-dialkyl substituted trioxanes.

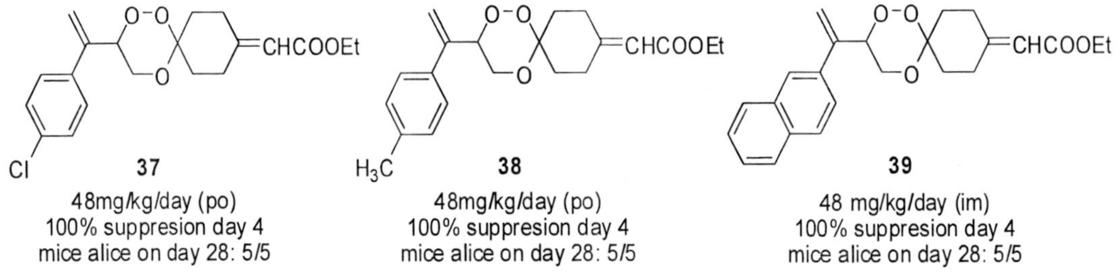

Figure 5.

Spiro 1,2,4-trioxanes were prepared using a similar strategy to forming natural compounds and geraniol was used as replacement for allyl alcohol[65,66] or steroidal ketones were used for cyclization,[67] nevertheless natural compounds did not augment the antimalarial activity that remained moderate. Improved results were obtained by using 3-aryl substituted allyl alcohols for ene-reaction followed by cyclization using 1,4-cyclohexadienone to obtain spiro-1,2,4-trioxane with the carbonyl function on the cyclohexane ring that served as an entry point for further transformation.[68] Screening for oral *in vivo* activity delivered two compounds 37 and 38 that were orally active with a profile that is better than β-arteether against *P. yoelii* (Figure 5). The authors claim that by changing phenyl ring into a naphthyl ring 39 they obtained a compound whose activity resembles that of artemisinin against rodent and simian malaria.

3.2. Tetraoxanes

3.2.1. 1,2,4,5-Tetraoxanes

1,2,4,5-Tetraoxanes differ from 1,2,4-trioxanes in having an additional oxygen atom in the six-membered ring. This means that they have two peroxide bonds per molecule instead of one. The use of symmetrical dispiro 1,2,4,5-tetraoxanes 42 as potent antimalarial agents was studied by the group of Vennerstrom. These compounds have been examined in detail and various substitution patterns and substituents studied to identify a candidate with the best antimalarial properties (Scheme 10).[69-72] Tetraoxane 42a prepared from 2-methylcyclohexanone 40a was selected as the most promising agent amongst the tetraoxanes having a good *in vitro* activity against *P. falciparum* K1 strain. While *in vivo* activity against *P. berghei* on a day 3 post-infection was similar to artemisinin, survival time of mice was better.[71,73] The synthesis of these compounds should be straightforward requiring only the acid-catalyzed peroxidation of cyclohexanones with aqueous hydrogen peroxide. However, selectivity depends on the structure of the ketone and various peroxides are formed, with hexaoxonane 43 being the main byproduct.[74-76] For this reason, the actual synthesis is mainly performed by ozonolysis of O-methyl oximes 41 or enol ethers.

The use of 30% aqueous hydrogen peroxide would be more desirable than ozone. Using fluorinated alcohols (2,2,2-trifluoroethanol - TFE or hexafluoro-*iso*-propanol – HFIP) as solvents for acid-catalyzed peroxidative cyclization tetraoxanes were formed selectively from a variety of ketones without the formation of other peroxides. [74,77] The use of fluorinated alcohols enabled 3,3,6,6-tetraalkyl-1,2,4,5-tetraoxanes 44 to be synthesized. These were studied because of the ease by which small variations in polarity and structure of tetraoxanes could be achieved by changing the length and branching of alkyl chains (Scheme 11).[78] The steric effects play an important role on the *in vitro* antimalarial activity as is indicated by the lower activity of tetraoxane that have the *i*-Pr 44b and *t*-Bu 44c groups on the tetraoxane scaffold in relation to 44a. Furthermore, polarity, as expressed by the length of alkyl chains, influences activity *i.e.*, smaller chains results in improved activity (44a *vs.* 44d), but the effect is not so clear as a significant difference in activity exists between methyl/ethyl substituted tetraoxane (44e/44d) albeit their polarity is similar.[78]

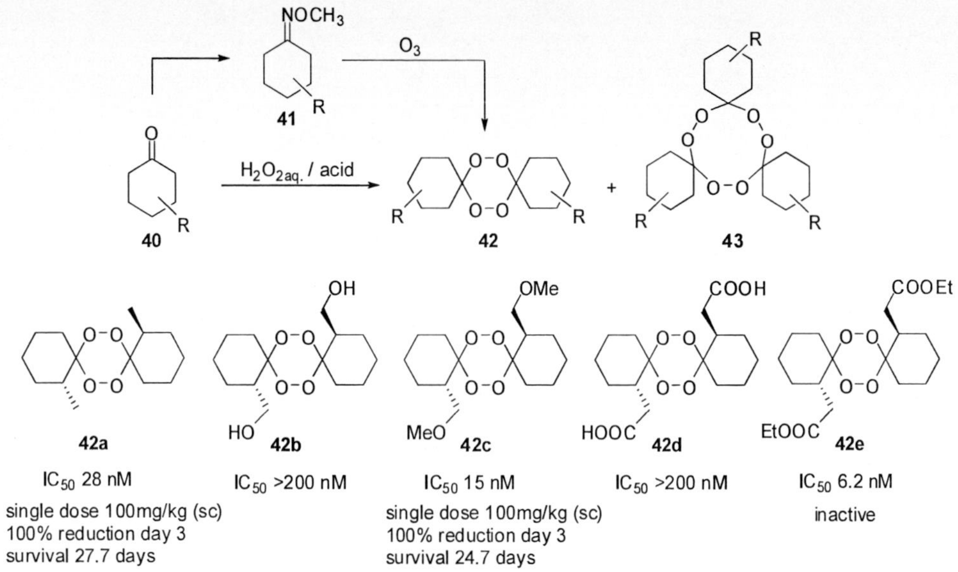

Scheme 10.

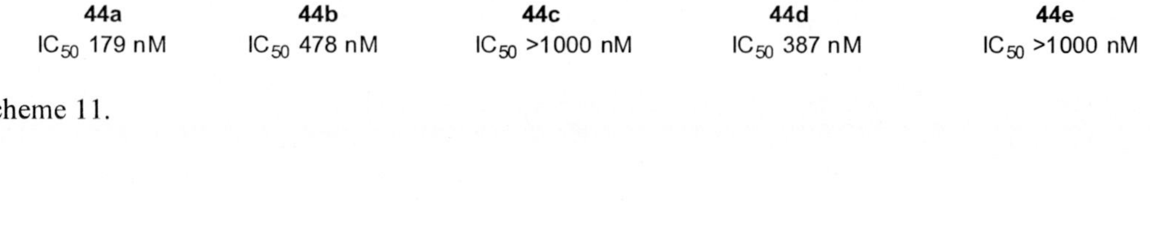

Scheme 11.

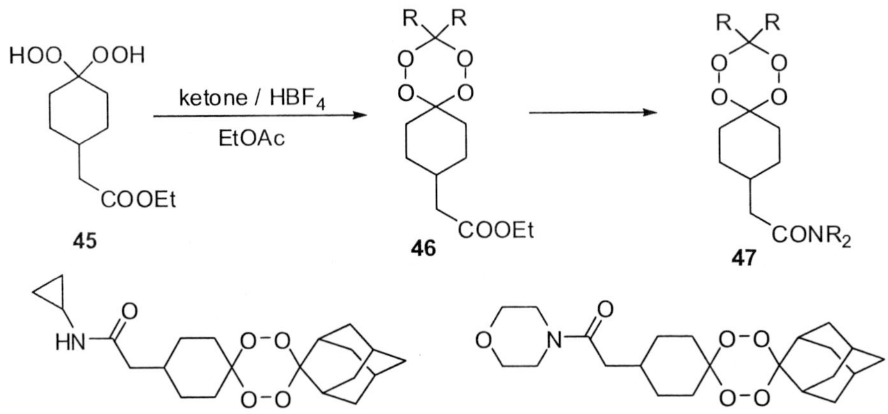

47a
IC_{50} 2.3 nM
ED_{50} 10.27 mg/kg

47b
IC_{50} 5.2 nM
ED_{50} 3.18 mg/kg

Scheme 12.

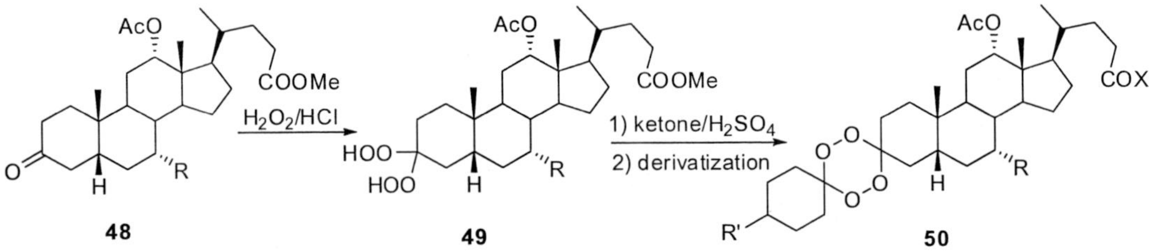

50a: R = H, R' = (R)-CH_3
X = $NHCH_2CH_2NMe_2$

IC_{50} 14.72 nM

160mg/kg/day
3/5 alive on day 31
survival 29 days

50b: R = OAc, R' = (R)-CH_3
X = $NHCH_2CH_2NMe_2$

IC_{50} 12.24 nM

40mg/kg/day
1/5 alive on day 31
survival 19 days

50c: R = OAc, R' = H
X = $NHCH_2CH_2NMe_2$

IC_{50} 19.65 nM

320mg/kg/day
5/5 alive on day 31
survival 31 days

40mg/kg/day
2/5 alive on day 31
survival 21 days

Scheme 13.

Further advances were made by O'Neill's group by developing a second generation of tetraoxanes with improved stability and oral bioavailability.[79] Important was the introduction of the amino and amido polar groups into the tetraoxane skeleton to increase water solubility and therefore bioavailability (Scheme 12). Different strategies were taken to obtain targeted compounds with excellent in vitro antimalarial activity against P. falciparum with the best of them having an activity better than that of artemisinin. Similar observations were also observed during in vivo experiments against P. berghei, where oral activity was enhanced by the presence of a cyclododecane or a adamantane ring in the dispiro tetraoxane structure. This confirms that dispiro tetraoxanes are more active than spiro analogues. Two particularly interesting facts are their relatively simple synthesis from simple starting compounds and that they are achiral. Nevertheless, it is that 47a and 47b have an activity equivalent to artemisinin should be emphasized.

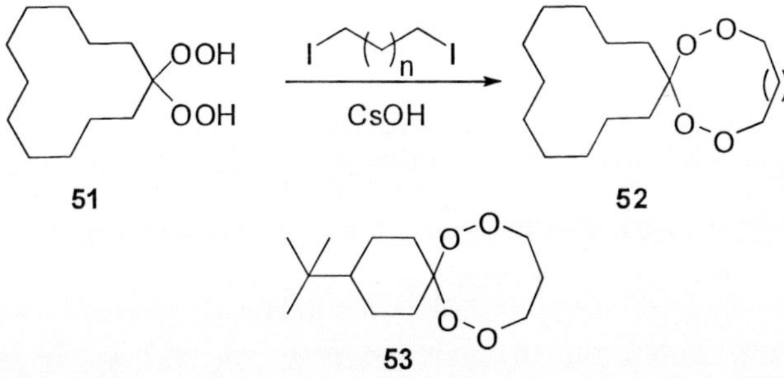

	EC_{50} (nM)	ED_{50} (ip) (mg/kg)
52a: n=1	25	12
52b: n=2	100	48
52c: n=3	280	
52d: n=4	10	160
52e: n=6	1700	
53	3	15

Scheme 14.

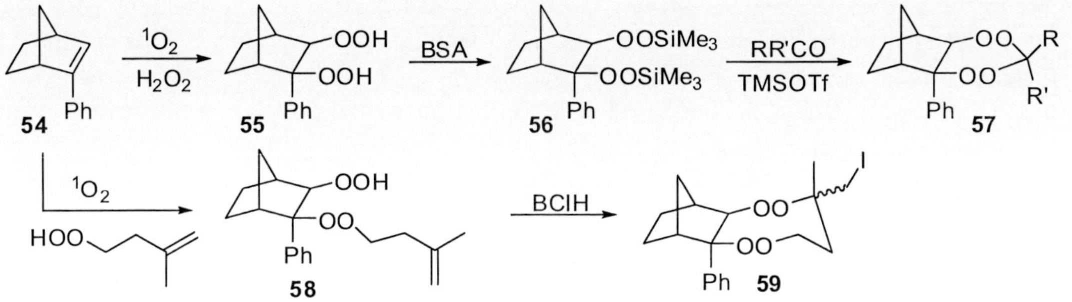

Scheme 15.

Non-symmetrical tetraoxanes offer the possibility to incorporate two different structures on each side of the tetraoxane ring. This makes it possible to obtain compounds with the desired properties that can be tuned by each of the substituents on the tetraoxane scaffold. Recently new methods for the preparation of non-symmetrical or mixed tetraoxanes have become available.[77,79-83] A group led by Solaja published an interesting approach for increasing the bioavailability of antimalarial tetraoxanes by attaching a tetraoxane ring to a natural steroid carrier. Moreover, by using cholic acid as a ketone, symmetrical steroidal tetraoxanes were synthesized with the carboxylic acid function on the side chain, which served as an entry point for further derivatization to various amides.[84,85] Their *in vitro* antimalarial activity against *P. falciparum* D6 clone matches almost that of artemisinin when the *n*-propylamido functional group was present on the side chain, while toxicity is low with the selectivity >10,000. Furthermore, these bis-steroidal tetraoxanes have antiproliferative activity and can induce apoptic cell death. This approach was upgraded to a second-generation of steroidal tetraoxanes consisting of mixed tetraoxanes to provide a selective and broad manipulation of the tetraoxane scaffold. The carrier properties of steroid was preserved, while additional activity was achieved by incorporating another smaller cyclic structure and at the same time the molecular mass of the active compound was reduced. [83,86,87] Synthesis consisted of preparing *gem*-dihydroperoxide of cholic or deoxycholic acid followed by acid-catalyzed cyclization with another ketone and the targeted derivatization of the ester in the cholic group (Scheme 13). Deoxycholic acid was introduced to increase metabolic stability, but despite this their antimalarial activity *in vitro* and *in vivo* was weaker than that of the corresponding cholic acid derivatives.[87] Derivatives with a secondary amido group on the steroid side chain have comparable *in vitro* activity to artemisinin against the multidrug resistant *P. falciparum* strain TM91C935. *In vivo* testing showed that compounds 50 are effective in curing mice infected with *P. berghei* without any observable toxic effects.

3.2.2. 1,2,4,5-Tetraoxacycloalkanes

gem-Dihydroperoxides are also the starting compounds for the preparation of 1,2,4,5-tetraoxacycloalkanes, macrocyclic analogues of a more elaborate group of 1,2,4,5-tetraoxanes. Silver(I) oxide or caesium hydroxide were used as a catalyst for the cyclization of *gem*-dihydroperoxides, prepared from ketones with hydrogen peroxide or ozonolysis of enol ethers, with 1,n-diiodoalkanes to obtain the 1,2,4,5-tetraoxacycloalkanes (Scheme 14). The influence of the peroxide ring structure on the activity is clear and can not be correlated just to the size of the ring. However, the *in vitro* activity of some compounds against *P. falciparum* is similar to artemisinin. The most active compound is 53. This compound together with its low toxicity has a selectivity index of 10,000. *In vivo* experiments against *P. berghei* pointed at 52a as the best candidate. Further tests have shown that 52a is also orally active and comparison with artemisinin proves it is a better agent.[88]

3.2.3. 1,2,5,6-Tetraoxacycloalkanes

Another variant of the peroxidic macrocycle *i.e.*, with four oxygen atoms is derived from the cyclization of *vic*-dihydroperoxide with a ketone or 1,*n*-diiodoalkane. *vic*-Dihydroperoxide 55 or 58 were prepared by a singlet oxygen oxygenation of 2-phenyl-norbornene 54 in the presence of hydrogen peroxide or alkenyl hydroperoxide, respectively (Scheme 15). Cyclization of 1,2-dihydroperoxides was performed according to Scheme 14 or alternatively both hydroperoxy groups were first trimethylsilylated and afterwards cyclized with a carbonyl compound catalyzed by a Lewis acid. The alkenyl perethers 58 were cyclized *via* the reaction with bis(collidine)iodine hexafluorophosphate (BCIH).[89] The antimalarial activity of these 1,2,5,6-tetraoxocycloalkanes is moderate.

3.3. Trioxolanes

Ozonolysis of alkenes forms 1,2,4-trioxolanes or secondary ozonides as intermediate products that are usually not stable. However, tetrasubstituted ozonides are stable due to the absence of α-hydrogen atoms. Grisbaum developed a method for the preparation of secondary ozonides by coozonolysis of O-alkyl oximes and carbonyl compounds.[90] The group of Vennerstrom used this to develop antimalarial 1,2,4-trioxolanes through an interesting approach.[91] They first selected a candidate that satisfied the essential characteristics for a trioxolane antimalarial drug: low product cost, good potency and pharmacokinetic properties for a maximum 3-day treatment regime with once-daily administration, low potential toxicity and no evidence for resistance development. To do this a candidate selection matrix was composed and subjected to further optimization and compound selection. Eventually, trioxolane 60 (OZ277 or Rbx-11160) was selected as a structurally simple drug development candidate with an economically feasible and scalable synthesis, with superior antimalarial activity and a biopharmaceutical profile (Figure 6). The development of this drug was taken further in a public-private partnership with the Medicines for Malaria Venture (MMV) and Ranbaxy Pharmaceuticals. According to a series of recommendation given by WHO for the usage of combined therapy, this drug is now undergoing clinical studies trials in combination with piperaquine. According to Ranbaxy Pharmaceuticals Inc. a new drug application for a combination of Rbx-11160 and PQP could be filed in early 2009.[92] This would be the first synthetic antimalarial peroxide on the market.

Ketone oxime ether 61 reacts with ozone at 0°C in the presence of a second ketone 62 and forms a stable tetrasubstituted secondary ozonide or 1,2,4-trioxolane 63.[91,93] Its antimalarial activity is derived from the trioxolane 63 structure, which incorporates a peroxide bond in the organic scaffold (Scheme 16). However, because the peroxide bond in diadamantyl 63a or dicyclohexyl 63c trioxolanes is sterically inaccessible or too exposed respectively, these compounds are inactive, while mixed trioxolane 63b has equilibrated both effects and is highly active. Evidence for the importance of the peroxide bond in determining the activity of the trioxolanes was confirmed by the synthesis of the inactive 1,3-dioxolane derivative.[94,95]

These substrates are however, too lipophilic and have low aqueous solubility and low oral bioavailability. Improved biopharmaceutical properties were expected from trioxolanes with polar functional groups. The best choice was deemed as having an adamantane structure on one side and a 4-substituted cyclohexane ring with functional groups that would allow further derivatization (R= OH, CH_2OH, COOH, CH_2COOH, NH_2, CH_2NH_2, =O) on the other side of the trioxolane ring. The structure of 65 excludes any problems associated with chirality and its diastereoselectivity is defined only by the substituent on the cyclohexanone ring. Ozonolysis occurs preferentially by forming the *cis*-isomer (Scheme 17).[93]

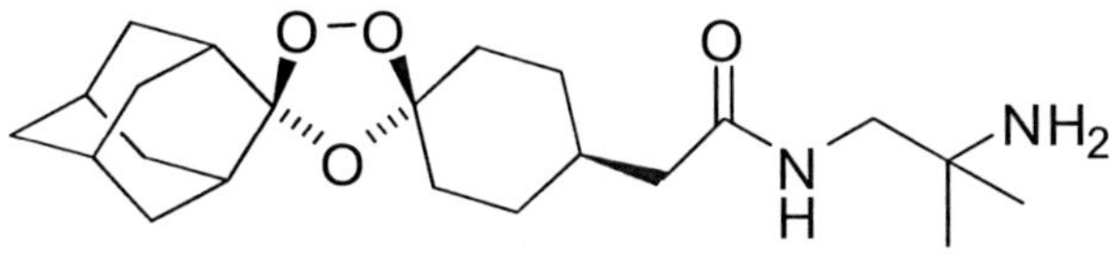

60, OZ227, Rbx-11160

Figure 6.

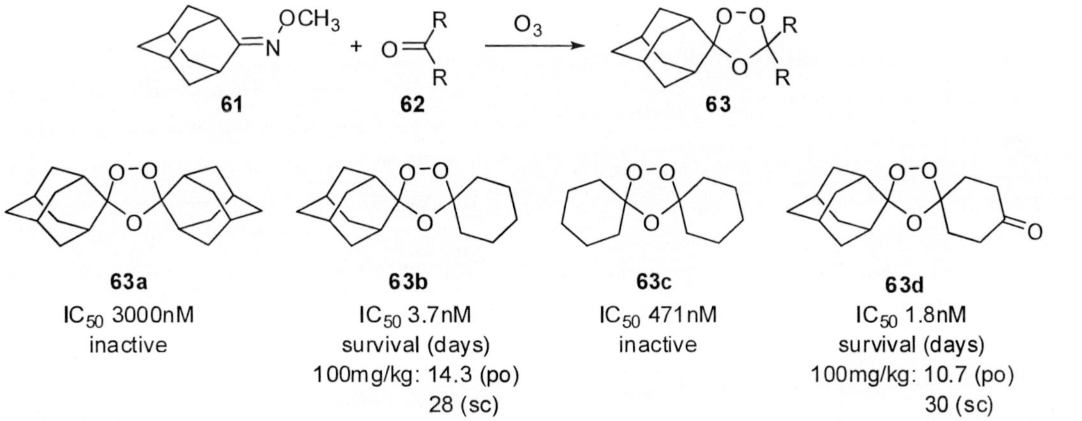

Scheme 16.

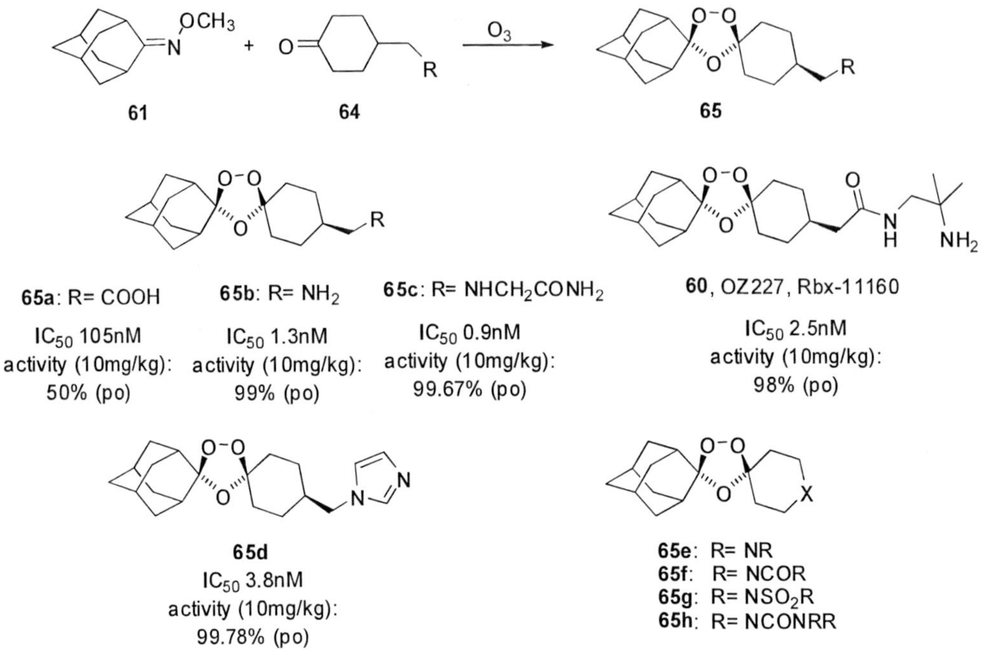

Scheme 17.

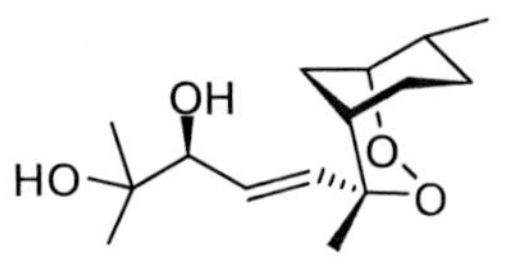

66, Yingzhaosu A

Figure 7.

Initially, three derivatives were targeted among which 60 had the best profile (OZ227, Rbx11160).[91,96] Later, new derivatives were synthesized to identify those substrates that have the optimum characteristics of lipophilicity and aqueous solubility and containing those functional groups linked with antimalarial properties. Analogues of 65 with a piperidine ring instead of cyclohexane and additionally functionalized into amines, amides, sulfonamides and ureas 65e-h possessed good *in vitro* activity but were less potent *in vivo*. Furthermore, products with acid functional groups are less potent than those with neutral or weak basic functional groups.[97,98] This led Vennerstrom and his team to prepare weak base trioxolanes with good oral activity, good biopharmaceutical properties and low toxicity. Targeted compounds 65b-d were derivatives with amino (alkylamino, amides, aminoxy, amino acid, guanidine) and azole functional groups.[99]

3.4. Endoperoxides

Yingzhaosu A 66 (Figure 7) is another active substance from the arsenal of traditional Chinese herbal medicine that has had a tremendous impact on the development of peroxidic antimalarial agents. Although it was isolated from a rare ornamental vine *Artabotrys uncinatus Lam.* and its structure determined in approximately the same time as that of artemisinin,[39,100] its biologically activity was only quantified recently.[101] One reason could be the scarcity of the herb and the difficulty in isolating the active compound. The total synthesis of yingzhaosu A consists of 14 steps and is economically unviable.[39,102] In 2005, however, a new total synthesis was reported consisting of only 8 steps starting with (S)-limonene and resulting in a in 7.3% overall yield of yingzhaosu A. This enabled the determination of its antimalarial activity and cytotoxicity.[101] Yingzhaosu A contains a 2,3-dioxo[3.3.1]nonane structure which has become a unique bicyclic scaffold for various synthetic endoperoxide antimalarials.

3.4.1. Bicyclic Endoperoxides

Since yingzhaosu A 66 itself was not available, various approaches were taken to synthesize substrates that mimics its structure. The leading method for the preparation of 2,3-dioxo[3.3.1]nonanes was developed by Hoffmann-La Roche and consists of reacting an isomer of carvone or its analogue 67 by a singlet oxygen ene reaction, followed by cyclization of the intermediate hydroperoxide 68 into the desired bicyclic yingzhaosu A analogue 69 (Scheme 18).[103-105] The structure of 69 was then transformed by incorporation of various alkyl chains, aryl structures or functional groups. The cyclic structure by itself 69a is active *in vivo* (sc) and it is surprisingly non-sensitive to stereochemical factors (ED_{50} of stereoisomers is 51 and 61mg/kg, respectively). Lipophilic groups 69b enhance the activity as well as aryl groups with electron-withdrawing substitutents. The difference in the activities of these endoperoxides and artemisinin is smaller *in vivo* than *in vitro*, while the dioxane structure is more stable leading to longer half-lives in plasma. Arteflene 69c was selected as the most potent drug candidate and entered into Phase II clinical trials, before it was abandoned. However, indepth studies on the mechanism of activity has provided important insights into the interaction of the peroxidic antimalarial agent with Fe(II) and the formation of secondary carbon radicals.[106,107]

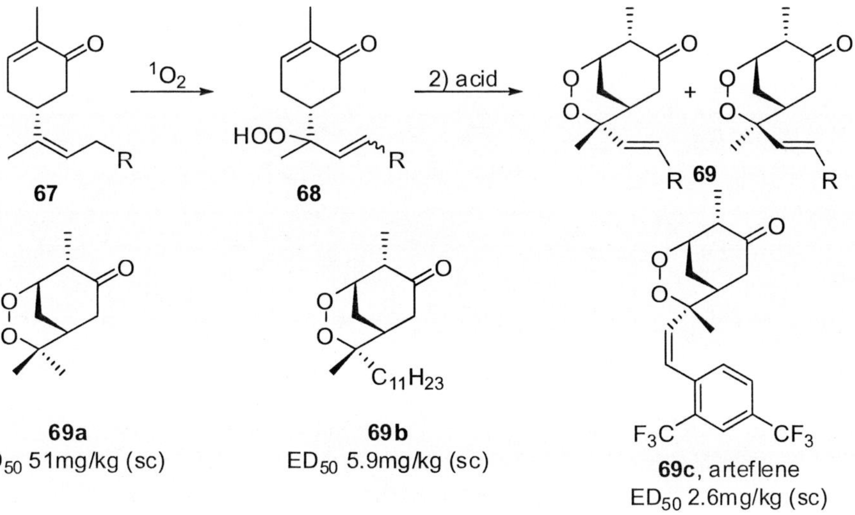

Scheme 18.

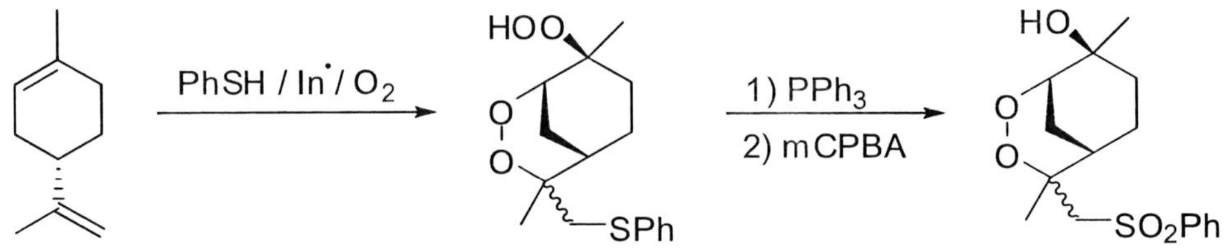

70 71 72

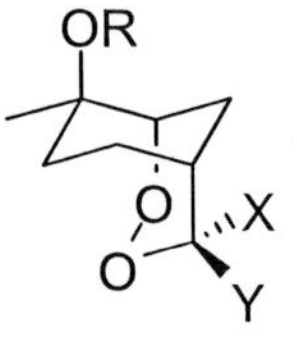

	R	X	Y	IC_{50} (nM)	ED_{50} (mg/kg)
72a	H	CH_2SO_2Ph	Me	55	>30
72b	Ac	CH_2SO_2Ph	Me	17	3.7
72c	Bn	CH_2SO_2Ph	Me	6.5	0.8
72d	Ac	Ph	CH_2SO_2Ph	42	

Scheme 19.

2,3-Dioxabicyclo[3.3.1]nonane was established as a pharmacophore and further synthetic strategies for its synthesis were initiated. An important group of bicyclic β-sulfonyl endoperoxides 72 was developed by thiol induced free radical cyclization of monoterpenes and oxygen (TOCO – Thiol Olefin Co-Oxygenation), where molecular oxygen is used to provide peroxide functionality as exemplified in the case of limonene 70 (Scheme 19).[108] Here, a free radical reaction is initiated by UV light, and AIBN or di-tert-butylperoxolate and an endoperoxide 71 is produced in a single step by forming five bonds simultaneously. The transformation into the sulfonyl derivative 72 involves the reduction of the hydroperoxy group into an hydroxy group followed by the oxidation of the sulfenyl group to a sulfonyl one. The hydroxy functional group in 72 presents an entry point for further transformations, while derivatives of monoterpenes, taken as starting substrates, can also broaden the structural scope. In this way, over fifty derivatives were synthesized and tested for their antimalarial activity.[109,110] Results were similar to arteflene derivatives regarding the insensitivity of their activity to stereoelectronic effects. Furthermore, the best endoperoxides have a better activity than the currently used artemisinin-based drugs, especially for in vivo values (sc). They also have a low toxicity.

Further indication of the importance of 2,3-dioxabicyclo[3.3.1]nonane pharmacophore emerged after the preparation of iodinated analogues of yingzhaosu A. Unsaturated hydroperoxides 73 were prepared by Co(II)-catalyzed autooxidation of alkenes, which is another method for introducing a peroxide functional group using molecular oxygen. Bis(collidine)iodine hexafluorophosphate (BCIH) was then used to construct the endoperoxide structure, where the reaction proceeds first by iodination of the double bond to form the cyclic iodonium ion that reacts with the hydroperoxy group as an internal nucleophile to form an endoperoxide bridge (Scheme 20).[111,112] The *in vitro* antimalarial activity of iodo-functionalized 2,3-dioxa[3.3.1]nonanes was high, however lower *in vivo* activity suggests that derivatives should be transformed to transfer this high activity from *in vitro* into *in vivo*.

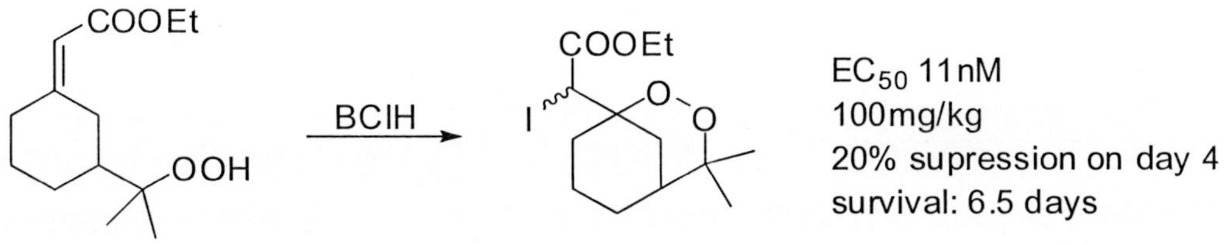

73 **74**

Scheme 20.

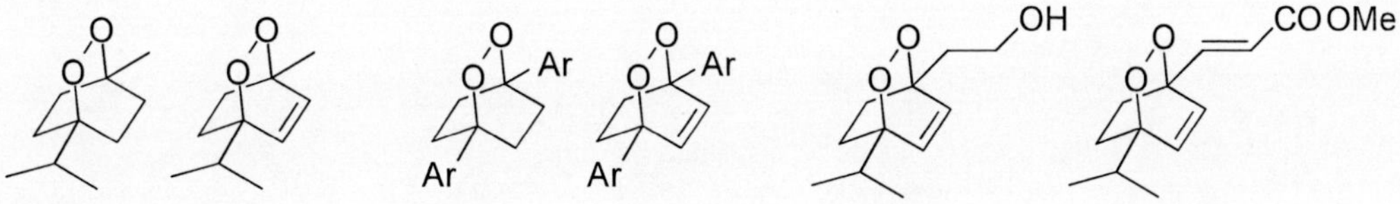

75 **76** **77** **78** **79** **80**

Scheme 21.

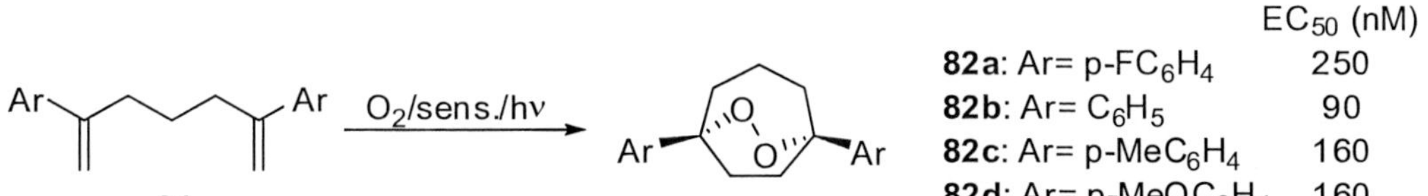

81

Scheme 22.

Monoterpenes are interesting starting compounds for the formation of various bicyclic [2.2.2]endoperoxides by the addition of singlet oxygen. Examples of these compounds are found in nature like ascaridol 75 and dihydroascaridol 76 (Scheme 21). A disadvantage is their volatile nature and lipophilicity. Diaryl substituted bicyclic [2.2.2] endoperoxides 77, 78 were first synthesized by the same method to study their antimalarial properties.[113,114] While more polar analogues of bicyclic [2.2.2] endoperoxides 79 and 80 were prepared from terpenoids with polar functional groups like nopol or perillyl alcohol by the addition of singlet oxygen.[115]

Another type of antimalarial bicyclic endoperoxide was synthesized from 2,6-diaryl-1,6-heptadienes 81 by photo-electron transfer oxygenation to give the bicyclic [3.2.2] endoperoxides 82 shown in Scheme 22. *In vitro* antimalarial activities against *P. falciparum* were moderate and dependent on the substituents. Interestingly, fluorine in 82a did not enhance the activity of the bicyclic endoperoxidic [3.2.2]nonanes. The mechanism of decomposition of these endoperoxides with Fe(II) salt has been studied and besides a common electron transfer reaction leading to the generation of oxygen and carbon radicals a Lewis acid pathway is also possible by generating the carbocation.[116]

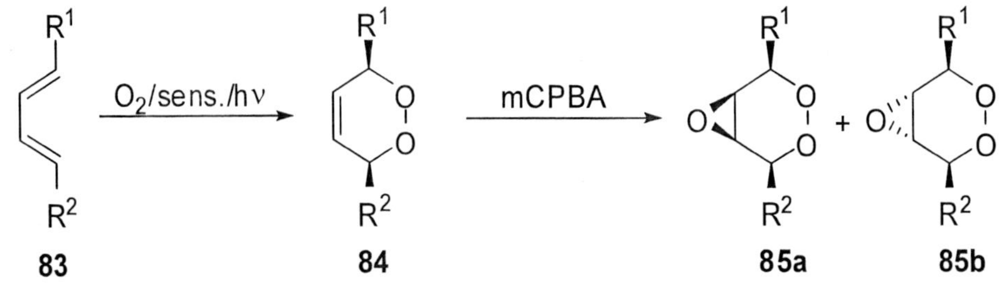

Scheme 23.

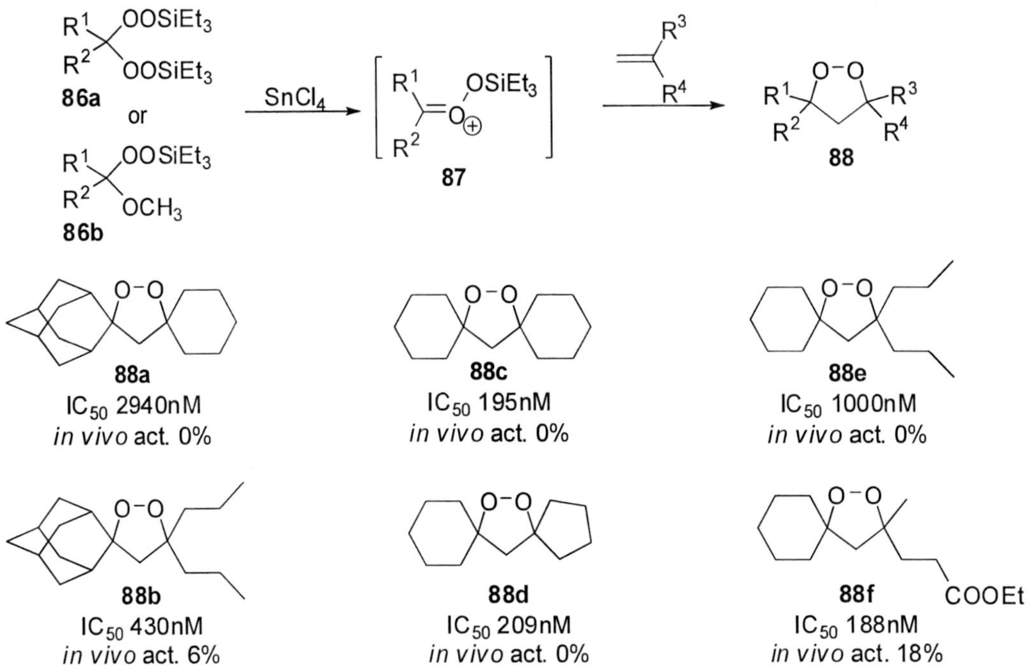

Scheme 24.

3.4.2. Dioxanes and Dioxolanes

There are only a few examples of simple dioxanes and dioxolanes as antimalarial agents. Unsaturated dioxanes 84 are produced by the cycloaddition of a singlet oxygen to 1,3-dienes 83 (Scheme 23). Another oxygen atom is introduced by epoxidation of the double bond to form an epoxy dioxane 85. This serves as a steric analogue to the cyclic oxygen atom in the 1,2,4-trioxane ring of artemisinin.[117] Although the reactivity of unsaturated and saturated dioxanes were moderate, research points to the importance of the interaction with Fe(II) atoms.

To better understand the effect that the peroxide bond has on the antimalarial action of organic peroxides Vennerstrom's group conducted research into the activity of various dioxolanes and compared their results with those obtained for trioxanes and tetraoxanes or trioxolanes, respectively.[118] They used a different synthetic strategy to prepare dioxolanes 88 involving the peroxycarbenium ion 87. This was formed from silylated *gem*-dihydroperoxides 86a or perketals 86b before being transformed by an annulation reaction with alkenes into dioxolanes 88. Their study provides an interesting insight into the structural features of peroxides relevant for good antimalarial activity. Dioxolanes exhibit a similar high influence of the steric effects on the antimalarial activity and in the case of dispiro dioxolanes, adamantane derivative 88a is less active than the cyclohexane derivative 88c and even less reactive that its spiro analogue 88b. A contrary effect was observed for cyclohexane analogues where dispiro dioxolanes 88c and 88d are more active than the spiro one 88e, while for *in vivo* activity a polar group is essential. The key element for good antimalarial activity of the cyclic peroxides is, according to the authors, the perketal structure. Therefore, the C-atom that bears the peroxide bond should have attached another oxygen atom because the β-scission of oxy radicals to form a carbon radical is faster in the perketal series than in the alkyl peroxides. In accordance with this observation 1,2,4-trioxanes and 1,2,4,5-tetraoxanes are more reactive than the corresponding dioxanes, while 1,2,4-trioxolanes are more reactive than 1,2-dioxolanes.[118]

3.4.3. Cyclic Perketals

This group of antimalarial peroxides share a common perketal structure with 1,2,4-trioxanes and 1,2,4-trioxolanes but only the peroxide group is a part of the cyclic structure.

3.4.3.1. Yingzhaosu A analogues

In accordance with the previous finding the perketal group is an important structural element of antimalarial drugs. In fact, it is only the yingzhaosu A and its analogues that does not have a perketal functional group, while all the others have this structural element included – 1,2,4-trioxanes, 1,2,4-trioxolanes and 1,2,4,5-tetraoxanes. However, yingzhaosu A analogues with a 2,3-dioxabicyclo[3.3.1]nonane scaffold and with a perketal structural element included were also prepared (Scheme 25). Synthesis starts by converting dihydrocarvone 89 via olefination and ozonolysis into an unsaturated hydroperoxy ketal 90 or hydroperoxide. The final step involves the cyclization into a bicyclic [3.3.1]nonane structure 91 by ozonolysis in 2,2,2-trifluoroethanol (TFE).[112,119]

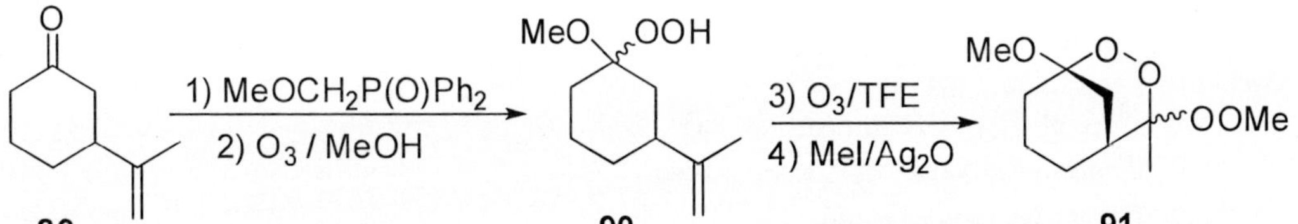

89 90 91

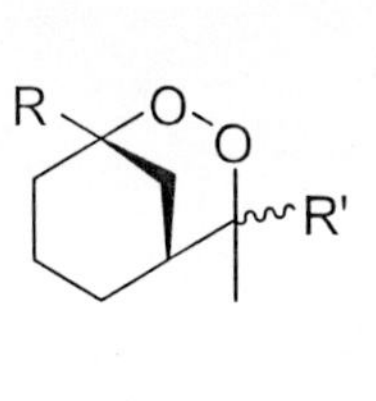

	R	R'	EC_{50} (M)
91a	Me	OOMe	2.9×10^{-5}
91b	OMe	OOMe	4.6×10^{-6}
91c endo	OOMe	OMe	1.7×10^{-5}
exo			1.8×10^{-6}
91d	OOMe	Me	7×10^{-5}

Scheme 25.

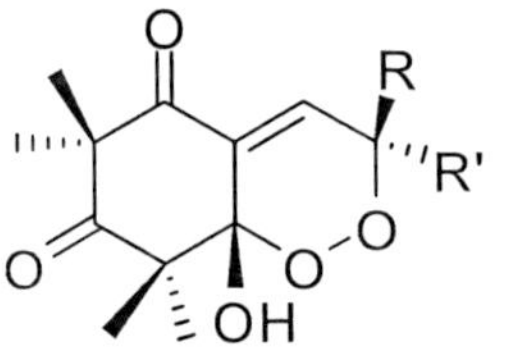

92
G1: R=Me, R'=Et
G2: R=Et, R'=Me
G3: R=Me, R'=Me

Figure 8.

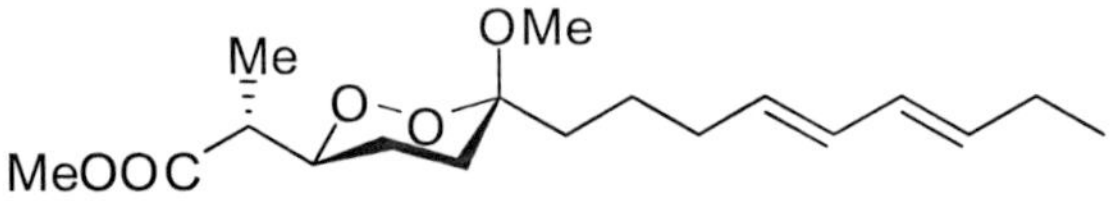

93a, Peroxyplacoric acid A_3 methyl ester

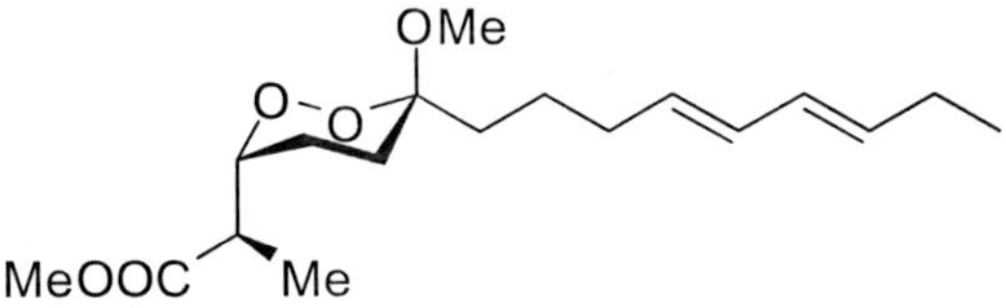

93b, Peroxyplacoric acid B_3 methyl ester

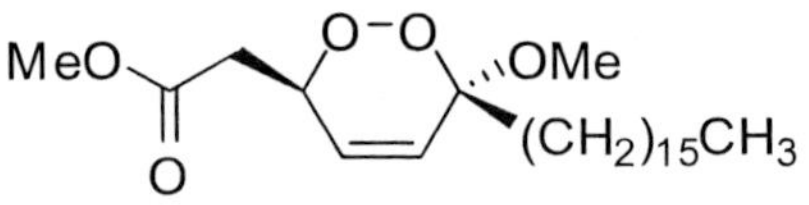

93c, Chondrillin

Figure 9.

3.4.3.2. G Factors

G factors 92 are phytohormones and growth regulators first isolated from *Eucalyptus grandis* (Figure 8). As they are toxic to the plant they are synthesized on demand. Synthesis for the preparation of G factor analogues begins with the syncarpic acid that is first transformed into ethylenic ketones that spontaneously react with oxygen to form a compound based on a G factor skeleton.[120,121] The G3 factor 92 does possess some antimalarial activity (IC_{50} 20μM) that can be improved by methylation of the hydroxy group to give a more active compound (IC_{50} 0.28μM). The activity still remains significantly lower than that of artemisinin, albeit they do have comparable electrochemical properties.[120]

The structure of G factors does allow some modification on the C-atom of the peroxide bond as well as through derivatization of the hydroxy group. It was shown that substitution of hydroxy group with fluorine, to produce a dioxane structure, completely inhibits activity. This is more evidence to the important role of the ketal structure. Alkylation of the hydroxy group in G3 increases its activity but the length of the alkyl chain does not have a big influence.[122-124]

3.4.3.3. Spongean Peroxides

Spongean metabolites have been investigated on account of their physiological activity. Two peroxidic metabolites were isolated: Plakorin 93a,b and Chondrillin 93c.[125,126] They share the same cyclic 3-methoxy-1,2-dioxane cyclic structure with a carbomethoxymethyl substituent in position 6 (Figure 9).

These natural derivatives possess some antimalarial activity with plakorins being more active than chondrillins. A simple method was developed to construct the basic 6-carbomethoxymethly-3-methoxy-1,2-dioxane scaffold that is necessary for the activity. It consists of a peroxyhemiacetalization of ketoester 94 with a urea-hydrogen peroxide complex to give 95 that is converted by an intramolecular Michael addition in fluorinated alcohol to yield 96 (Scheme 26).[127] To fine tune the antimalarial activity, different 3-alkyl substituents were added to adjust lipophilicity.[128] However, the alkyl group does not influence only the lipophilicity but the stability of its radical also needs to be taken into account. The pentyl radical is unstable and the parent 96b is only moderately active. Alternatively, the alkenyl chain can form a radical that is stabilized and therefore longer-lived to react with the parasite's proteins.[129] Although reported activities were better, these substrates remain inactive *in vivo* due to the hydrolysis of the ester group. The methyl ester was transformed into amide that is not hydrolysed by the organism and therefore possess *in vivo* activity.[130]

3.4.3.4. Other Cyclic Perketals

A similar synthetic route was applied to various cyclic perketals with photooxygenation being replaced by intramolecular addition of an hydroperoxy group obtained by hydroperoxidation of the ketone using hydrogen peroxide.

Scheme 27 presents a typical synthesis of the simplified analogue of artemisinin 99 with a 7-membered endoperoxide structure and cyclic perketal oxygen atom incorporated into the tricyclic system.[131]

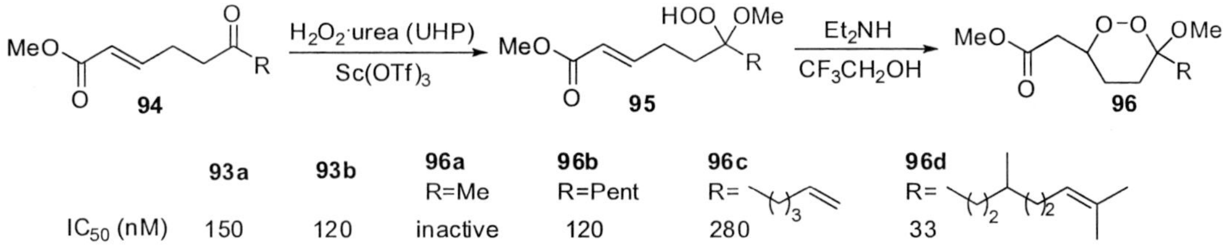

Scheme 26.

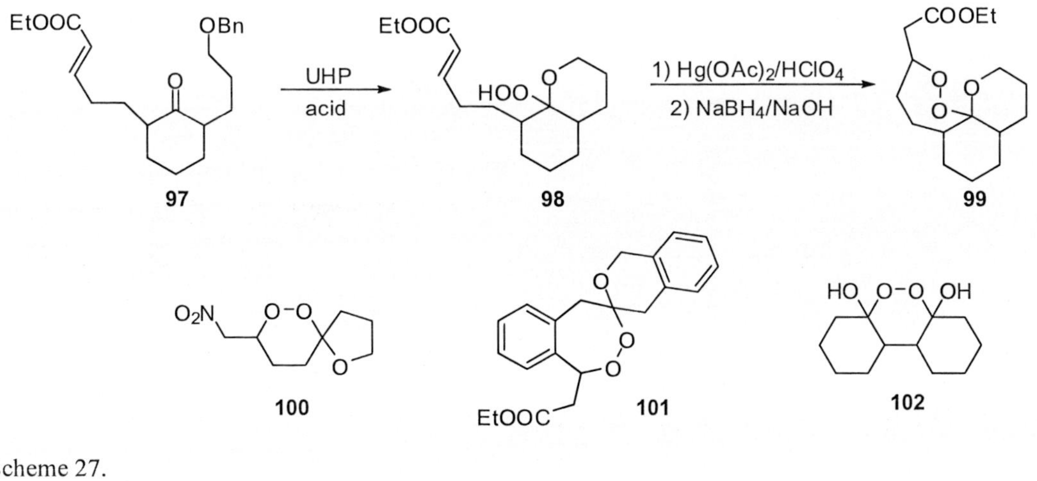

Scheme 27.

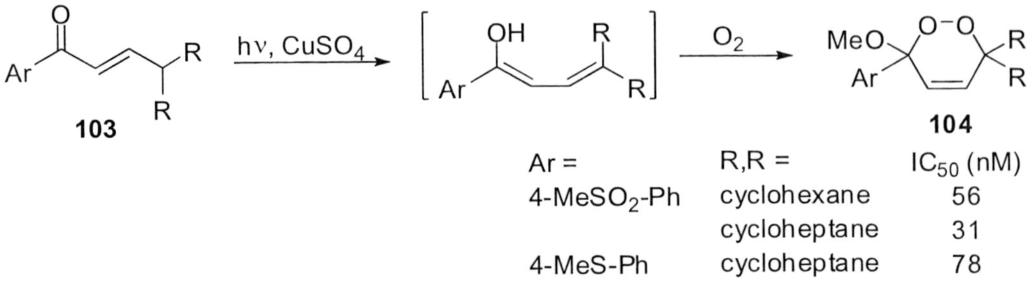

Scheme 28.

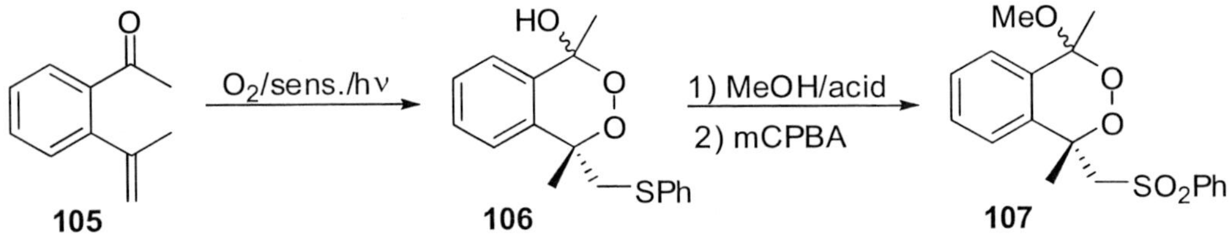

Scheme 29.

Spiroperoxides 100 and 101, analogues of 99 with an endoperoxide bond in the 6- or 7-membered cycle were also prepared. When a benzene ring was fused to the peroxy ring, formation of seven-membered derivatives is facilitated, while a nitro group can replace the carboethoxy one. Although in this case due to a possible interference of the nitro group in radical processes, the *in vitro* antimalarial activity is reduced. The activity of spiroperketals with the carboethoxy group is in the order of the peroxyplakoric acid derivatives.[132] The perketal 102, prepared from bis(cyclohexyl)-2,2'-dione and hydrogen peroxide, also shows an antimalarial activity.[133]

The majority of synthetic procedures still utilize the singlet oxygen as a source of peroxide functionality. Posner's group has synthesized 3-aryl-3-methoxy-1,2-dioxene derivatives 104 as a special case of cyclic peroxy ketals *via* the photooxygenation of unsaturated ketones 103 (Scheme 28).[134] Although mechanistic debates were focused on the possibility of a single oxygen addition or radical oxygenation[42] this does not effect their potent antimalarial activity with the most active substrates prepared having a quarter of artemisinin's activity. Interestingly, the activity of the spiro perketals is higher with the spiro cycloheptane ring than with the cyclohexane one.

The TOCO protocol (Scheme 19), used to construct 1,2-dioxanes, was modified to obtain the 3-alkoxy-1,2-dioxane scaffold.[135] The unsaturated ketone 105 (2'-isopropenyl acetophenone) was taken as the starting substrate and transformed by thiol-olefin-co-oxygenation into a cyclic benzo-fused peroxy hemiketal 106. Finally, the hydroxy group was methylated, while the thiol group was oxidized into a sulfonyl one (Scheme 29) to obtain the desired benzo-fused ketal 107, however it has a relatively weak antimalarial activity.

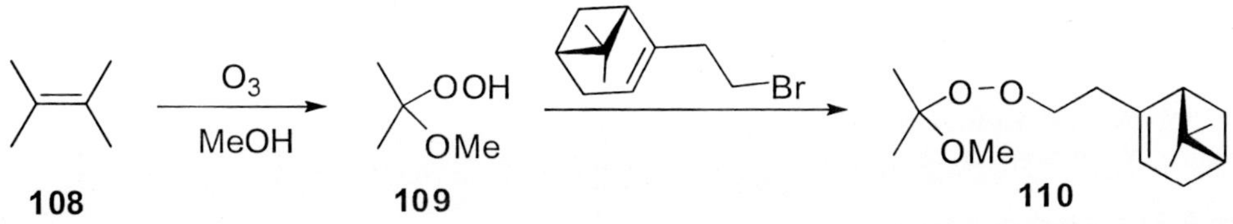

Scheme 30.

3.4. Perketals and Bis-Perethers

Even a simple peroxide compound like hydrogen peroxide and *tert*-butyl hydroperoxide possess some antimalarial activity. In the series of cyclic peroxides it is shown that an additional oxygen atom bound to a C-atom at the peroxide bond significantly increases the antimalarial activity.[118] Trioxanes and trioxolanes are therefore more reactive than dioxolanes and this is true for tetraoxanes *vs*. dioxanes. It is reasonable to believe that among acylic derivatives the same observation is valid and that perketals and perethers are more active than hydroperoxides or peroxides. Recent years have witnessed considerable development in the field of α-heteroatom substituted peroxides to allow the targeted preparation of perketals and perethers.[42]

The most common route to synthesize perketals is ozonolysis of the double bonds in the presence of alcohol to form a hydroperoxy ketal that is then transformed into a more stable perketal.[136] This method was used for the preparation of various 2-methoxyprop-2-yl peroxides 110 by the ozonolysis of 2,3-dimethyl-2-butene 108 in methanol followed by the alkylation of the hydroperoxy ketal 109 with the addition of desired alkyl bromide (Scheme 30).[137] However, their antimalarial activity remains moderate to weak with the highest *in vitro* IC_{50} activity of 370nM recorded for 110. Interestingly, the activity diminishes by introducing polar groups like ethers, ciano or esters, while a pinene structure augments the activity.

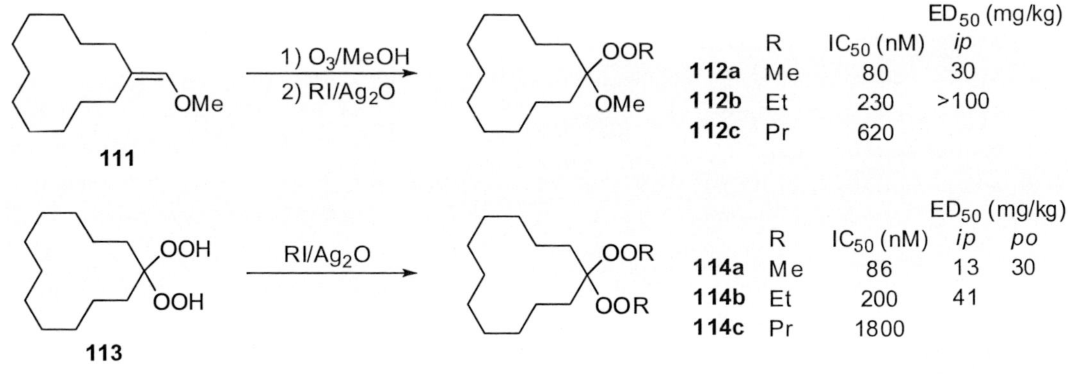

Scheme 31.

The research on 1,2,4,5-tetraoxacycloalkanes reveal that the cyclododecane ring is a potent structure for antimalarial agents. Therefore, the antimalarial activity of open analogues were studied, these included *gem*-bisperethers, together with the open analogues of 1,2,4-

trioxanes and 1,2,4-trioxolanes, namely peroxy ketals.[138] The starting point for both products was cyclododecanone and for the synthesis of perketals it was first methoxymethylated into 111 before being converted into the hydroperoxy ketal by ozonolysis in methanol. The hydroperoxide group was then alkylated to obtain the targeted perketals 112. The *in vitro* antimalarial activity was tested on *P. falciparum* chloroquine sensitive FCR-3 strain. The compounds possess potent antimalarial activity that is sensitive to the structure of the alkyl groups and that of the organic core (Scheme 31). The highest activity belonged to a dimethylated cyclododecyl perketal 112a, while prolongation of alkyl chain on the peroxide oxygen rapidly reduces activity. Again, cyclic perketals 112 are more active than dialkyl analogues with the dibutyl derivative having an IC_{50} of 8600nM.

A similar relationship between activity and the peroxide structure is observed in the bis-perether series. The preparation of perethers includes the transformation of a ketone into the *gem*-dihydroperoxide 113 with hydrogen peroxide or ozonolysis followed by alkylation of the hydroperoxide groups (Scheme 31). Results from the *in vitro* screening of 114 produced results similar to those for perketals 112 *i.e.*, the methylated derivatives are the most active.

Four days *in vivo* studies against *P. berghei* revealed large differences between perketals and bis-perethers and the latter being more active. Surprisingly, 114a is also orally active and on *ip* administration of 20mg/kg dose 4 day suppressive test failed to show any parasites in the blood.

4. Chimeric Compounds

Peroxides are fast acting antimalarial agents but require an additional longer acting agent to clear the host of malaria parasites. Furthermore, during combination therapy parasites have more trouble in developing resistance. In chimeric compounds this approach of combination therapy is united in a single molecule in which two different pharmacophores are covalently joined.

The first such compound was published by the group led by Meunier who synthesized "trioxaquines" as agents for "covalent bitherapy".[139-141] The group was able to join a 4-aminoquinoline structure (facilitates the penetration of an agent) to a trioxane structural element (a potent alkylating agent essential for good antimalarial activity) to obtain trioxaquines *i.e.*, 115 (Figure 10). Research on the mechanism of action of peroxidic antimalarial suggests that the alkylating property might not be the only factor governing antimalarial activity, nevertheless trioxaquines are potent antimalarial drugs that act against the asexual and sexual (gametocytes) stages of the parasite cycle.[142] The most active among this group is di-citrate trioxaquine DU1302 115 having a potent oral *in vivo* activity against *P. yoelli nigeriensis* higher than artemisinin and artemether and similar to artesunate. Furthermore, with an oral dose of 50mg/kg/day trioxaquine 115 cured all infected mice without any recrudescence at 60 days. Another chimeric compound 116 was prepared from a spiro 1,2,4-trioxane ring and 4-aminoquinoline.[143] A comparison of the oral activity of each separate element of the hybrid molecule and that of 116 was made to determine the *in vivo* antimalarial activity against multidrug resistant *P. yoelii.* Some improvement was observed over the trioxane part or quinoline, however neither of the three drugs resulted in a cure after 28 days.

Following a similar approach, mefloquine and artemisinin were joined in two ways: directly or with a divisible linker that can be hydrolysed *in vivo*.[144] Chimera 117 proved potent *in vitro* against various strains of *P. falciparum* with an IC_{50} of 2.4 to 17.2nM, an activity comparable to artemether but better than mefloquine. Chimera 117 was more effective *in vivo* (ip, 30mg/kg) in controlling the parasitemia than the parent compounds, the CF_3-artemisinin and artemether, in completely inhibiting the growth of parasites. Another report presents the hybrid antimalarial agent 118 as a combination of quinine and dihydroartemisinin linked by an ester linkage.[145] Its potency has been compared to each individual component as well as to a mixture of both. Results show that the hybrid is more potent (IC_{50} against chloroquine resistant strain FcB1 was 9.6nM), while artemisinin and quinine have an IC_{50} of 50.0 and 96.8nM, respectively and the mixture of quinine and artemisinin had an IC_{50} of 27.9nM. Interesting is that the chimera 118 is more active than a mixture of both components and each component individually, which could indicate some kind of synergetic action when both constituents are covalently bound. This observation requires a more complete understanding of the exact mechanism of action of chimeras.

A masked combination therapy is an innovative approach based on a directly acting pharmacophore from the peroxide family and a latent pharmacophore of the chalcone family. Once inside the parasite, the peroxide pharmacophore damages the parasite either by alkylation or some other action. In this case the peroxide is reduced to the ketone function to form another pharmacophore in the form of a chalcone that is a cysteine protease inhibitor that further damages the parasite.[146] This approach is akin to a "Trojan horse". When composing such an agent 119, a yingzhaosu A analogue similar to arteflene is used as a peroxide pharmacophore, while substituents on position 4 of the endoperoxide ring were modulated to form a chalcone during decomposition. Indeed, the reaction of 119 with an Fe(II) salt reveals that the reaction with the heme should produce a secondary carbon radical 121 that is toxic to the parasite's proteins and the targeted chalcone 120 as a second agent for inhibiting cysteine protease (Scheme 32). Activities of these chimeras were determined only *in vitro* against *P. falciparum* chloroquine resistant strain K1 but show that substrates have an activity higher than that of the parent arteflene.

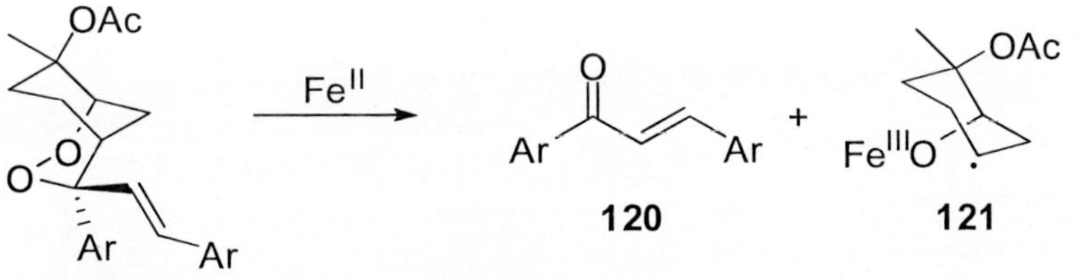

119

	IC_{50} (nM)
119a: R = C_6H_5	34
119b: R = *p*-FC_6H_4	29
119c: R = *p*-ClC_6H_4	23

Scheme 32.

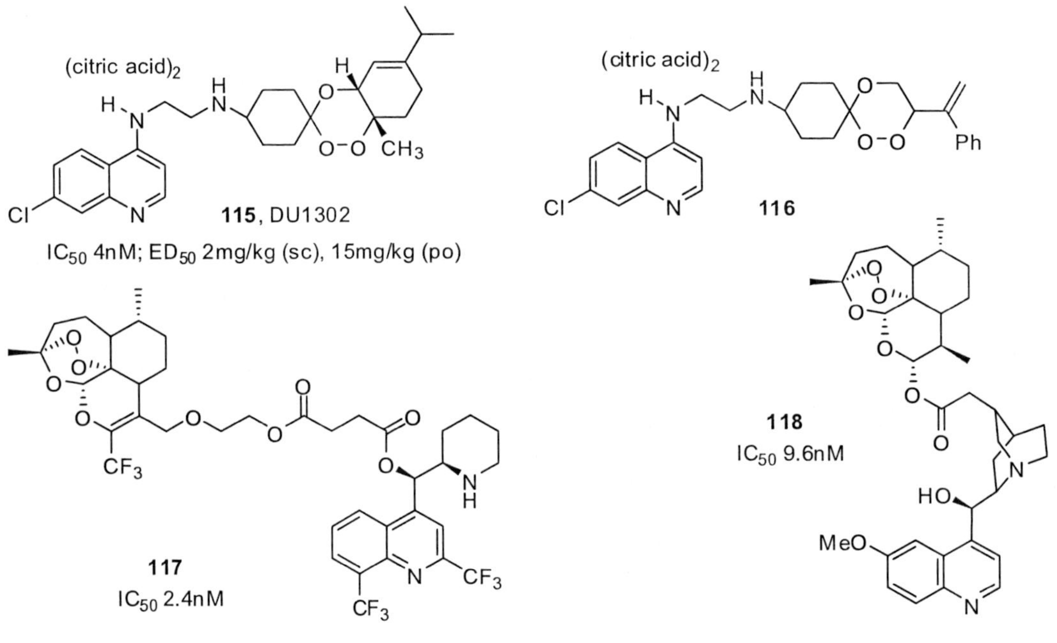

Figure 10.

5. Conclusion

The field of organic peroxides has witnessed immense progress after the recognition of the value of artemisinin in the current chemotherapy of malaria. New synthetic methods have been developed to incorporate a peroxide bond into an organic scaffold that enables the preparation of organic peroxides deemed to have antimalarial activity. Until recently, the main methods involved using singlet oxygen and ozone, however their use has drawbacks when the synthesis is scaled up. Alternative solutions proposed in recent years include radical reactions with molecular oxygen and the use of aqueous hydrogen peroxide, which is a much safer reagent. Improvement in the synthesis of organic peroxides has increased the spectrum of available organic peroxides that provide a greater understanding of the mechanism of action of antimalarial peroxides. With this knowledge it will be possible to define better antimalarial agents and transfer good ideas from theory to practice.

References

[1] Joy, D. A.; Feng, X.; Mu, J.; Furuya, T.; Chotivanich, K.; Krettli, A. U.; Ho, M.; Wang, A.; White, N. J.; Suh, E.; Beerli, P.; Su, X. Z.; Early Origin and Recent Expansion of *Plasmodium Falciparum*. *Science*, 2003, *300*, 318-321.

[2] Snow, R. W.; Guerra, C. A.; Noor, A. M.; Myint, H. Y.; Hay, S. I.; The Global Distribution of Clinical Episodes of *Plasmodium Falciparum* Malaria. *Nature*, 2005, *434*, 214-217.

[3] Breman, J. G.; The Ears of the Hippopotamus: Manifestations, Determinants, and Estimates of the Malaria Burden. *Am. J. Trop. Med. Hyg.*, 2001, *64*, 1-11.
[4] Meshnick, S. R.; Dobson, M. J. The History of Antimalarial Drugs. In *Antimalarial Chemotherapy. Mechanisms of Action, Resistance, and New Directions in Drug Discovery*; Rosenthal, P. J., Editor; Humana Press: Totowa, 2001; Chapter 2, pp 15-25.
[5] Martin, P. H.; Lefebvre, M. G.; Malaria and Climate: Sensitivity of Malaria Potential Transmission to Climate. *Ambio*, 1995, *24*, 200-207.
[6] Rogers, D. J.; Randolph, S. E.; The Global Spread of Malaria in a Future, Warmer World. *Science*, 2000, *289*, 1763-1766.
[7] Hopkins, A. L.; Witty, M. J.; Nwaka, S.; Mission Possible. *Nature*, 2007, *449*, 166-169.
[8] Nwaka, S.; Hudson, A.; Innovative Lead Discovery Strategies for Tropical Diseases. *Nat. Rev. Drug Discovery*, 2006, *5*, 941-955.
[9] Nwaka, S.; Ridley, R. G.; Virtual Drug Discovery and Development for Neglected Diseases through Public-Private Partnerships. *Nat. Rev. Drug Discovery*, 2003, *2*, 919-928.
[10] Tripathi, R. P.; Mishra, R. C.; Dwivedi, N.; Tewari, N.; Verma, S. S.; Current Status of Malaria Control. *Curr. Med. Chem.*, 2005, *12*, 2643-2659.
[11] Mital, A.; Recent Advances in Antimalarial Compounds and their Patents. *Curr. Med. Chem.*, 2007, *14*, 759-773.
[12] Jana, S.; Paliwal, J.; Novel Molecular Targets for Antimalarial Chemotherapy. *Int. J. Antimicrob. Agents*, 2007, *30*, 4-10.
[13] Padmanaban, G.; Nagaraj, V. A.; Rangarajan, P. N.; Drugs and Drug Targets against Malaria. *Curr. Sci.*, 2007, *92*, 1545-1555.
[14] O'Neill, P. M.; Ward, S. A.; Berry, N. G.; Jeyadevan, J. P.; Biagini, G. A.; Asadollaly, E.; Park, B. K.; Bray, P. G.; A Medicinal Chemistry Perspective on 4-Aminoquinoline Antimalarial Drugs. *Curr. Top. Med. Chem.*, 2006, *6*, 479-507.
[15] Biagini, G. A.; O'Neill, P. M.; Nzila, A.; Ward, S. A.; Bray, P. G.; Antimalarial Chemotherapy: Young Guns or Back to the Future? *Trends Parasitol.*, 2003, *19*, 479-487.
[16] Hsu, E.; Reflections on the 'Discovery' of the Antimalarial Qinghao. *Br. J. Clin. Pharmacol.*, 2006, *61*, 666-670.
[17] Mueller, M. S.; Runyambo, N.; Wagner, I.; Borrmann, S.; Dietz, K.; Heide, L.; Randomized Controlled Trial of a Traditional Preparation of Artemisia Annua L. (Annual Wormwood) in the Treatment of Malaria. *Trans. R. Soc. Trop. Med. Hyg.*, 2004, *98*, 318-321.
[18] Rath, K.; Taxis, K.; Walz, G.; Gleiter, C. H.; Li, S. M.; Heide, L.; Pharmacokinetic Study of Artemisinin after Oral Intake of a Traditional Preparation of *Artemisia Annua L.* (Annual Wormwood). *Am. J. Trop. Med. Hyg.*, 2004, *70*, 128-132.
[19] Jansen, F. H.; The Herbal Tea Approach for Artemisinin as a Therapy for Malaria? *Trans. R. Soc. Trop. Med. Hyg.*, 2006, *100*, 285-286.
[20] Heide, L.; Artemisinin in Traditional Tea Preparations of *Artemisia Annua*. *Trans. R. Soc. Trop. Med. Hyg.*, 2006, *100*, 802.
[21] http://www.who.int/mediacentre/news/releases/2006/pr02/en/
[22] Krishna, S.; Woodrow, C. J.; Staines, H. M.; Haynes, R. K.; Mercereau-Puijalon, O.; Re-Evaluation of How Artemisinins Work in Light of Emerging Evidence of *in Vitro* Resistance. *Trends Mol. Med.*, 2006, *12*, 200-205.

[23] Jefford, C. W.; New Developments in Synthetic Peroxidic Drugs as Artemisinin Mimics. *Drug Discovery Today*, 2007, *12*, 487-495.

[24] Haynes, R. K.; From Artemisinin to New Artemisinin Antimalarials: Biosynthesis, Extraction, Old and New Derivatives, Stereochemistry and Medicinal Chemistry Requirements. *Curr. Top. Med. Chem.*, 2006, *6*, 509-537.

[25] O'Neill, P. M.; The Therapeutic Potential of Semi-Synthetic Artemisinin and Synthetic Endoperoxide Antimalarial Agents. *Expert Opin. Investig. Drugs*, 2005, *14*, 1117-1128.

[26] Begue, J.-P.; Bonnet-Delpon, D.; The Future of Antimalarials: Artemisinins and Synthetic Endoperoxides. *Drugs Fut.*, 2005, *30*, 509-518.

[27] O'Neill, P. M.; Posner, G. H.; A Medicinal Chemistry Perspective on Artemisinin and Related Endoperoxides. *J. Med. Chem.*, 2004, *47*, 2945-2964.

[28] Borstnik, K.; Paik, I. H.; Posner, G. H.; Malaria: New Chemotherapeutic Peroxide Drugs. *Mini-Rev. Med. Chem.*, 2002, *2*, 573-583.

[29] Singh, C.; Chaudhary, S.; Puri, S. K.; New Orally Active Derivatives of Artemisinin With High Efficacy against Multidrug-Resistant Malaria in Mice. *J. Med. Chem.*, 2006, *49*, 7227-7233.

[30] Posner, G. H.; Paik, I. H.; Chang, W.; Borstnik, K.; Sinishtaj, S.; Rosenthal, A. S.; Shapiro, T. A.; Malaria-Infected Mice are Cured by a Single Dose of Novel Artemisinin Derivatives. *J. Med. Chem.*, 2007, *50*, 2516-2519.

[31] Paik, I. H.; Xie, S.; Shapiro, T. A.; Labonte, T.; Narducci Sarjeant, A. A.; Baege, A. C.; Posner, G. H.; Second Generation, Orally Active, Antimalarial, Artemisinin-Derived Trioxane Dimers with High Stability, Efficacy, and Anticancer Activity. *J. Med. Chem.*, 2006, *49*, 2731-2734.

[32] Haynes, R. K.; Fugmann, B.; Stetter, J.; Rieckmann, K.; Heilmann, H. D.; Chan, H. W.; Cheung, M. K.; Lam, W. L.; Wong, H. N.; Croft, S. L.; Vivas, L.; Rattray, L.; Stewart, L.; Peters, W.; Robinson, B. L.; Edstein, M. D.; Kotecka, B.; Kyle, D. E.; Beckermann, B.; Gerisch, M.; Radtke, M.; Schmuck, G.; Steinke, W.; Wollborn, U.; Schmeer, K.; Romer, A.; Artemisone - A Highly Active Antimalarial Drug of the Artemisinin Class. *Angew. Chem., Int. Ed.*, 2006, *45*, 2082-2088.

[33] Magueur, G.; Crousse, B.; Ourevitch, M.; Bonnet-Delpon, D.; Begue, J.-P.; Fluoro-Artemisinins: When a Gem-Difluoroethylene Replaces a Carbonyl Group. *J. Fluorine Chem.*, 2006, *127*, 637-642.

[34] http://www.rbm.who.int/wmr2005/html/3-2.htm#3_2_3

[35] Ferreira, J. F. S.; Laughlin, J. C.; Delabays, N.; de Magalhaes, P. M.; Cultivation and Genetics of *Artemisia Annua L.* for Increased Production of the Antimalarial Artemisinin. *Plant Genet. Resour.*, 2005, *3*, 206-229.

[36] Ro, D. K.; Paradise, E. M.; Ouellet, M.; Fisher, K. J.; Newman, K. L.; Ndungu, J. M.; Ho, K. A.; Eachus, R. A.; Ham, T. S.; Kirby, J.; Chang, M. C. Y.; Withers, S. T.; Shiba, Y.; Sarpong, R.; Keasling, J. D.; Production of the Antimalarial Drug Precursor Artemisinic Acid in Engineered Yeast. *Nature*, 2006, *440*, 940-943.

[37] Avery, M. A.; Jennings-White, C.; Chong, W. K. M.; The Total Synthesis of (+)-Artemesinin and (+)-9-Desmethylartemesinin. *Tetrahedron Lett.*, 1987, *28*, 4629-4632.

[38] Avery, M. A.; Chong, W. K. M.; Jennings-White, C.; Stereoselective Total Synthesis of (+)-Artemisinin, the Antimalarial Constituent of *Artemisia Annua L. J. Am. Chem. Soc.*, 1992, *114*, 974-979.

[39] Zhou, W. S.; Xu, X. X.; Total Synthesis of the Antimalarial Sesquiterpene Peroxide *Qinghaosu* and *Yingzhaosu A*. *Acc. Chem. Res.*, 1994, *27*, 211-216.

[40] Tang, Y.; Dong, Y.; Vennerstrom, J. L.; Synthetic Peroxides as Antimalarials. *Med. Res. Rev.*, 2004, *24*, 425-448.

[41] Masuyama, A.; Wu, J. M.; Nojima, M.; Kim, H. S.; Wataya, Y.; 1,2,4,5-Tetraoxacycloalkanes: Synthesis and Antimalarial Activity. *Mini-Rev. Med. Chem.*, 2005, *5*, 1035-1043.

[42] Zmitek, K.; Zupan, M.; Iskra, J.; Alpha-Substituted Organic Peroxides: Synthetic Strategies for a Biologically Important Class of Gem-Dihydroperoxide and Perketal Derivatives. *Org. Biomol. Chem.*, 2007, *5*, 3895-3908.

[43] Eckstein-Ludwig, U.; Webb, R. J.; Van Goethem, I. D. A.; East, J. M.; Lee, A. G.; Kimura, M.; O'Neill, P. M.; Bray, P. G.; Ward, S. A.; Krishna, S.; Artemisinins Target the SERCA of *Plasmodium Falciparum*. *Nature*, 2003, *424*, 957-961.

[44] Krishna, S.; Uhlemann, A. C.; Haynes, R. K.; Artemisinins: Mechanisms of Action and Potential for Resistance. *Drug Resist. Updates*, 2004, *7*, 233-244.

[45] Dockrell, H. M.; Playfair, J. H. L.; Killing of Blood-Stage Murine Malaria Parasites by Hydrogen-Peroxide. *Infect. Immun.*, 1983, *39*, 456-459.

[46] Clark, I. A.; Cowden, W. B.; Butcher, G. A.; Free Oxygen Radical Generators as Anti-Malarial Drugs. *Lancet*, 1983, *1*, 234.

[47] Jefford, C. W.; Velarde, J. A.; Bernardinelli, G.; Bray, D. H.; Warhurst, D. C.; Milhous, W. K.; Synthesis, Structure, and Antimalarial Activity of Tricyclic 1,2,4-Trioxanes Related to Artemisinin. *Helv. Chim. Acta*, 1993, *76*, 2775-2788.

[48] Cazelles, J.; Camuzat-Dedenis, B.; Provot, O.; Robert, A.; Mayrargue, J.; Meunier, B.; Alkylating Properties of Synthetic Trioxanes Related to Artemisinin. *J. Chem. Soc. Perkin Trans. 1*, 2000, 1265-1270.

[49] Posner, G. H.; Cumming, J. N.; Woo, S. H.; Ploypradith, P.; Xie, S.; Shapiro, T. A.; Orally Active Antimalarial 3-Substituted Trioxanes: New Synthetic Methodology and Biological Evaluation. *J. Med. Chem.*, 1998, *41*, 940-951.

[50] Posner, G. H.; Oh, C. H.; Wang, D.; Gerena, L.; Milhous, W. K.; Meshnick, S. R.; Asawamahasadka, W.; Mechanism-Based Design, Synthesis, and *in Vitro* Antimalarial Testing of New 4-Methylated Trioxanes Structurally Related to Artemisinin: The Importance of a Carbon-Centered Radical for Antimalarial Activity. *J. Med. Chem.*, 1994, *37*, 1256-1258.

[51] Posner, G. H.; Oh, C. H.; Gerena, L.; Milhous, W. K.; Extraordinarily Potent Antimalarial Compounds: New, Structurally Simple, Easily Synthesized, Tricyclic 1,2,4-Trioxanes. *J. Med. Chem.*, 1992, *35*, 2459-2467.

[52] Posner, G. H.; Jeon, H. B.; Ploypradith, P.; Paik, I. H.; Borstnik, K.; Xie, S.; Shapiro, T. A.; Orally Active, Water-Soluble Antimalarial 3-Aryltrioxanes: Short Synthesis and Preclinical Efficacy Testing in Rodents. *J. Med. Chem.*, 2002, *45*, 3824-3828.

[53] Posner, G. H.; Jeon, H. B.; Parker, M. H.; Krasavin, M.; Paik, I. H.; Shapiro, T. A.; Antimalarial Simplified 3-Aryltrioxanes: Synthesis and Preclinical Efficacy/Toxicity Testing in Rodents. *J. Med. Chem.*, 2001, *44*, 3054-3058.

[54] Water-Soluble Trioxanes as Potent and Safe Antimalarial Agents. *Expert Opin. Ther. Patents*, 2001, *11*, 1351-1354.

[55] O'Neill, P. M.; Rawe, S. L.; Borstnik, K.; Miller, A.; Ward, S. A.; Bray, P. G.; Davies, J.; Oh, C. H.; Posner, G. H.; Enantiomeric 1,2,4-Trioxanes Display Equivalent *in Vitro*

Antimalarial Activity versus *Plasmodium Falciparum* Malaria Parasites: Implications for the Molecular Mechanism of Action of the Artemisinins. *ChemBioChem*, 2005, *6*, 2048-2054.

[56] Posner, G. H.; Maxwell, J. P.; O'Dowd, H.; Krasavin, M.; Xie, S.; Shapiro, T. A.; Antimalarial Sulfide, Sulfone, and Sulfonamide Trioxanes. *Bioorg. Med. Chem.*, 2000, *8*, 1361-1370.

[57] Jefford, C. W.; Rossier, J. C.; Milhous, W. K.; The Structure and Antimalarial Activity of Some 1,2,4-Trioxanes, 1,2,4,5-Tetroxanes, and Bicyclic Endoperoxides. Implications for the Mode of Action. *Heterocycles*, 2000, *52*, 1345-1352.

[58] Jefford, C. W.; Kohmoto, S.; Jaggi, D.; Timari, G.; Rossier, J. C.; Rudaz, M.; Barbuzzi, O.; Gerard, D.; Burger, U.; Kamalaprija, P.; Mareda, J.; Bernardinelli, G.; Manzanares, I.; Canfield, C. J.; Fleck, S. L.; Robinson, B. L.; Peters, W.; Synthesis, Structure, and Antimalarial Activity of Some Enantiomerically Pure, *Cis*-Fused Cyclopenteno-1,2,4-trioxanes. *Helv. Chim. Acta*, 1995, *78*, 647-662.

[59] Peters, W.; Robinson, B. L.; Rossier, J. C.; Jefford, C. W.; The Chemotherapy of Rodent Malaria. XLVIII. The Activities of Some Synthetic 1,2,4-Trioxanes against Chloroquine-Sensitive and Chloroquine-Resistant Parasites. Part 1: Studies Leading to the Development of Novel *Cis*-Fused Cyclopenteno Derivatives. *Ann. Trop. Med. Parasitol.*, 1993, *87*, 1-7.

[60] Griesbeck, A. G.; Blunk, D.; El-Idreesy, T. T.; Raabe, A.; Bicyclic Peroxides and Perorthoesters with 1,2,4-Trioxane Structures. *Angew. Chem., Int. Ed.*, 2007, *46*, 8883-8886.

[61] O'Neill, P. M.; Pugh, M.; Davies, J.; Ward, S. A.; Park, B. K.; Regioselective Mukaiyama Hydroperoxysilylation of 2-Alkyl- or 2-Aryl-prop-2-en-1-ols: Application to a New Synthesis of 1,2,4-Trioxanes. *Tetrahedron Lett.*, 2001, *42*, 4569-4571.

[62] Griesbeck, A. G.; Bartoschek, A.; El-Idreesy, T. T.; Hoinck, L.-O.; Miara, C.; Type II Photooxygenation in Polymer Matrices for the Synthesis of New Antimalarial Peroxides. *J. Mol. Catal. A - Chem.*, 2006, *251*, 41-48.

[63] Griesbeck, A. G.; El-Idreesy, T. T.; Hoinck, L. O.; Lex, J.; Brun, R.; Novel Spiroanellated 1,2,4-Trioxanes with High *in Vitro* Antimalarial Activities. *Bioorg. Med. Chem. Lett.*, 2005, *15*, 595-597.

[64] Griesbeck, A. G.; El-Idreesy, T. T.; Fiege, M.; Brun, R.; Synthesis of Antimalarial 1,2,4-Trioxanes *via* Photooxygenation of a Chiral Allylic Alcohol. *Org. Lett.*, 2002, *4*, 4193-4195.

[65] Singh, C.; Gupta, N.; Puri, S. K.; Geraniol-Derived 1,2,4-Trioxanes with Potent *in-Vivo* Antimalarial Activity. *Bioorg. Med. Chem. Lett.*, 2003, *13*, 3447-3450.

[66] Singh, C.; Gupta, N.; Puri, S. K.; Photo-Oxygenation of Geraniol: Synthesis of a Novel Series of Hydroxy-Functionalized Anti-malarial 1,2,4-Trioxanes. *Bioorg. Med. Chem. Lett.*, 2002, *12*, 1913-1916.

[67] Singh, C.; Sharma, U.; Saxena, G.; Puri, S.; Orally Active Antimalarials: Synthesis and Bioevaluation of a New Series of Steroid-Based 1,2,4-Trioxanes Against Multi-Drug Resistant Malaria in Mice. *Bioorg. Med. Chem. Lett.*, 2007, *17*, 4097-4101.

[68] Singh, C.; Malik, H.; Puri, S. K.; Orally Active 1,2,4-Trioxanes: Synthesis and Antimalarial Assessment of a New Series of 9-Functionalized 3-(1-Arylvinyl)-1,2,5-trioxaspiro[5.5]undecanes against Multi-drug-resistant *Plasmodium Yoelii Nigeriensis* in Mice. *J. Med. Chem.*, 2006, *49*, 2794-2803.

[69] Vennerstrom, J. L.; Dong, Y.; Andersen, S. L.; Ager, A. L. Jr.; Fu, H. N.; Miller, R. E.; Wesche, D. L.; Kyle, D. E.; Gerena, L.; Walters, S. M.; Wood, J. K.; Edwards, G.; Holme, A. D.; McLean, W. G.; Milhous, W. K.; Synthesis and Antimalarial Activity of Sixteen Dispiro-1,2,4,5-Tetraoxanes: Alkyl-Substituted 7,8,15,16- Tetraoxadispiro [5.2.5.2]hexadecanes. *J. Med. Chem.*, 2000, *43*, 2753-2758.

[70] McCullough, K. J.; Wood, J. K.; Bhattacharjee, A. K.; Dong, Y.; Kyle, D. E.; Milhous, W. K.; Vennerstrom, J. L.; Methyl-Substituted Dispiro-1,2,4,5-tetraoxanes: Correlations of Structural Studies with Antimalarial Activity. *J. Med. Chem.*, 2000, *43*, 1246-1249.

[71] Dong, Y.; Matile, H.; Chollet, J.; Kaminsky, R.; Wood, J. K.; Vennerstrom, J. L.; Synthesis and Antimalarial Activity of 11 Dispiro-1,2,4,5-tetraoxane Analogues of WR 148999. 7,8,15,16-Tetraoxadispiro[5.2.5.2]hexadecanes Substituted at the 1 and 10 Positions With Unsaturated and Polar Functional Groups. *J. Med. Chem.*, 1999, *42*, 1477-1480.

[72] Vennerstrom, J. L.; Fu, H. N.; Ellis, W. Y.; Ager, A. L. Jr.; Wood, J. K.; Andersen, S. L.; Gerena, L.; Milhous, W. K.; Dispiro-1,2,4,5-tetraoxanes: a New Class of Antimalarial Peroxides. *J. Med. Chem.*, 1992, *35*, 3023-3027.

[73] Vennerstrom, J. L.; Ager, A. L. Jr.; Andersen, S. L.; Grace, J. M.; Wongpanich, V.; Angerhofer, C. K.; Hu, J. K.; Wesche, D. L.; Assessment of the Antimalarial Potential of Tetraoxane WR 148999. *Am. J. Trop. Med. Hyg.*, 2000, *62*, 573-578.

[74] Zmitek, K.; Stavber, S.; Zupan, M.; Bonnet-Delpon, D.; Iskra, J.; Fluorinated Alcohol Directed Formation of Dispiro-1,2,4,5-tetraoxanes by Hydrogen Peroxide under Acid Conditions. *Tetrahedron*, 2006, *62*, 1479-1484.

[75] Dong, Y.; Vennerstrom, J. L.; Differentiation Between 1,2,4,5-Tetraoxanes and 1,2,4,5,7,8-Hexaoxonanes Using H-1 and C-13 NMR Analyses. *J. Heterocycl. Chem.*, 2001, *38*, 463-466.

[76] Dong, Y.; Vennerstrom, J. L.; Dispiro-1,2,4,5-tetraoxanes *via* Ozonolysis of Cycloalkanone O-Methyl Oximes: A Comparison with the Peroxidation of Cycloalkanones in Acetonitrile-Sulfuric Acid Media. *J. Org. Chem.*, 1998, *63*, 8582-8585.

[77] Iskra, J.; Bonnet-Delpon, D.; Begue, J. P.; One-Pot Synthesis of Non-Symmetric Tetraoxanes with the H_2O_2/MTO/Fluorous Alcohol System. *Tetrahedron Lett.*, 2003, *44*, 6309-6312.

[78] Zmitek, K.; Stavber, S.; Zupan, M.; Bonnet-Delpon, D.; Charneau, S.; Grellier, P.; Iskra, J.; Synthesis and Antimalarial Activities of Novel 3,3,6,6-Tetraalkyl-1,2,4,5-tetraoxanes. *Bioorg. Med. Chem.*, 2006, *14*, 7790-7795.

[79] Amewu, R.; Stachulski, A. V.; Ward, S. A.; Berry, N. G.; Bray, P. G.; Davies, J.; Labat, G.; Vivas, L.; O'Neill, P. M.; Design and Synthesis of Orally Active Dispiro 1,2,4,5-tetraoxanes; Synthetic Antimalarials with Superior Activity to Artemisinin. *Org. Biomol. Chem.*, 2006, *4*, 4431-4436.

[80] Kim, H. S.; Tsuchiya, K.; Shibata, Y.; Wataya, Y.; Ushigoe, Y.; Masuyama, A.; Nojima, M.; McCullough, K. J.; Synthetic Methods for Unsymmetrically-Substituted 1,2,4,5-Tetroxanes and of 1,2,4,5,7-Pentoxocanes. *J. Chem. Soc., Perkin Trans. 1*, 1999, 1867-1870.

[81] Opsenica, I.; Terzic, N.; Opsenica, D.; Milhous, W. K.; Solaja, B. A.; 7,8,15,16-Tetraoxa-dispiro[5.2.5.2]hexadecane-3-carboxylic Acid Derivatives and their Antimalarial Activity. *J. Serb. Chem. Soc.*, 2004, *69*, 919-922.

[82] Terent'ev, A. O.; Kutkin, A. V.; Starikova, Z. A.; Antipin, M. Y.; Ogibin, Y. N.; Nikishina, G. I.; New Preparation of 1,2,4,5-Tetraoxanes. *Synthesis*, 2004, 2356-2366.

[83] Solaja, B. A.; Terzic, N.; Pocsfalvi, G.; Gerena, L.; Tinant, B.; Opsenica, D.; Milhous, W. K.; Mixed Steroidal 1,2,4,5-Tetraoxanes: Antimalarial and Antimycobacterial Activity. *J. Med. Chem.*, 2002, *45*, 3331-3336.

[84] Opsenica, D.; Angelovski, G.; Pocsfalvi, G.; Juranic, Z.; Zizak, Z.; Kyle, D.; Milhous, W. K.; Solaja, B. A.; Antimalarial and Antiproliferative Evaluation of Bis-Steroidal Tetraoxanes. *Bioorg. Med. Chem.*, 2003, *11*, 2761-2768.

[85] Opsenica, D.; Pocsfalvi, G.; Juranic, Z.; Tinant, B.; Declercq, J. P.; Kyle, D. E.; Milhous, W. K.; Solaja, B. A.; Cholic Acid Derivatives as 1,2,4,5-Tetraoxane Carriers: Structure and Antimalarial and Antiproliferative Activity. *J. Med. Chem.*, 2000, *43*, 3274-3282.

[86] Opsenica, I.; Terzic, N.; Opsenica, D.; Angelovski, G.; Lehnig, M.; Eilbracht, P.; Tinant, B.; Juranic, Z.; Smith, K. S.; Yang, Y. S.; Diaz, D. S.; Smith, P. L.; Milhous, W. K.; Dokovic, D.; Solaja, B. A.; Tetraoxane Antimalarials and their Reaction with Fe(II). *J. Med. Chem.*, 2006, *49*, 3790-3799.

[87] Terzic, N.; Opsenica, D.; Milic, D.; Tinant, B.; Smith, K. S.; Milhous, W. K.; Solaja, B. A.; Deoxycholic Acid-Derived Tetraoxane Antimalarials and Antiproliferatives. *J. Med. Chem.*, 2007, *50*, 5118-5127.

[88] Kim, H. S.; Nagai, Y.; Ono, K.; Begum, K.; Wataya, Y.; Hamada, Y.; Tsuchiya, K.; Masuyama, A.; Nojima, M.; McCullough, K. J.; Synthesis and Antimalarial Activity of Novel Medium-Sized 1,2,4,5-Tetraoxacycloalkanes. *J. Med. Chem.*, 2001, *44*, 2357-2361.

[89] Kim, H. S.; Begum, K.; Ogura, N.; Wataya, Y.; Nonami, Y.; Ito, T.; Masuyama, A.; Nojima, M.; McCullough, K. J.; Antimalarial Activity of Novel 1,2,5,6-Tetraoxacycloalkanes and 1,2,5-Trioxacycloalkanes. *J. Med. Chem.*, 2003, *46*, 1957-1961.

[90] Griesbaum, K.; Liu, X.; Kassiaris, A.; Scherer, M.; Ozonolyses of O-Alkylated Ketoximes in the Presence of Carbonyl Groups: a Facile Access to Ozonides. *Liebigs Ann. Recl.*, 1997, 1381-1390.

[91] Vennerstrom, J. L.; Arbe-Barnes, S.; Brun, R.; Charman, S. A.; Chiu, F. C. K.; Chollet, J.; Dong, Y.; Dorn, A.; Hunziker, D.; Matile, H.; McIntosh, K.; Padmanilayam, M.; Santo Tomas, J.; Scheurer, C.; Scorneaux, B.; Tang, Y.; Urwyler, H.; Wittlin, S.; Charman, W. H.; Identification of an Antimalarial Synthetic Trioxolane Drug Development Candidate. *Nature*, 2004, *430*, 900-904.

[92] http://www.ranbaxyusa.com/newsdisp12132005.aspx

[93] Tang, Y.; Dong, Y.; Karle, J. M.; DiTusa, C. A.; Vennerstrom, J. L.; Synthesis of Tetrasubstituted Ozonides by the Griesbaum Coozonolysis Reaction: Diastereoselectivity and Functional Group Transformations by Post-Ozonolysis Reactions. *J. Org. Chem.*, 2004, *69*, 6470-6473.

[94] Dong, Y.; Chollet, J.; Matile, H.; Charman, S. A.; Chiu, F. C. K.; Charman, W. N.; Scorneaux, B.; Urwyler, H.; Santo Tomas, J.; Scheurer, C.; Snyder, C.; Dorn, A.; Wang, X.; Karle, J. M.; Tang, Y.; Wittlin, S.; Brun, R.; Vennerstrom, J. L.; Spiro and

Dispiro-1,2,4-trioxolanes as Antimalarial Peroxides: Charting a Workable Structure-Activity Relationship Using Simple Prototypes. *J. Med. Chem.*, 2005, *48*, 4953-4961.

[95] Tang, Y.; Dong, Y.; Wang, X.; Sriraghavan, K.; Wood, J. K.; Vennerstrom, J. L.; Dispiro-1,2,4-trioxane Analogues of a Prototype Dispiro-1,2,4-trioxolane: Mechanistic Comparators for Artemisinin in the Context of Reaction Pathways with Iron(II). *J. Org. Chem.*, 2005, *70*, 5103-5110.

[96] Osorio, L.; Murillo, C.; Aponte, S.; Mayagaya, V.; Scheurer, C.; Brun, R.; Matile, H.; Wittlin, S.; *In Vitro* Susceptibility of *P. Falciparum* Populations from Colombia and Tanzania to a New Synthetic Peroxide (OZ277). *Am. J. Trop. Med. Hyg.*, 2007, *76*, 1024-1026.

[97] Padmanilayam, M.; Scorneaux, B.; Dong, Y.; Chollet, J.; Matile, H.; Charman, S. A.; Creek, D. J.; Charman, W. N.; Santo Tomas, J.; Scheurer, C.; Wittlin, S.; Brun, R.; Vennerstrom, J. L.; Antimalarial Activity of N-Alkyl Amine, Carboxamide, Sulfonamide, and Urea Derivatives of a Dispiro-1,2,4-trioxolane Piperidine. *Bioorg. Med. Chem. Lett.*, 2006, *16*, 5542-5545.

[98] Dong, Y.; Tang, Y.; Chollet, J.; Matile, H.; Wittlin, S.; Charman, S. A.; Charman, W. N.; Santo Tomas, J.; Scheurer, C.; Snyder, C.; Scorneaux, B.; Bajpai, S.; Alexander, S. A.; Wang, X.; Padmanilayam, M.; Cheruku, S. R.; Brun, R.; Vennerstrom, J. L.; Effect of Functional Group Polarity on the Antimalarial Activity of Spiro and Dispiro-1,2,4-trioxolanes. *Bioorg. Med. Chem.*, 2006, *14*, 6368-6382.

[99] Tang, Y.; Dong, Y.; Wittlin, S.; Charman, S. A.; Chollet, J.; Chiu, F. C. K.; Charman, W. N.; Matile, H.; Urwyler, H.; Dorn, A.; Bajpai, S.; Wang, X.; Padmanilayam, M.; Karle, J. M.; Brun, R.; Vennerstrom, J. L.; Weak Base Dispiro-1,2,4-trioxolanes: Potent Antimalarial Ozonides. *Bioorg. Med. Chem. Lett.*, 2007, *17*, 1260-1265.

[100]Liang, X. T.; Yu, D. Q.; Wu, W. L.; Deng, H. C.; *Huaxue Xuebao (Acta Chimica Sinica)*, 1979, *37*, 215.

[101]Szpilman, A. M.; Korshin, E. E.; Rozenberg, H.; Bachi, M. D.; Total Syntheses of Yingzhaosu A and of its C(14)-Epimer Including the First Evaluation of their Antimalarial and Cytotoxic Activities. *J. Org. Chem.*, 2005, *70*, 3618-3632.

[102]Xu, X. X.; Zhu, J.; Huang, D. Z.; Zhou, W. S.; Total Synthesis of (+)-Yingzhaosu-A. *Tetrahedron Lett.*, 1991, *32*, 5785-5788.

[103]Hofheinz, W.; Burgin, H.; Gocke, E.; Jaquet, C.; Masciadri, R.; Schmid, G.; Stohler, H.; Urwyler, H.; Ro42-1611 (Arteflene), a New Effective Antimalarial: Chemical Structure and Biological Activity. *Trop. Med. Parasitol.*, 1994, *45*, 261-265.

[104]Jaquet, C.; Stohler, H. R.; Chollet, J.; Peters, W.; Antimalarial Activity of the Bicyclic Peroxide Ro42-1611 (Arteflene) in Experimental Models. *Trop. Med. Parasitol.*, 1994, *45*, 266-271.

[105]Weidekamm, E.; Dumont, E.; Jaquet, C.; Tolerability and Pharmacokinetics of Ro42-1611 (Arteflene) in Man. *Trop. Med. Parasitol.*, 1994, *45*, 278-283.

[106]O'Neill, P. M.; Bishop, L. P. D.; Searle, N. L.; Maggs, J. L.; Storr, R. C.; Ward, S. A.; Park, B. K.; Mabbs, F.; Biomimetic Fe(II)-Mediated Degradation of Arteflene (Ro-42-1611). The First EPR Spin-Trapping Evidence for the Previously Postulated Secondary Carbon-Centered Cyclohexyl Radical. *J. Org. Chem.*, 2000, *65*, 1578-1582.

[107]Cazelles, J.; Robert, A.; Meunier, B.; Characterization of the Main Radical and Products Resulting From a Reductive Activation of the Antimalarial Arteflene (Ro 42-1611). *J. Org. Chem.*, 1999, *64*, 6776-6781.

[108]Korshin, E. E.; Hoos, R.; Szpilman, A. M.; Konstantinovski, L.; Posner, G. H.; Bachi, M. D.; An Efficient Synthesis of Bridged-Bicyclic Peroxides Structurally Related to Antimalarial Yingzhaosu A Based on Radical Co-Oxygenation of Thiols and Monoterpenes. *Tetrahedron*, 2002, *58*, 2449-2469.

[109]O'Neill, P. M.; Verissimo, E.; Ward, S. A.; Davies, J.; Korshin, E. E.; Araujo, N.; Pugh, M. D.; Cristiano, M. L. S.; Stocks, P. A.; Bachi, M. D.; Diels-Alder/Thiol-Olefin Co-Oxygenation Approach to Antimalarials Incorporating the 2,3-Dioxabicyclo[3.3.1]nonane Pharmacophore. *Bioorg. Med. Chem. Lett.*, 2006, *16*, 2991-2995.

[110]Bachi, M. D.; Korshin, E. E.; Hoos, R.; Szpilman, A. M.; Ploypradith, P.; Xie, S.; Shapiro, T. A.; Posner, G. H.; A Short Synthesis and Biological Evaluation of Potent and Nontoxic Antimalarial Bridged Bicyclic Beta-Sulfonyl-Endoperoxides. *J. Med. Chem.*, 2003, *46*, 2516-2533.

[111]Kim, H. S.; Begum, K.; Ogura, N.; Wataya, Y.; Tokuyasu, T.; Masuyama, A.; Nojima, M.; McCullough, K. J.; Antimalarial Activity of Yingzhaosu A Analogues. *J. Med. Chem.*, 2002, *45*, 4732-4736.

[112]Tokuyasu, T.; Masuyama, A.; Nojima, M.; McCullough, K. J.; Kim, H. S.; Wataya, Y.; Yingzhaosu A Analogues: Synthesis by the Ozonolysis of Unsaturated Hydroperoxides, Structural Analysis and Determination of Anti-Malarial Activity. *Tetrahedron*, 2001, *57*, 5979-5989.

[113]Kamata, M.; Satoh, C.; Kim, H. S.; Wataya, Y.; Iron(II)-Promoted Rearrangement of 1,4-Diaryl-2,3-dioxabicyclo[2.2.2]oct-5-enes: A Mechanism Distinct from that Postulated Previously. *Tetrahedron Lett.*, 2002, *43*, 8313-8317.

[114]Posner, G. H.; Tao, X.; Cumming, J. N.; Klinedinst, D.; Shapiro, T. A.; Antimalarially Potent, Easily Prepared, Fluorinated Endoperoxides. *Tetrahedron Lett.*, 1996, *37*, 7225-7228.

[115]Hatzakis, E.; Opsenica, I.; Solaja, B. A.; Stratakis, M.; Synthesis of Novel Polar Derivatives of the Antimalarial Endoperoxides Ascaridole and Dihydroascaridole. *Arkivoc*, 2007, *2007*, 124-135.

[116]Kamata, M.; Ohta, M.; Komatsu, K. I.; Kim, H. S.; Wataya, Y.; Synthesis, Fe(II)-Induced Degradation, and Antimalarial Activities of 1,5-Diaryl-6,7-dioxabicyclo[3.2.2]nonanes: Direct Evidence for Nucleophilic O-1,2-Aryl Shifts. *Tetrahedron Lett.*, 2002, *43*, 2063-2067.

[117]Taylor, D. K.; Avery, T. D.; Greatrex, B. W.; Tiekink, E. R. T.; Macreadie, I. G.; Macreadie, P. I.; Humphries, A. D.; Kalkanidis, M.; Fox, E. N.; Klonis, N.; Tilley, L.; Novel Endoperoxide Antimalarials: Synthesis, Heme Binding, and Antimalarial Activity. *J. Med. Chem.*, 2004, *47*, 1833-1839.

[118]Wang, X.; Dong, Y.; Wittlin, S.; Creek, D.; Chollet, J.; Charman, S. A.; Santo Tomas, J.; Scheurer, C.; Snyder, C.; Vennerstrom, J. L.; Spiro- and Dispiro-1,2-dioxolanes: Contribution of Iron(II)-Mediated One-Electron *vs* Two-Electron Reduction to the Activity of Antimalarial Peroxides. *J. Med. Chem.*, 2007, *50*, 5840-5847.

[119]Tokuyasu, T.; Masuyama, A.; Nojima, M.; Kim, H. S.; Wataya, Y.; Synthesis and Anti-Malarial Activity of Yingzhaosu A Analogues From Unsaturated Hydroperoxy Acetals. *Tetrahedron Lett.*, 2000, *41*, 3145-3148.

[120]Najjar, F.; Baltas, M.; Gorrichon, L.; Moreno, Y.; Tzedakis, T.; Vial, H.; Andre-Barres, C.; Synthesis and Electrochemical Studies of New Antimalarial Endoperoxides. *Eur. J. Org. Chem.*, 2003, 3335-3343.

[121]Gavrilan, M.; Andre-Barres, C.; Baltas, M.; Tzedakis, T.; Gorrichon, L.; Bicyclic Peroxides in the G Factors Series: Synthesis and Electrochemical Studies. *Tetrahedron Lett.*, 2001, *42*, 2465-2468.

[122]Najjar, F.; Andre-Barres, C.; Baltas, M.; Lacaze-Dufaure, C.; Magri, D. C.; Workentin, M. S.; Tzedakis, T.; Electrochemical Reduction of G3-Factor Endoperoxide and its Methyl Ether: Evidence for a Competition Between Concerted and Stepwise Dissociative Electron Transfer. *Chem. Eur. J.*, 2007, *13*, 1174-1179.

[123]Najjar, F.; Gorrichon, L.; Baltas, M.; Andre-Barres, C.; Vial, H.; Alkylation of Natural Endoperoxide G3-Factor. Synthesis and Antimalarial Activity Studies. *Org. Biomol. Chem.*, 2005, *3*, 1612-1614.

[124]Najjar, F.; Gorrichon, L.; Baltas, M.; Vial, H.; Tzedakis, T.; Andre-Barres, C.; Crucial Role of the Peroxyketal Function for Antimalarial Activity in the G-Factor Series. *Bioorg. Med. Chem. Lett.*, 2004, *14*, 1433-1436.

[125]Kobayashi, M.; Kondo, K.; Kitagawa, I.; Antifungal Peroxyketal Acids from an Okinawan Marine Sponge of *Plakortis Sp. Chem. Pharm. Bull.*, 1993, *41*, 1324-1326.

[126]Quinoa, E.; Kho, E.; Manes, L. V.; Crews, P.; Bakus, G. J.; Heterocycles from the Marine Sponge *Xestospongia Sp. J. Org. Chem.*, 1986, *51*, 4260-4264.

[127]Murakami, N.; Kawanishi, M.; Itagaki, S.; Horii, T.; Kobayashi, M.; Facile Construction of 6-Carbomethoxymethyl-3-methoxy-1,2-dioxane, a Core Structure of Spongean Anti-Malarial Peroxides. *Tetrahedron Lett.*, 2001, *42*, 7281-7285.

[128]Murakami, N.; Kawanishi, M.; Itagaki, S.; Horii, T.; Kobayashi, M.; New Readily Accessible Peroxides with High Anti-Malarial Potency. *Bioorg. Med. Chem. Lett.*, 2002, *12*, 69-72.

[129]Kawanishi, M.; Kotoku, N.; Itagaki, S.; Horii, T.; Kobayashi, M.; Structure-Activity Relationship of Anti-Malarial Spongean Peroxides Having a 3-Methoxy-1,2-dioxane Structure. *Bioorg. Med. Chem.*, 2004, *12*, 5297-5307.

[130]Murakami, N.; Kawanishi, M.; Mostaqul, H. M.; Li, J.; Itagaki, S.; Horii, T.; Kobayashi, M.; New Anti-Malarial Peroxides with *in Vivo* Potency Derived from Spongean Metabolites. *Bioorg. Med. Chem. Lett.*, 2003, *13*, 4081-4084.

[131]Zhang, Q.; Wu, Y.; Simplified Analogues of Qinghaosu (Artemisinin). *Tetrahedron*, 2007, *63*, 10407-10414.

[132]Jin, H. X.; Zhang, Q.; Kim, H. S.; Wataya, Y.; Liu, H. H.; Wu, Y.; Design, Synthesis and *in Vitro* Antimalarial Activity of Spiroperoxides. *Tetrahedron*, 2006, *62*, 7699-7711.

[133]Howarth, J.; Wilson, D.; 1,4-Dihydroxy-2,3-dioxatricyclo[8.4.0.0]tetradecane and Derivatives with *in Vitro* Activity against *Plasmodium Falciparum*, *Trypanasoma b Brucei*, *Trypanasoma Cruzi*, and *Leishmaniasis Infantum*. *Bioorg. Med. Chem. Lett.*, 2003, *13*, 2013-2015.

[134]Posner, G. H.; O'Dowd, H.; Ploypradith, P.; Cumming, J. N.; Xie, S.; Shapiro, T. A.; Antimalarial Cyclic Peroxy Ketals. *J. Med. Chem.*, 1998, *41*, 2164-2167.

[135]Kim, J.; Li, H. B.; Rosenthal, A. S.; Sang, D.; Shapiro, T. A.; Bachi, M. D.; Posner, G. H.; Ground State Oxygen in Synthesis of Cyclic Peroxides. Part 1: Benzo Fused Ketals. *Tetrahedron*, 2006, *62*, 4120-4127.

[136]Dussault, P.; Sahli, A.; 2-Methoxyprop-2-yl Hydroperoxide - A Convenient Reagent for the Synthesis of Hydroperoxides and Peracids. *J. Org. Chem.*, 1992, *57*, 1009-1012.

[137]Cointeaux, L.; Berrien, J. F.; Peyrou, V.; Provot, O.; Ciceron, L.; Danis, M.; Robert, A.; Meunier, B.; Mayrargue, J.; Synthesis and Antimalarial Activity of 2-Methoxyprop-2-yl Peroxides Derivatives. *Bioorg. Med. Chem. Lett.*, 2003, *13*, 75-77.

[138]Hamada, Y.; Tokuhara, H.; Masuyama, A.; Nojima, M.; Kim, H. S.; Ono, K.; Ogura, N.; Wataya, Y.; Synthesis and Notable Antimalarial Activity of Acyclic Peroxides, 1-(Alkyldioxy)-1-(methyldioxy)cyclododecanes. *J. Med. Chem.*, 2002, *45*, 1374-1378.

[139]Dechy-Cabaret, O.; Benoit-Vical, F.; Loup, C.; Robert, A.; Gornitzka, H.; Bonhoure, A.; Vial, H.; Magnaval, J. F.; Seguela, J. P.; Meunier, B.; Synthesis and Antimalarial Activity of Trioxaquine Derivatives. *Chem. Eur. J.*, 2004, *10*, 1625-1636.

[140]Basco, L. K.; Dechy-Cabaret, O.; Ndounga, M.; Meche, F. S.; Robert, A.; Meunier, B.; *In Vitro* Activities of DU-1102, a New Trioxaquine Derivative, against *Plasmodium Falciparum* Isolates. *Antimicrob. Agents Chemother.*, 2001, *45*, 1886-1888.

[141]Dechy-Cabaret, O.; Benoit-Vical, F.; Robert, A.; Meunier, B.; Preparation and Antimalarial Activities of "Trioxaquines", New Modular Molecules with a Trioxane Skeleton Linked to a 4-Aminoquinoline. *ChemBioChem*, 2000, *1*, 281-283.

[142]Benoit-Vical, F.; Lelievre, J.; Berry, A.; Deymier, C.; Dechy-Cabaret, O.; Cazelles, J.; Loup, C.; Robert, A.; Magnaval, J. F.; Meunier, B.; Trioxaquines are New Antimalarial Agents Active on all Erythrocytic Forms, Including Gametocytes. *Antimicrob. Agents Chemother.*, 2007, *51*, 1463-1472.

[143]Singh, C.; Malik, H.; Puri, S. K.; Synthesis and Antimalarial Activity of a New Series of Trioxaquines. *Bioorg. Med. Chem.*, 2004, *12*, 1177-1182.

[144]Grellepois, F.; Grellier, P.; Bonnet-Delpon, D.; Begue, J.-P.; Design, Synthesis and Antimalarial Activity of Trifluoromethylartemisinin- Mefloquine Dual Molecules. *ChemBioChem*, 2005, *6*, 648-652.

[145]Walsh, J. J.; Coughlan, D.; Heneghan, N.; Gaynor, C.; Bell, A.; A Novel Artemisinin-Quinine Hybrid with Potent Antimalarial Activity. *Bioorg. Med. Chem. Lett.*, 2007, *17*, 3599-3602.

[146]O'Neill, P. M.; Stocks, P. A.; Pugh, M. D.; Araujo, N. C.; Korshin, E. E.; Bickley, J. F.; Ward, S. A.; Bray, P. G.; Pasini, E.; Davies, J.; Verissimo, E.; Bachi, M. D.; Design and Synthesis of Endoperoxide Antimalarial Prodrug Models. *Angew. Chem., Int. Ed.*, 2004, *43*, 4193-4197.

In: New Developments in Medicinal Chemistry
Editors: Marta P. Ortega and Irene C. Gil
ISBN 978-1-60456-810-3

Chapter 7

CHEMICAL ECOLOGY AND MEDICINAL CHEMISTRY OF MARINE NF-κB INHIBITORS*

F. Folmer, M. Schumacher, M. Jaspars, M. Dicato and M. Diederich
Laboratoire de Biologie Moléculaire et Cellulaire du Cancer, Luxembourg

ABSTRACT

NF-κB is an inducible transcription factor found in virtually all types of vertebrate cells, as well as in some invertebrate cells. While normal activation of NF-κB is required for cell survival and immunity, its deregulated expression is characteristic of cancer, inflammation, and numerous other diseases. Hence, NF-κB has recently become one of the major targets in drug discovery.

Several marine organisms use NF-κB (or analogues thereof), NF-κB inducers, or NF-κB inhibitors as chemical defence mechanisms, for parasitic invasion, for symbiosis, or for larval development. In particular, a wide range of marine natural products have been reported to possess NF-κB inhibitory properties, and some of these marine metabolites are currently in clinical trials as anticancer or anti-inflammatory drugs.

In the present review, we discuss the role of NF-κB inhibitors in marine chemical ecology, as well as in biomedicine. We also describe synthetic modifications that have been made to a range of highly promising marine NF-κB inhibitors, including the macrolide bryostatin 1 isolated from the bryozoan *Bugula neritina*, the lactone-γ-lactam salinosporamide A isolated from the actinomycete *Salinispora tropica*, the alkaloid hymenialdisine isolated from various sponges, the sesquiterpenoid hydroquinone avarol isolated from the sponge *Dysidea avara*, and the sesterterpene lactone cacospongonolide B isolated from the sponge *Fasciospongia cavernosa*, to increase their bioactivity and bioavailability, to decrease their level of toxicity or to lower the risk of other detrimental side-effects, and to increase the sustainability of their pharmaceutical production by facilitating their chemical synthesis.

---

* A version of this chapter was also published in *Aquatic Ecosystem Research Trends*, edited by George H. Nairne published by Nova Science Publishers, Inc. It was submitted for appropriate modifications in an effort to encourage wider dissemination of research.

Abbreviations

CDK	cyclin dependent kinase;
CK	casein kinase;
DAG	1,2-diacyl-*sn*-glycerol;
GFP	green fluorescent protein;
GSK-3b	glycogen synthase kinase;
HIF-1	hypoxia-inducible transcription factor-1;
HIV-1	human immuno-deficiency virus-1;
IkB	inhibitor of kappa B
IKK	kinase of IκB;
IL-1β	interleukin-1β;
K_i	kinase inhibitory activity;
NF-κB	nuclear factor-κB;
PKC	protein kinase C;
PMA	phorbol 12-myristate 13-acetate;
RBL	rat basophilic leukemia;
ROS	reactive oxygen species;
SAR	structure-activity relationship;
TNF-α	tumour necrosis factor-α;
V-ATP-ase	vacuolar-type H^+-ATPase.

1. Introduction

The sea covers seventy percent of the earth's surface, and it is an even more important ecological entity in terms of the total volume of habitable space than terrestrial habitats [1]. Therefore, it is not surprising that the oceans have drawn the attention of numerous scientists from a wide range of disciplines, and that new discoveries about ecology and biodiversity of marine organisms are made every day. Even though a vast majority of the oceans remains unexplored and our knowledge about marine ecology is mostly restricted to the littoral, which is a relatively narrow section of the oceans at the land-sea interface, marine scientists are well aware that the oceans form a unique source of very high biodiversity [2-8]. This high level of biodiversity can be attributed in part to the sheer complexity of the marine ecosystem [3, 4, 8]. Marine habitats cover a wide range of temperatures, air exposure levels, UV exposure levels, pH, dissolved gas concentrations, sun light availability, nutrient availability, pressure levels, current prevalence, turbulence levels, and sediment compositions [5]. Along with the latter abiotic factors, a wide range of biotic factors also govern the marine ecosystem. These include predation, parasitism, fouling, competition for resources, and competition for space [1, 3, 4, 6] (figure 1).

To cope with the biotic and abiotic factors reigning in the marine realm, marine organisms have developed various physical and chemical defence mechanisms. Physical defences include locomotion, camouflage, and the development of a hard external shell [1], while chemical defences rely on the production of toxic or deterrent natural products [3, 4, 8, 9].

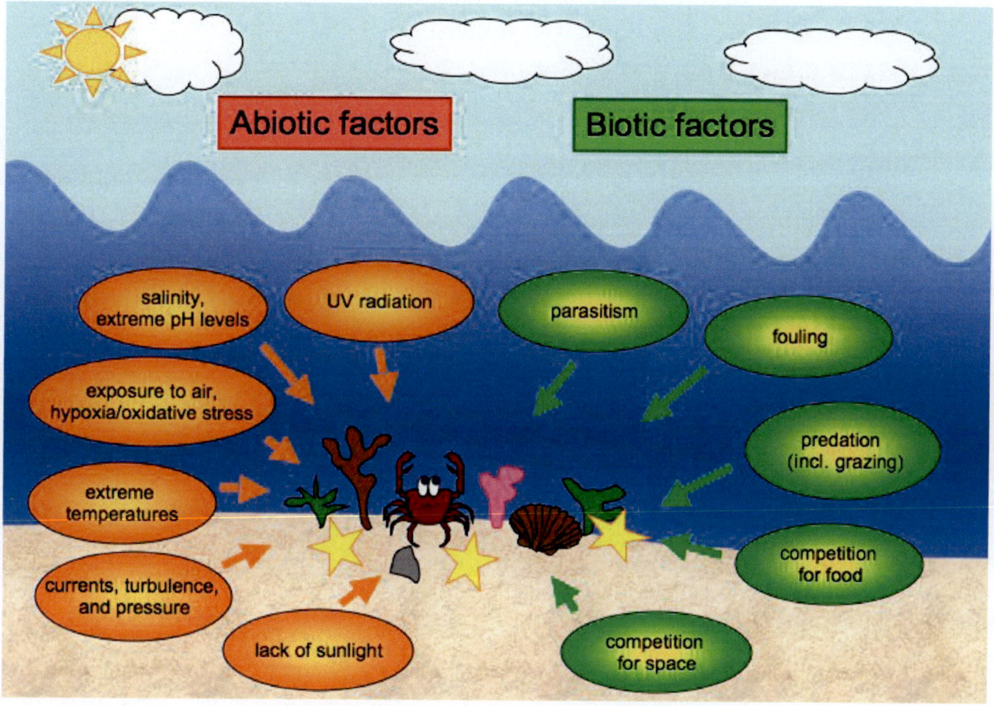

Figure 1. The biotic and abiotic factors governing the marine ecosystem.

In the 1940's, a new branch of ecology termed *marine chemical ecology* was established based on the observation that sessile, soft-bodied marine plants and animals rely heavily on natural products as a defence mechanism, as well as for communication in the ocean [3, 4, 8-11]. Heterotrophs can either produce the secondary metabolites used as chemical defences themselves or acquire them through dietary intake or symbiosis. Algae, which are sessile autotrophs, do not have any dietary intake of secondary metabolites, and the opportunities of evolving symbiotic relationships that might provide defence are quite limited. Furthermore, their sessile lifestyle prevents them from moving away from grazers or fouling organisms or spatial competitors. Hence, the autonomous production of natural products as chemical defences is a particularly important mechanism in algae [6].

The probing of the biomedical properties of marine natural products started in the 1960's [5]. This research avenue, which is often referred to as the search for "*Drugs from the Seas*", has led to the recognition of the extraordinary biomedical potential of marine natural products [7, 12-14].

In the present review, we discuss the ecological and biomedical roles of marine natural products that inhibit the nuclear transcription factor-κB (NF-κB), which has recently been recognized as a key target in cancer- and inflammation-related drug discovery [15-17]. Unfortunately, the natural production of bioactive secondary metabolites by wild or cultured marine organisms is generally unsustainable for pharmaceutical applications, and chemical synthesis is, in most cases, a necessity in order to complete clinical trials [14, 18, 19]. In the final section of the present review, we describe how the synthesis of a range of highly promising marine NF-κB inhibitors and their analogues has provided invaluable information about the structure-activity relationship of marine natural products targeting NF-κB.

2. The Transcription Factor NF-κB: Its Pathological Implications in Cancer Development and Inflammation, and Its Activation Pathway

The transcription of DNA into RNA is the first stage in the expression of genes. Transcription is orchestrated by a group of proteins called transcription factors, which bind to DNA on specific binding sites within the genes' promoter regions and initiate the expression of the genes [20, 21]. The transcription factor nuclear factor-κB (NF-κB), which was discovered in 1986 by Sen and Baltimore, has been shown to be implicated in the regulation of over 150 different human genes [22-24]. Although NF-κB is required for both innate and adaptive immunity [25-27], most of the genes regulated by NF-κB encode for proteins that play critical roles in cancer development, inflammation, and other pathologies [23, 24, 28]. Up-regulated activity of NF-κB has been observed primarily in patients with various types of lymphomas [29-35], with several types of solid cancers [36, 37], with asthma [38, 39], with rheumatoid arthritis [40-42], with various inflammatory bowel diseases [43], with multiple sclerosis [44, 45], and with osteoporosis [46]. Furthermore, aberrant NF-κB activation has been recognized as one of the major resistance factors to chemotherapy and radiotherapy, as it impairs the ability of damaged, malfunctioning, or cancerous cells to undergo apoptosis [47-54]. Because of its numerous pathological implications in cancer development and inflammation, NF-κB has become a major target in drug discovery, and the activation pathway of the transcription factor has been studied in depth in several laboratories around the world [15, 29, 55-58].

NF-κB is a dimer of proteins belonging to the Rel family, which includes RelA (p65), RelB, c-Rel, p50 (NF-κB1), and p52 [59, 60]. All five Rel family proteins contain a well-conserved Rel homology domain (RHD) responsible for the dimerization of NF-κB, for the interactions of NF-κB with its cytoplasmic inhibitory protein IκB, and for the binding of NF-κB to DNA [60, 61].

The nuclear localization signal (NLS) at the C-terminus of the RHD plays an important role in the nuclear translocation of activated NF-κB [59, 60]. p65, RelB, and c-Rel also contain a terminal transactivation domain (TAD) at their C-terminal end which is required for the activation of transcription [60]. p50 and p52 lack a terminal transactivation domain and are therefore transcriptionally inactive [60] (figure 2). The present paper focuses on the inhibition of the heterodimer p50/p65, which is the most common form of NF-κB [62-64].

NF-κB is normally found in the cytoplasm, in an inactive form as it is bound to its cytoplasmic inhibitor IκB [60, 61, 65]. IκB is a family of a handful of closely related molecules, the most common ones being IκBα and IκBβ. All known IκB proteins contain a 30-33 amino acid long sequence called ankyrin repeat, which binds, via non-covalent interactions, and with extremely high affinity, to the RHD of NF-κB [65], and which masks the NLS of NF-κB [61], thereby preventing the latter's nuclear translocation [59] (figure 3). The N-terminal region of IκB proteins is known to play a major role in the signal-dependent degradation of IκB by the proteasome, whereas the C-terminus domain is implicated in the regulation of NF-κB-DNA binding, and in the nuclear export of NF-κB [61].

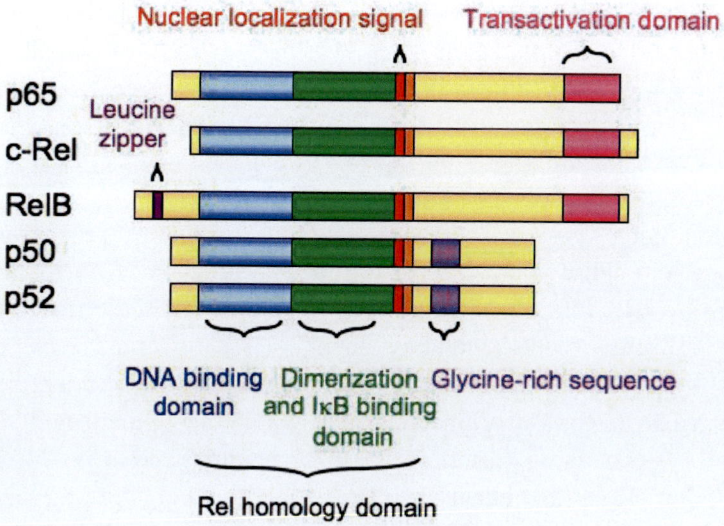

Figure 2. Schematic representation of the five Rel family proteins RelA (p65), c-Rel, RelB, p50, and p52, showing the highly conserved Rel homology domain (RHD) including a DNA binding domain and a dimerization and IκB binding domain on the N-terminus of each protein. The transactivation domain (TAD) of RelA, c-Rel, and RelB, as well as the Leucine zipper (LZ) in the case of RelB are required for the transcriptional activation. p50 and p52, which lack a transactivation domain, are transcriptionally inactive. The glycine-rich region towards the C-terminus in p50 and p52 is involved in the generation of the latter from their precursors (p105 and p100, respectively) [60]. (figure modified from Keutgens *et al.* [60]).

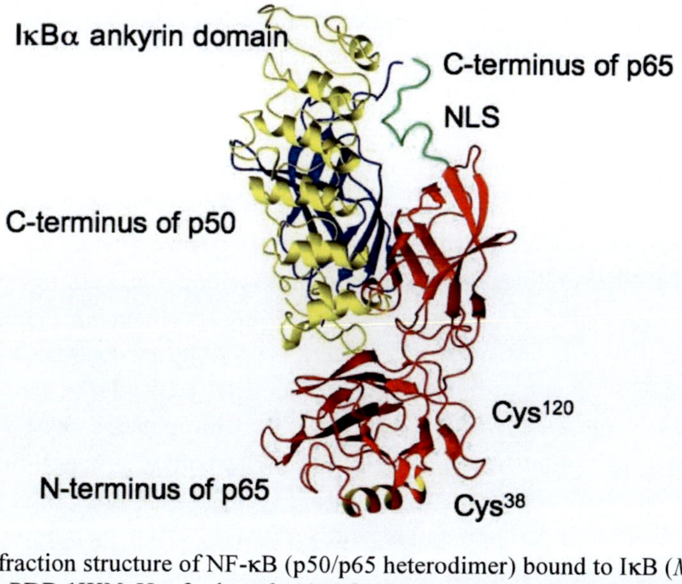

Figure 3. X-ray diffraction structure of NF-κB (p50/p65 heterodimer) bound to IκB (*Mus musculus*, synthetic construct, PDB 1IKN, Huxford *et al.*, 1998 [66]). Cartoon shows sheet strands (arrows pointing towards the C-terminus end of the chains) and α-helices. The red chain represents the N-terminus of p65. The C-terminus of p65 is shown in green. The blue chain represents the C-terminus of p50. The N-terminus of p50 is not shown in this figure. The ankyrin domain of IκBα (all α-helices) is shown in yellow. The N-terminus and C-terminus regions of IκBα are not shown in this figure. The image was generated using the programme MOLMOL [67].

In most cases, the activation of NF-κB is triggered by extra-cellular stimulations including ionizing radiation, oxidative stress, various toxins, and signalling through pro-inflammatory cytokines such as interleukin-1β (IL-1β), tumour necrosis factor-α (TNF-α), or receptor activator for nuclear factor-κB ligand (RANKL) [68-71]. RANKL, which binds to RANK (receptor activator for nuclear factor-κB) to induce the NF-κB activation pathway, a membrane protein expressed on the surface of osteoclasts, is, itself, activated by the ATP-dependent proton pump vacuolar-type H^+-ATPase (V-ATPase) [72]. The extra-cellular stimuli culminate, through a complex upstream signalling cascade, in the activation of the IκB kinase complex IKK [59, 60, 73-75]. IKK, which is composed of two catalytic subunits (IKKα and IKKβ) and of one regulatory subunit (IKKγ, which is also called NEMO), phosphorylates IκBα [59, 74, 75]. The phosphorylation of the N-terminal serines Ser^{32} and Ser^{36} of IκBα leads to conformational changes within the protein and to a consequent exposure of the Lys-containing region of IκBα for ubiquitination [59, 74, 75]. The Lys^{48}-linked polyubiquitin chains that attach to IκBα at Lys^{21} and/or Lys^{22} target IκBα for rapid degradation by the 26S proteasome [59, 76-78]. Once freed from IκB, NF-κB translocates into the nucleus, where it activates its target genes [79]. The NF-κB activation pathway described above, which is generally referred to as the "classical" or "canonical" NF-κB activation pathway, is illustrated in figure 4. It is the most common NF-κB activation pathway [62, 73, 80], and the only one discussed in the present paper.

The activation of NF-κB can be inhibited at different levels along the activation cascade. The major targets of NF-κB inhibitors are the binding of NF-κB to its DNA binding sites, the degradation of IκB by the 26S proteasome, and the phosphorylation of IκB by IKK (figure 4).

The binding of NF-κB to DNA can be inhibited by molecules that mask either the NF-κB binding sites on DNA or the DNA binding domain on NF-κB [81]. However, because of the large interaction surface mediating the binding of NF-κB to DNA, it is unusual to find natural products that are large enough to block the binding of NF-κB to DNA [81]. The majority of natural products reported to date to interfere directly with NF-κB-DNA binding are terrestrial sesquiterpene lactones [81, 82]. Sesquiterpene lactones possessing α,β-unsaturated carbonyl groups are thought to interfere with the binding of NF-κB to its binding sites on DNA by undergoing Michael-type conjugate additions to the nucleophilic cysteine sulfhydryl groups Cys^{38} and Cys^{120} in the p65 monomer of NF-κB [82-84]. Helenalin (1) isolated from the terrestrial plant *Arnica* sp. is one of the first natural sesquiterpene lactones reported to inhibit NF-κB [85]. Another approach to inhibit the binding of NF-κB to DNA involves the creation of steric hindrance in the DNA binding region of p50 through the formation of hydrogen bonds with amino acids in that region. One example of a natural product interfering with the binding of NF-κB to DNA in that way is gallic acid (2) isolated from garlic. Gallic acid 2 has been shown to form strong hydrogen bonds with Ser^{66} in the DNA-binding region of p50.

Further upstream, NF-κB activation can be inhibited by compounds targeting the proteasomal degradation of IκB [86-88]. Proteasome inhibitors include the peptide boronate Velcade® (3) (also known as bortezomib or PS-341) , which is already on the market as an anti-cancer drug [86, 88], and the marine γ-lactam-β-lactone salinosporamide A (4)[89], which is currently in clinical trials [86].

A further step upstream, NF-κB activation can be inhibited by compounds interfering with the kinase activity of IKK [81, 90].

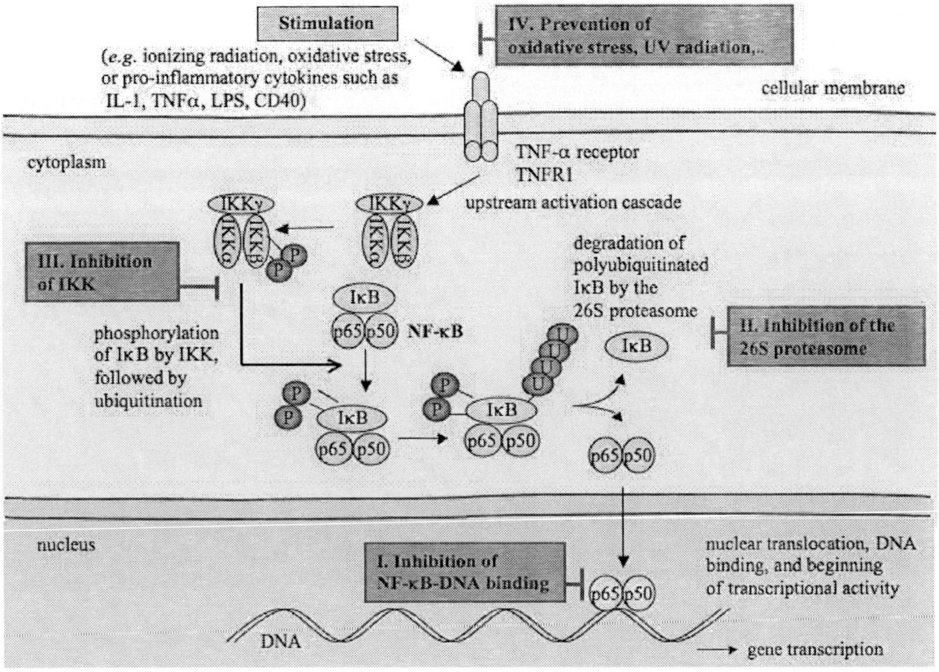

Figure 4. The NF-κB activation pathway ("classical pathway"). Upon stimulation by pro-inflammatory cytokines, ionizing radiation, or oxidative stress, a complex upstream activation cascade leads to the phosphorylation and activation of the IκB kinase complex IKK. IKK phosphorylates IκB. Once phosphorylated, IκB is polyubiquitinated, and the polyubiquitin chains target IκB for degradation by the 26S proteasome. After the degradation of IκB, NF-κB is free to translocate into the nucleus, and to induce the transcription of its target genes (modified after Keutgens *et al.* (2006))[60]. The major targets for NF-κB inhibitors are highlighted in green: NF-κB inhibitors can target the binding of NF-κB to DNA (I), the degradation of polyubiquitinated IκB by the 26S proteasome (II), or the phosphorylation of IκB by IKK (III). NF-κB activation can also be slowed down by compounds that protect against UV radiation or that prevent oxidative stress which would otherwise induce the NF-κB activation cascade.

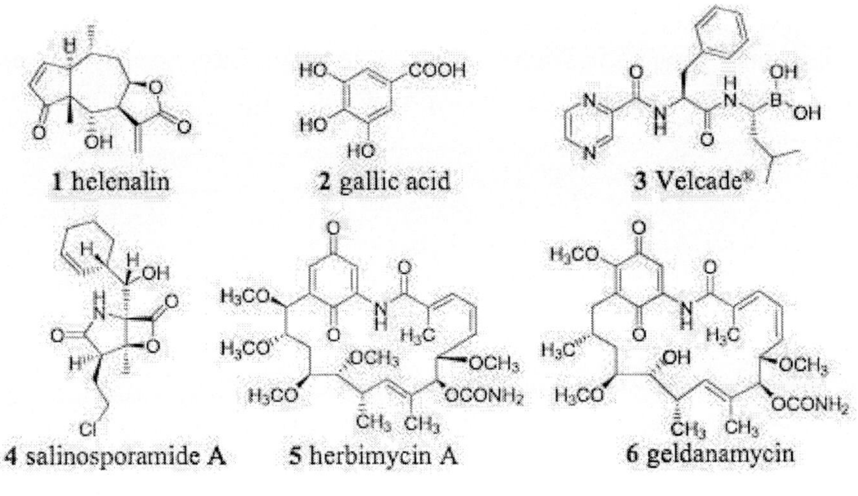

Scheme 1.

The inhibition of IKK is viewed as the most specific way of interfering with the NF-κB activation pathway, as IKK does not phosphorylate any protein outside the NF-κB signalling cascade, while the proteasome degradation of proteins and the binding of transcription factors to DNA are implicated in a wide range of biological processes [90]. Natural products specifically targeting the kinase activity of IKK include the two benzoquinones herbimycin A (5) [91] and geldanamycin (6) [92] isolated from bacteria.

Finally, NF-κB activation can be slowed down by antioxidants such as vitamins C (ascorbic acid) and E (α-tocopherol), co-enzyme Q10, and a variety of polyphenolics [81, 90, 93-95]. Antioxidants reduce reactive oxygen species (ROS) which could otherwise activate NF-κB [75, 81, 90, 93, 96].

3. The Role of NF-κB In Marine Chemical Ecology

As described in the introduction, marine chemical ecology is the field of science that investigates the ecological role of natural products in the marine ecosystem [3, 4, 8-11]. Although there is a coherent link between the anti-predatory effects of toxic natural products released by sessile soft-bodied marine organisms and the anti-cancer activity of the former, in many cases, there is no easy explanation for the production of secondary metabolites by a particular organism. If the ecological reason behind the production of a secondary metabolite of interest can be understood, this can be of tremendous value for the large-scale production of the latter. For instance, once the ecological role of the compound of interest is understood, bioprospecting for the compound can be enhanced by culturing the source organism under conditions that trigger the production of the compound. The chemical ecology behind the production of a natural product can also provide valuable cues about various environments or organisms which can potentially serve as an alternative source of the compound of interest, or of analogues thereof. In this way, the sustainability of the large-scale production of the compound can be increased, and the natural biodiversity of the source organisms can be protected. It is generally accepted that natural products have evolved under the pressure of natural selection to bind to specific receptors in ecological targets, and that evolutionary pressures may have compelled marine organisms to produce substances that cause growth inhibition or mortality in their competitors [6]. In the case of NF-κB inhibition, there are several potential evolutionary and ecological explanations for the finding of NF-κB inhibiting natural products in marine organisms (figure 5).

3.1. Evolution of the Presence of NF-κB and NF-κB Analogues in Marine Organisms

From an evolutionary point of view, one interesting potential explanation for the finding of NF-κB inhibitors in marine organisms is the fact that marine invertebrates and fish, no matter how distantly related to us they appear to be, possess, in many cases, NF-κB or closely related analogues. As a matter of fact, the NF-κB/IκB cascade has a very ancient evolutionary origin reaching all the way back to the "living fossil" and most ancient arthropod

Carcinoscorpius rotundicauda (horseshoe crab) [97, 98]. Members of the Rel/NF-κB family have been reported in the sea hare *Aplysia californica* [99], in the sea urchin *Strongylocentrotus purpuratus* [100], in the mussel *Mytillus galloprovincialis* [101], in the crab *Chasmagnathus granulatus* [102], in ascidians [103], in the oyster *Crassostera gigas* [104], in the hydrothermal vent mussel *Bathymodiolus azoricus* [105], in marine cnidarians [106], and in sponges [106, 107]. In marine bivalves, including the mussels *Mytillus galloprovincialis* and *Bathymodiolus azoricus* and the oyster *Crassostera gigas,* NF-κB appears to function primarily as an inducer of the innate immune system, as it is involved in the transcription of genes encoding for antibacterial molecules such as mytilin [105].

Oysters (*Crassostera gigas* and *Pinctada fucata*) have been shown to possess an IKK-like protein oIKK that share structural and functional properties with their mammalian homologues. When transfected into human cells, oIKK has been shown to activate NF-κB-controlled reporter genes [108, 109].

Another mollusc, the Hawaiian squid *Euprymna scolopes,* has also been shown to possess analogues of several elements of the NF-κB activation pathway, including Rel, IκB, IKKγ, and TRAF6 [110]. Goodson *et al.* (2005) [110] showed that the NF-κB activation pathway of is modulated by the squid's beneficial symbiont *Vibrio fischeri.*

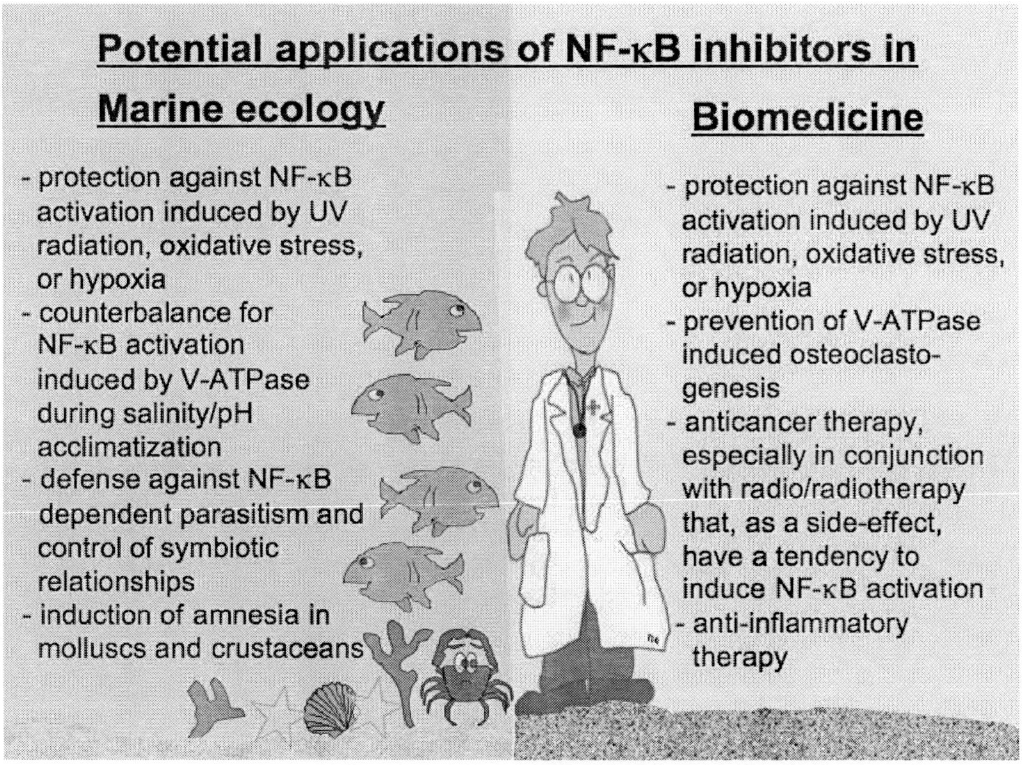

Figure 5. Potential applications of NF-kB inhibitors in marine ecology and in biomedicine.

In the interesting case of ascidians, the NF-κB proteins As-rel1 and As-rel2 have been shown to be involved in the regulation of the formation and degradation of the notochord [103]. Ascidians are unique in the sense that they start their larval life as vertebrates, bearing, like all other chordates, a notochord at the centre of the larval tail, and then gradually losing their notochord to become invertebrate adults [103].

With the presence of NF-κB or NF-κB analogues in marine organisms [97, 99, 100, 102, 111-115], it can be hypothesized that the same organisms also produce antagonists of the activation of NF-κB. Such antagonists might play critical roles in the switching-off of positive feed-back loops within the NF-κB activation pathway, or in the halting of aberrant NF-κB production.

3.2. Potential Ecological Reasons for the Production of NF-κB Inhibitors by Marine Organisms

3.2.1. Defence Against UV Radiation, Oxidative Stress, and Hypoxia

Ultraviolet (UV) radiation mainly affects intertidal plants and animals, but planktonic and subtidal benthic organisms, and sessile coral reefs organisms in particular, are also exposed to harmful levels of UV radiation in surface waters and at shallow depths [3]. Absorption of UV photons causes organic molecules to undergo conformational changes that can interfere with their vital metabolic functions, and it is now widely accepted that short wavelength UV radiation induces NF-κB activation [116-118]. In addition to it's own effects on the NF-κB activation cascade, UV radiation mediates the formation of reactive oxygen species (ROS) including the superoxide anion radical peroxides, ($^{\cdot}O_2^{-}$), the hydroxyl radical ($^{\cdot}OH$), hydrogen peroxide (H_2O_2), and other free radicals that cause NF-κB-inducing oxidative stress [117, 118]. As a chemical defence against UV radiation, many marine organisms produce potent UV-absorbing sunscreens and antioxidants capable of quenching photo-oxidative reactions [3, 119]. Marine sunscreens, the most common examples of which include mycosporine-like amino acids (MAAs), scytonemin (7), phlorotannins, coumarins, and polyphenolics, have been documented in depth by Karentz (2001) [119]. Marine antioxidants include several carotenoids, tocopherols, phycocyanins, and anthocyanins [95, 119, 120]. Amongst the strongest marine antioxidants are the algal derivatives cymopol (8) and avrainvilleol (9) and the spongean metabolite puupehenone (10) [95].

Antioxidants play a particularly important role in sessile or slow-moving intertidal organisms. As a matter of fact, the latter are not just exposed to ROS formed by UV radiation. They are also continuously experiencing aerobiosis-anaerobiosis transitions, and the abrupt reintroduction of oxygen into their metabolism during the return from an anoxic state back to an oxygen-rich state often leads to a rapid generation of ROS [121].

Oxidative stress is not the only oxygen-related environmental condition that has been associated with NF-κB induction. At the other extreme, hypoxia, which impairs energy availability and hence cell viability [122], is also linked to NF-κB activation [123]. In humans, it has recently been discovered that NF-κB activation is necessary for the transcription of *Hif1α* mRNA under hypoxic conditions [123, 124]. Hypoxia-inducible transcription factor-1 (HIF-1) upregulates pro-angiogenic factors to restore nutrient, oxygen, and energy supply in order to maintain tissue integrity and homeostasis [123]. HIF-1 activation has also been associated with the innate immune response against bacterial infection [123]. However, aberrant expression and activation of HIF-1α can be detrimental [125, 126], and it can be hypothesized that marine organisms have developed antagonists to the expression of HIF-1, including compounds targeting the NF-κB-induced transcription of the protein. Several marine natural products, including the diterpenoid laurenditerpenol (11) isolated from the tropical alga *Laurencia intricata* [127], the triterpenoids sodwanone A (12) and yardenone A (13) isolated from the South African sponge *Axinella* sp. [128], norsesterterpene peroxides from the marine sponge *Diacarnus levii* [125], various benzochromenones isolated from the marine crinoid *Comantheria rotula* [129], the phenolic pyrrole 7-hydroxyneolamellarin A (14) isolated from the sponge *Dendrilla nigra* [130], various strongylophorines isolated from the sponge *Petrosia (Strongylophora) strongylata* [131], and the macrolide latrunculin A (15) isolated from the Red Sea sponge *Negombata magnifica* [132] have been reported as HIF-1 inhibitors.

3.2.2. Osmoregulation and Adaptation to Changes in pH

Marine organisms are frequently exposed to salinity or pH changes, and in order to acclimatize to those changes, they rely on enzymes such as Na^+, K^+-ATPase and vacuolar-type H^+-ATPase (V-ATPase) [133]. V-ATPase is found in vacuoles, lysosomes, and plasma membranes of many different types of cells. Plasma membrane V-ATPase is generally associated with pH homeostasis. In the intercalated cells of kidneys, V-ATPase is implicated in the pumping of protons into the urine, to allow the reabsorption of bicarbonate into the blood [134, 135].

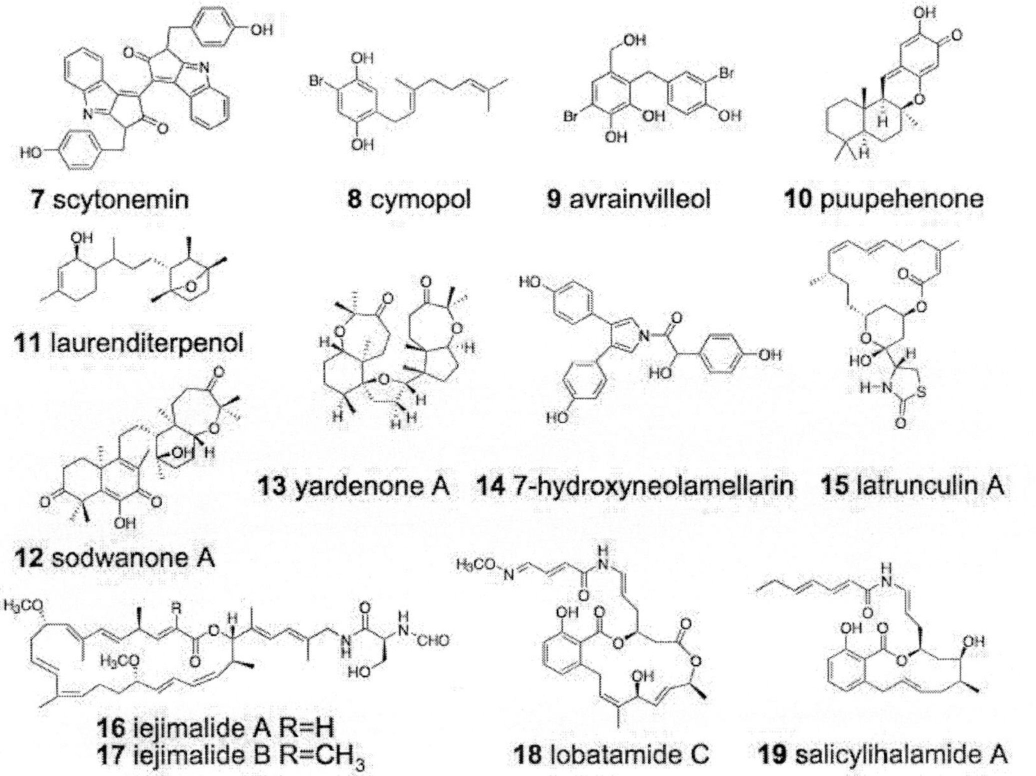

Scheme 2.

As mentioned in the introduction, V-ATPase-mediated acidosis activates RANKL in osteoclasts [72]. The binding of RANKL to RANK on the membrane surface of osteoclasts induces the NF-κB activation pathway and promotes NF-κB-dependent osteoclastogenesis [72]. Again, it can be hypothesized that marine organisms are likely to produce V-ATPase inhibitors to counteract aberrant activation of V-ATPase. Alternatively, they may produce compounds that target NF-κB activity resulting from V-ATPase activation induced during the process of osmoregulation or of adaptation to changes in pH. Marine natural products inhibiting V-ATPase include the macrolides iejimalide A (16) and B (17) isolated from the tunicate *Eudistoma rigida*, lobatamide C (18) isolated from the tunicate *Aplidium lobatum* [134], and salicylihalamide A (19) isolated from the sponge *Haliclona* sp. [134, 135].

Examples of marine sunscreens, antioxidants, HIF-1 inhibitors, and V-ATPase inhibitors are presented in scheme 2.

3.2.3. Defence against parasites and interference with learning processes and memory in other marine organisms

The natural response of cells infected by parasites is often to commit apoptosis and deprive the parasite of an opportunity to proliferate within the infected organism. Although this particular aspect of ecology has, so far, been mainly investigated amongst terrestrial organisms and there is hardly any literature covering marine examples, there are many reports of viral, bacterial, and protozoan parasites taking advantage of the anti-apoptotic pathways to escape host defence mechanisms [136-140]. Parasites reported to induce the activation of NF-κB include human immunodeficiency virus type 1 (HIV-1), adenovirus 5, human T cell lymphotropic virus type 1 (HTLV-1), Epstein-Barr virus (EBV), hepatitis B virus (HBV), hepatitis C virus (HCV), the influenza virus, the tick *Theileria* sp., and the protista *Leishmania major*, *Cryptosporidium parvum* and *Toxoplasma gondii* [136, 137, 139, 141]. The tick *Theileria* sp. has been shown to depend on the constitutive activation of NF-κB for survival in its host. The constitutive activation of NF-κB is achieved through the recruitment of IKK into large IKK signalosome assemblies at the surface of the parasite [142]. IKK has also been shown to be constitutively activated in HIV-1, adenovirus 5, or HTLV-1 infected cells [137, 139, 143]. EBV, on the other hand, produces an oncoprotein, LMP1, that interacts with the TNF-α receptor associated factor 2 (TRAF2) [139], while the polypeptide HBx produced by HBV is directly implicated in the degradation of IκBα [139]. Noteworthy, it is now well established that constitutive activation of NF-κB as a consequence of viral infection often leads to malignancies in humans [137] (figure 5).

Some parasites, including the bacteria *Salmonella* sp., *Yersinia* sp., and *Pseudomonas* sp., are also producing NF-κB inhibitors, in order to attenuate inflammatory responses that would otherwise lead to their clearance from the host cells [138, 144] (figure 5).

One of the earliest reports of the presence of NF-κB analogues in marine invertebrates describes the role of ApNF-κB, an axoplasmic protein found in the mollusc *Aplysia californica*. ApNF-κB is involved in learning processes of the sea hare and is rapidly inactivated after nerve injury [99]. Similarly, the crab *Chasmagnathus granulatus* has been reported to possess an NF-κB analogue that plays a crucial role in long-term memory [102]. The inhibition of NF-κB in *Chasmagnathus granulatus* by inhibitors of IKK induces amnesia in the crab [102, 145] (figure 5). To date, there is no reported evidence for ecological roles of interferences with learning processes or long-term memory amongst marine organisms.

4. Marine Inhibitors of NF-κB

Several marine natural products with NF-κB inhibitory properties have been described during the last decade. However, the exact targets and mechanisms of action of most marine NF-κB inhibitors remain poorly understood [81, 146, 147].

Marine natural products inhibiting the 26S proteasome include the γ-lactam-β-lactone salinosporamide A (4) isolated from the actinomycete *Salinispora tropica* [148] and the spongean metabolites agosterol C (20) and secomucalolide A (21) [149] (scheme3). The macrolide bryostatin 1 (22) isolated from the marine bryozoan *Bugula neritina* and from its γ–proteobacterial symbiont *Candidatus endobugula sertula* [150] has never been reported as a direct inhibitor of NF-κB, but it has potent synergetic anti-cancer effects with several chemotherapeutical drugs, and with the prototypical proteasome inhibitor lactacystin in particular [151]. The major target of bryostatin 1 (22), which is currently in phase I and phase II clinical trials as an anti-cancer drug, is the serine/threonine protein kinase C (PKC) [152, 153]. The PKC signalling pathway, which occurs upstream of the NF-κB activation cascade, has been shown to be essential for the activation of NF-κB in various cell-lines [62, 154-156].

Inhibitors of IKK include the two naphtopyrones 6-methoxy-comaparvin (23) and 6-methoxy-comaparvin-5-methyl ether (24) isolated from the crinoid *Comanthus parvicirrus* [157] (scheme 3). The carotenoid astaxanthin (25) isolated from various bacteria and algae by [158]. and the indole-alkaloid sunscreen scytonemin (7) isolated from cyanobacteria [159], which have antioxidative and UV-absorbing properities, respectively, are known to inhibit NF-κB upstream of the activation pathway (scheme 3).

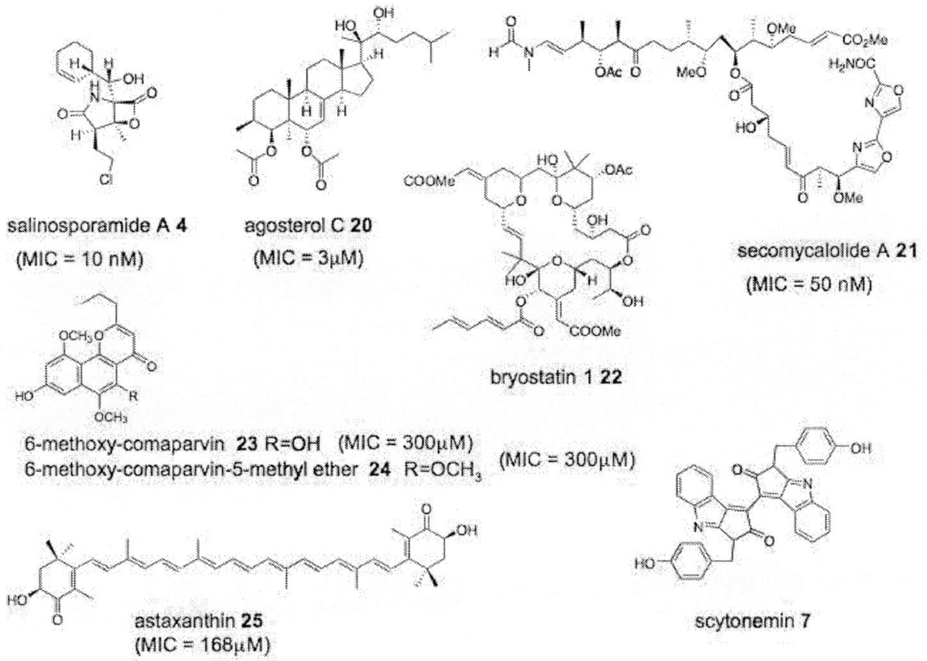

Scheme 3. Marine NF-kB inhibitors with identified targets. Where available, the MIC values are provided.

Marine natural products with unidentified or poorly established molecular targets include the bacterial secondary metabolites cycloprodigiosin hydrochloride (26) [160, 161] and streptochlorin (27) [162], as well as the fungal secondary metabolite verracurin (28) [163] and several natural products isolated from sponges. NF-κB inhibitors isolated from marine sponges include the sesterterpene lactones cacospongiolide B (29) [164], petrosaspongiolide M (30) [165], and cyclolinteinone (31) [166], the sesquiterpene hydroquinone avarol (32) [167], the bromopyrrol alkaloid hymenialdisin (33) [168-170], the sesquiterpene benzoquinone ilimaquinone (34) [171], and the diterpene cycloamphilectene (35) [172] (scheme 4).

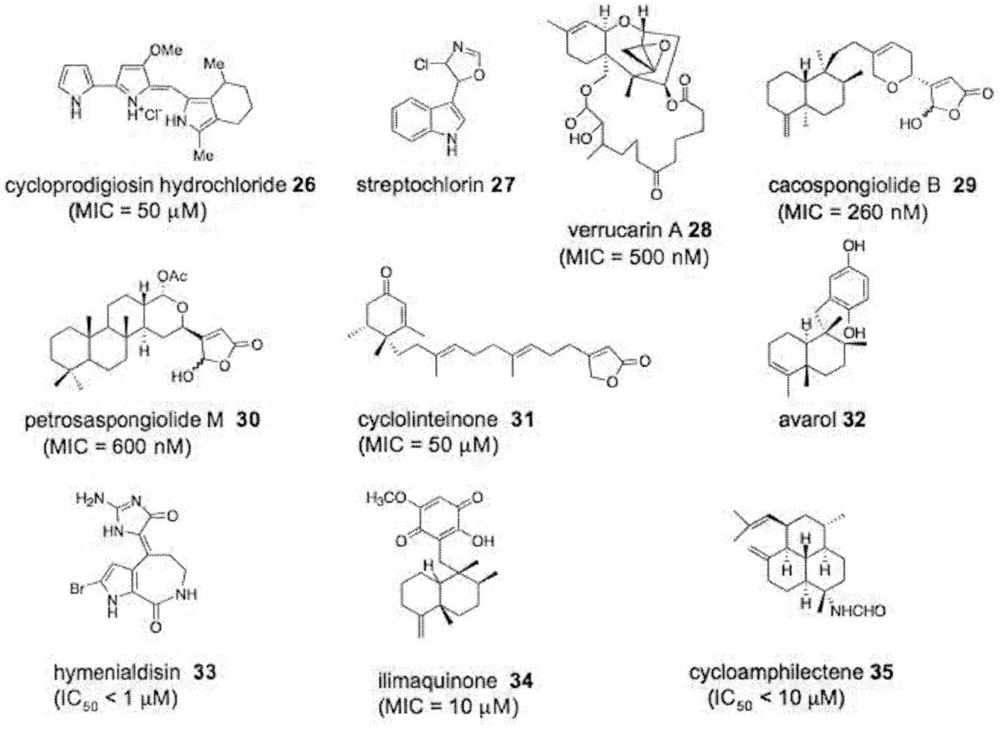

Scheme 4. Marine NF-kB inhibitors without identified targets. Where available, the MIC or the IC_{50} values are provided.

5. Examples of Synthetic Modifications that Have Been Attempted on Marine Natural Products in the Context of NF-κB Inhibitory Activity

Marine natural products are now widely recognized as highly promising drug candidates and, as mentioned in the introduction, several marine natural products are currently in clinical trials [2, 14, 173-177]. With the number of marine natural products identified as NF-κB inhibitors steadily increasing, marine metabolites will doubtlessly gain further importance in the field of anti-cancer and anti-inflammatory drug discovery. However, marine natural products are, in general, an unsustainable source of druggable compounds [13, 19, 146, 178,

179]. The natural abundance of the source organisms is, in most cases, too low to yield sufficient material for clinical trials without destroying the marine ecosystem, and the access to the collection sites is often very limited. Aquaculture of marine invertebrates and algae for pharmaceutical purposes has been revealed to be extremely challenging, and the large-scale fermentation of marine microbes, although more successful than other strategies used in the production of marine drugs, can be troublesome as well [7, 11, 13, 58, 179-184]. For these reasons, chemical synthesis is, in most cases, an unavoidable step in the large-scale production of marine drugs. In addition to serving as an approach to get around the hurdles associated with the harvest of marine drugs from their natural sources, the chemical synthesis of bioactive marine natural products offers the possibility to produce analogues of the natural products, and to use the former for structure-activity relationship (SAR) studies. SAR studies on analogues of lead molecules play a crucial role in the drug development, as they can lead to the discovery of more potent, less toxic, and more bioavailable drug candidates [185].

Five marine natural products involved in the inhibition of NF-κB activation have been synthesized and have become the subject of detailed SAR studies. These are the proteasome inhibitor salinosporamide A (4) isolated from the marine actinomycete *Salinospora tropica*, the protein kinase C (PKC) inhibitor bryostatin 1 (22) isolated from the bryozoan *Bugula neritina*, and the three sponge-derived natural products hymenialdisine (33), avarol (32), and cacospongionide B (29). A brief survey of the strategies used for the chemical synthesis of these five compounds and their analogues, and the results obtained from the NF-κB-related SAR studies are presented below.

5.1. Chemical Synthesis and NF-κB-Related SAR of Salinosporamide A and Its Analogues

The γ-lactam-β-lactone salinosporamide A (4) isolated from the marine actinomycete *Salinospora tropica* is by far the most widely documented marine lactone with NF-κB inhibitory properties [148, 186-188]. The compound, which was discovered at the beginning of this decade by William Fenical *et al.* at the Scripps Institution of Oceanography (University of California, San Diego (U.S.A.)) [189] and by Barbara Potts *et al.* at Nereus Pharmaceuticals in San Diego (U.S.A.) [148], targets the degradation of IκB by the 26S proteasome [148, 186-188] and it has been reported as a suppressor of RANKL-induced osteoclastogenesis [188]. Salinosporamide A (4) entered phase I clinical trials in May of 2006, initially against solid tumors and leukemia, and in April of 2007 another phase I trial against multiple myeloma was initiated [190]. Salinosporamide A (4) is still in anti-cancer clinical trials at the moment[190].

The first total syntheses of salinosporamide A (4) and of some of its analogues were achieved by Corey *et al.* at Harvard University (U.S.A.) [191, 192] and by Danishefsky *et al.* at Columbia University (U.S.A.) [193] in 2004-2005. One of the major challenges for both research groups was the stereo-controlled assembly of the γ-lactam group. Another challenge was the water-sensitive β-lactone moiety [191, 193]. Corey *et al.* started the synthesis from L-threonine methyl ester [191], while Danishefsky *et al.* started from a bicyclo derivative of L-glutamic acid [193]. In Corey's synthetic route, the L-threonine methyl ester underwent condensation and silyation before cyclizing into a *cis*-fused ring under Stork conditions.

Finally, the addition of cyclohexenyl zinc led to the diastereo-controlled formation of salinosporamide A (4) [191]. The first step of Danishefsky's synthetic route involved the formation of the γ-lactam frame through conjugate addition and alkylation of the starting block. After two cyclization steps, the synthesis of salinosporamide A (4) was completed through the addition of cyclohexenyl zinc as in Corey's synthetic route [193]. Following their initial synthesis of salinosporamide A (4) in 2004, Corey *et al.* have developed a new, slightly simpler synthetic route for the compound [194]. Recently, an asymmetric synthetic pathway has been described for (-)-salinosporamide A [195], and several derivatives of salinosporamide A (4) have been synthesized over the last three years [186, 192, 194]. Additionally, the large-scale fermentation of the source organism *Salinispora tropica* under various culture conditions has led to the isolation of several novel salinosporamide derivatives [186, 187].

The SAR of salinosporamide A (4) analogues has been studied in depth by Barbara Potts *et al.* at Nereus Pharmaceuticals (San Diego, U.S.A.). The results, which have been published by Macherla *et al.* (2005) [186] and by Reed *et al.* (2007) [187], are summarized in scheme 5.

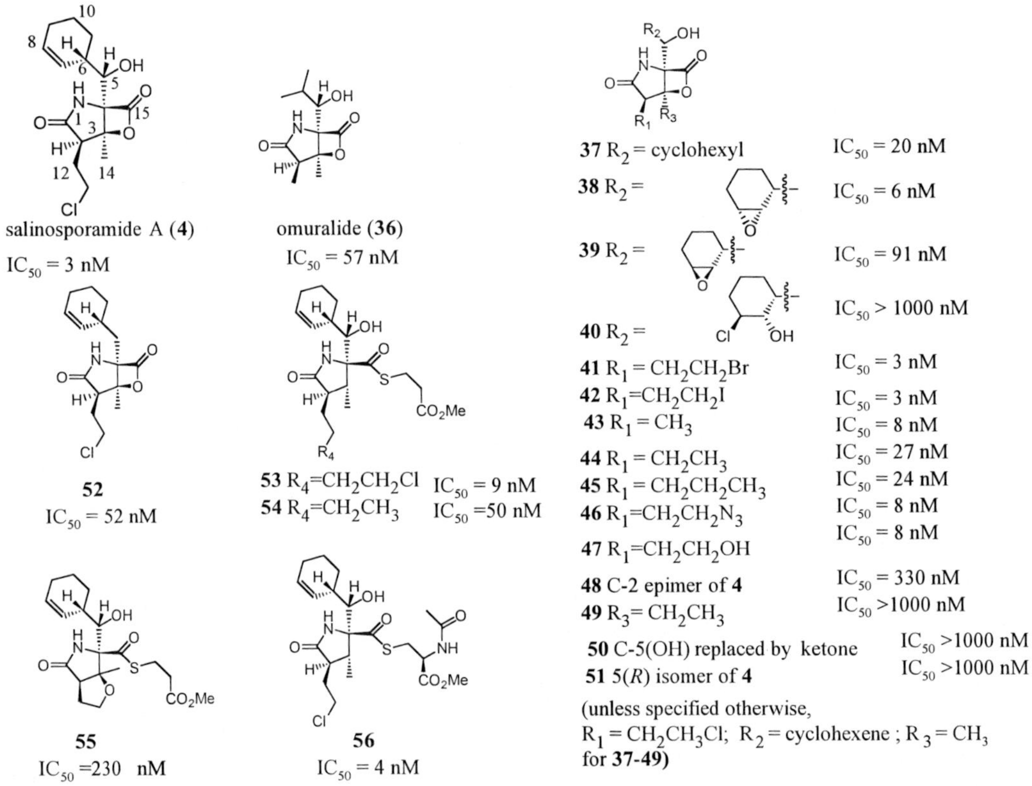

Scheme 5. Analogues of salinosporamide A (**4**) investigated by Macherla *et al.* (2005) [186]for their effects on the proteolytic activity of the 26S proteasome. IC_{50} values (in nM) refer to the inhibiton of the enzymatic activity of the chymotrypsin-like (CT-L) subunit of the 26S proteasome in rabbit (*in vitro*).

Amongst the salinosporamide A (4) analogues included in the SAR study performed by Macherla et al. (2005) [186], the strongest inhibitors of the proteolytic activity of the chymotrypsin-like (CT-L) catalytic unit of the 26S proteasome where salinosporamide A (4) itself and the analogues 38, 41, and 42 (scheme 5). These compounds were also the most potent inhibitors of NF-κB activation in luciferase reporter gene assays performed by Macherla *et al.* (2005) (with IC_{50} values ≤ 34 nM) [186].

The results of the SAR study performed by Potts *et al.* [186] show that modification to the cyclohexene ring is only moderately tolerated in terms of NF-κB inhibition potential and 20S proteasome inhibition activity. In the SAR studies reported by Macherla *et al.* (2005) [186], the hydrogenation of the cyclohexene ring in (37) resulted in up to 10-fold loss of activity. Furthermore, the (7*S*,8*S*)-epoxide (39) of salinosporamide A (4) was less active than its (7*R*,8*R*) diastereomer (38). This observation and the fact that chlorohydrin (40) was shown to be 3-log units less active than salinosporamide A (4) proved to Macherla *et al.* that steric hindrance at either side of the cylcohexene moiety has a negative effect on bioactivity. However, halogen exchange such as in 41 and 42 resulted in equivalent bioactivity, underlining the crucial role of the chlorine leaving group in salinosporamide A (4). The inversion of the C-2 or C-5 stereocenters (48; 51) resulted in a dramatic loss of the observed activity, providing further support for the hypothesis that the chlorine leaving group of salinosporamide A (4) plays a critical role in the bioactivity of the natural product. The role of the chlorine leaving group and of other halogenated leaving groups in the proteasome inhibitory activity of salinosporamide (4) analogues has been investigated by Brad Moore *et al.* at the Scripps Institution of Oceanography (University of California, San Diego, U.S.A.) [196, 197], and the results have revealed the opening up of the γ-lactam β-lactone ring and the covalent tethering of salinosporamide analogues to the proteasome is a mechanism that is reversible if fluoride is present in the leaving group, but irreversible if chloride is present in the leaving group [148, 196, 197]. The oxidation of the C-5 alcohol, which has been identified as a pharmacophore involved in the binding of salinosporamide A (4) to the threonine residue Thr^{21} of the chymotrypsin-like catalytic subunit of the 26S proteasome [186], to a ketone in (50), was shown by Macherla et al. (2005) [186] to result in a significant loss of activity. Finally, the substitution of the C-3 methyl group by an ethyl in (49) was accompanied by a significant loss of activity. This loss of activity was attributed by Macherla *et al.* to an increase in steric hindrance [186]. As described in Reed *et al.* [187], the presence of the thioester group is very likely to be responsible for the weaker 26S proteasomal inhibition potential of the salinosporamide A (4) analogues 53-56 [187].

The conclusions drawn from the SAR studies performed by Barbara Potts *et al.* at Nereus Pharmaceuticals (San Diego. U.S.A) and by Brad Moore *et al.* at the Scripps Institution of Oceanography (University of California, San Diego, U.S.A.) are presented in scheme 6. The results of the SAR studies, together with crystallography studies published by Groll *et al.* (2006) [148], have resulted in the identification of the pharmacophores of salinosporamide A (4) [186, 187], which are highlighted in blue and in magenta in scheme 6. The SAR and crystallography studies revealed that the mode of action of salinosporamide A (4) involves covalent ester linkages between the 26S proteasome and salinosporamide A (4) resulting from the nucleophilic addition of the threonine residue Thr^1 within the catalytic subunits of the proteasome to the carbonyl group of the β-lactone and from the opening of the lactone ring. The opening of the lactone ring is followed by the intramolecular nucleophilic addition of C-3O to the chloroethyl group of salinosporamide A (4), which leads to the formation of a cyclic ether [148, 186].

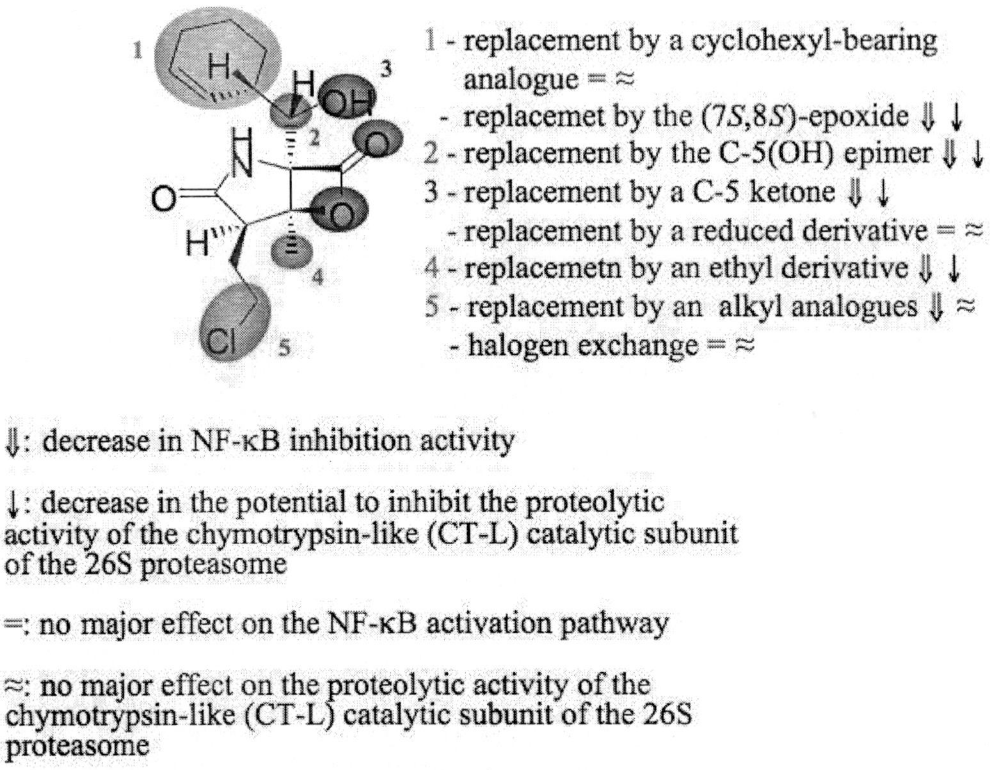

Scheme 6. Summary of the results obtained by Macherla *et al.* (2005) [186] in their SAR study of effects of salinosporamide A (**4**) analogues on the proteolytic activity of 26S proteasome and on the activation of NF-κB. The pharmacophores identified during the SAR study and through crystallography studies by Groll *et al.* (2006) [148] are highlighted in colour.

The protonation of the amine group of Thr[1] by C3-OH severely hinders the deacylation activity of the enzyme [148, 198]. The γ-lactam ring fixes the position of C-3O in the salinosporamide A (4) to the proteasome complex and precludes the regeneration of the β-lactone ring and the elimination of the compound from the proteasome [148, 186]. The alcohol function of C-5OH is involved in the formation of hydrogen bonds between salinosporamide A (4) and the amine of the proteasomal threonine residue Thr[21]. The amide of the γ-lactam ring binds to the glycine residue Gly[47]. The binding of salinosporamide to the proteasomal residues Thr[1] and Gly[47] hinders the enzymatic activity of the 26S proteasome by disturbing the proton shuttling through water molecules present in the vicinity of the latter residues in absence of salinosporamide A (4). The cyclohexenyl ring of salinosporamide A (4) offers additional hydrophobic interactions to the proteasome [148].

5.2. Chemical Synthesis and NF-κB-related SAR of Bryostatin 1 and its Analogues

Bryostatin 1 (22) is a macrolide isolated from the bryozoan *Bugula neritina*, and produced by the latter's γ-proteobacterial symbiont *Candidatus endobugula sertula* [150]. Bryostatin 1 (22) is a potent anti-neoplastic agent with activity mainly in leukaemic tumours

[199, 200]. The macrolide is currently in phase I and II clinical trials as an anti-cancer drug, but as part of a combination with other drug rather than as a single entity. Although bryostatin 1 (22) does not directly inhibit NF-κB, it has been shown to have synergetic anti-cancer effects together with the prototypical proteasome inhibitor lactacystin [151], as well as with other chemotherapeutic agents, including paclitaxel, and vincristine [184]. The mode of action of bryostatin 1 (22) is related to the macrolide's strong affinity for the regulatory domain of various kinases, and of the serine/threonine protein kinase C (PKC) in particular [152, 153]. In the context of the present paper, it is noteworthy that the PKC signalling pathway, which occurs upstream of the NF-κB activation cascade, has been shown to be essential for the activation of NF-κB in various cell-lines [62, 154-156].

Bryostatin 1 (22) and its analogues of it have been synthesized by several research groups around the world, in order to meet the clinical need for multi-gram quantities of compound [201-204]. However, due to the large number of synthetic steps and the low overall yield, total synthesis has been considered to be inappropriate for the supply of bryostatin 1 (22) for clinical trials. To this date, clinical material is mainly provided through aquaculture carried out by Mendola *et al.* at CalBioMarine (U.S.A.) [205]. Culturing the symbiont of the bryozoan, γ-proteobacteria *Candidatus Endobugula sertula* could dramatically increase the supply of bryostatin1 (22) [206], but *Candidatus Endobugula sertula* remains, to date, unculturable [207].

SAR studies of the inhibition of the kinase activity of PKC by bryostatin 1 (22) and its analogues have revealed that that the epoxidation of the C13-C30 double bond does not alter the binding activity, in contrast to the epoxidation of the two C16-C17 and C21-C34 alkenes which results in a lower activity [208]. The *R*-configuration is required for binding activity to PKC as the chemically modified 26-*epi*-analogue of bryostatin 1 (22) has been shown to exhibit a 25-fold lower activity, and the acetylating of the C26 hydroxyl has lead to a significant reduction in the binding capability of bryostatin 1 (22) to the C1 unit of PKC [208]. The absence of the C19 hydroxyl function in the naturally occurring bryostatin analogues has resulted in a 2 orders of magnitude reduction in activity [184].

Based on these observations, Wender *et al.* developed a computer model for the comparison of the calculated low-energy molecule of 1,2-diacyl-*sn*-glycerol (DAG) (57), the x-ray structures of phorbol 12-myristate 13-acetate (PMA) (58) and bryostatin 1 (22) [209, 210] (scheme 7). The computer model proposed an important role of the C19 and C26 hydroxyl functionalities and the C1 carboxyl group of bryostatin 1 (22), which corroborate spatially with the C4, C9, C20 oxygen atoms of PMA. Based on the study by Wender *et al.*, the C19, C26 and C1 oxygen groups are required for binding to the C1 domain of PKC, whereas the lipophilic regions influences the binding orientation of the molecule and the partition into the membrane [211]. Moreover, the region of the two rings A and B, referred to as "spacer domain", is needed for the control of the conformation of the recognition domain (C4-C16) [210, 212] (scheme 8).

Wender *et al.* pursued their studies by designing two derivatives with simplified A- and B-rings, an intact C-ring, except for a deleted methyl group in analogue 2 (59) [210]. The hydropyranyl (B-ring) was substituted by a dioxane moiety, allowing a simplified synthesis. The coupling of the recognition domain and spacer domain, though macroacetalization, prior to lactonization, led to a convergent synthesis resulting in a reduced number of steps, an improved overall yield and the possibility to easily introduce synthetic modifications.

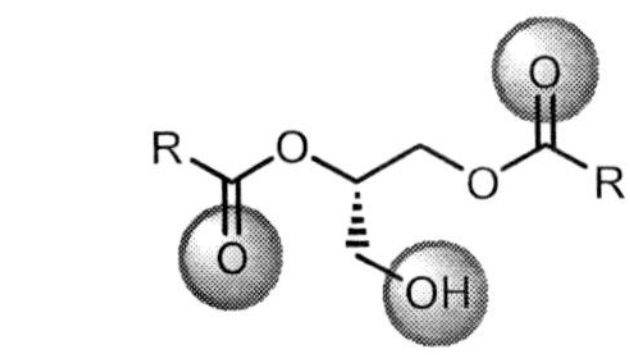

1,2-diacyl-*sn*-glycerol (DAG) **57**

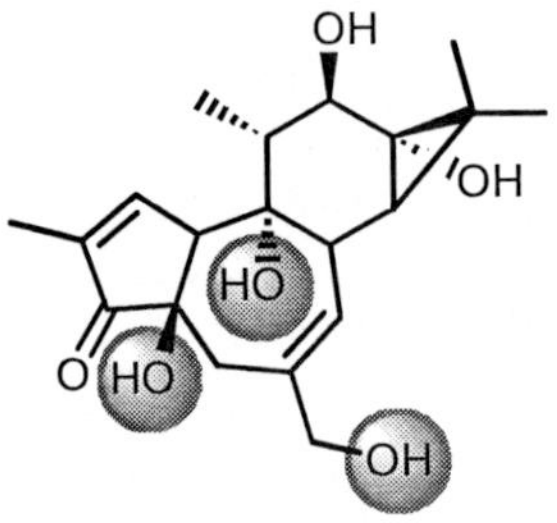

phorbol 12-myristate 13-acetate (PMA) **58**

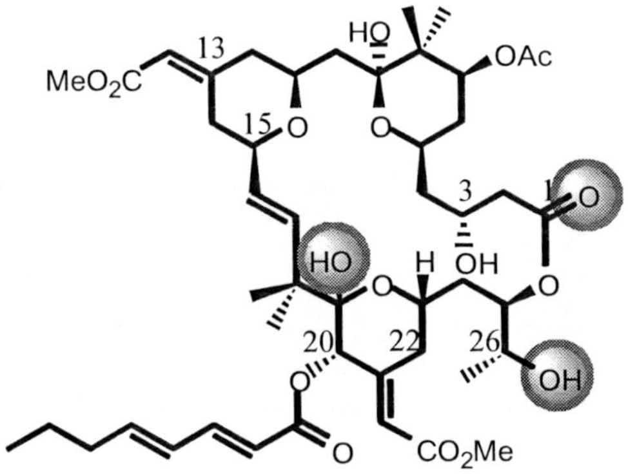

bryostatin 1 (**22**)

Scheme 7. Chemical structures of bryostatin 1 (22), DAG (57), and PMA (58) used in computer model-based SAR studies of the inhibition of PKC by Wender *et al.* (1998) [210]. The pharmacophores are highlighted in blue.

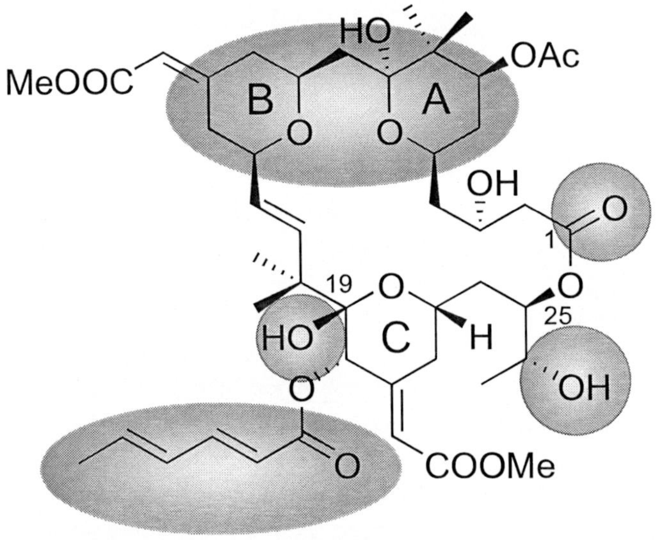

bryostatin 1 (**22**)

Scheme 8. Pharmacophores of bryostatin 1 (22) as identified by Wender *et al.* (1998) [210].

Step economy is one of the hallmarks of the research group of Wender *et al.* permitting the economic supply of sufficient bryostatin analogues (also referred to as "bryologs") for future clinical trials [210, 212, 213] (schemes 9 and 10).

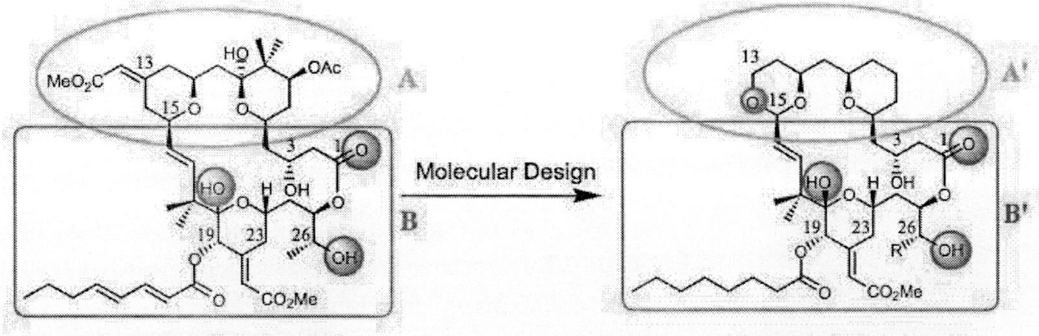

bryostatin 1 (**22**)

R = Me Analogue 1 (**59**)

R = H Analogue 2 (**60**)

Scheme 9. The first bryostatin 1 analogues 59 and 60 synthesized by Wender *et al.* in 1998 [210]. In comparison with the natural product bryostatin 1 (22), the functionalities along C7-C13 have been eliminated, and the carbon atom in position C14 has been replaced by an oxygen atom.

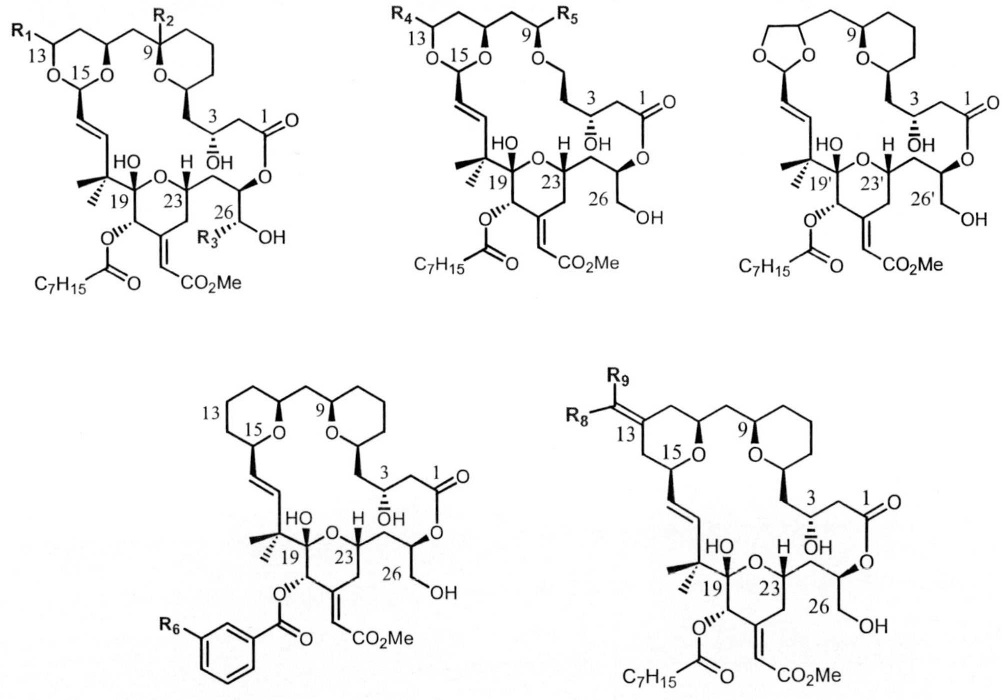

Scheme 10. Chemical structures of the "bryolog" analogues of bryostatin 1 (22) synthesized by Wender *et al.* (1998) [210-217, 220-226]. R_1-R_9 are listed in tables 1 to 5.

The two bryostatin 1 (22) analogues 59 and 60 were assessed by Wender *et al.* (1998-2008) [210, 214-216] and by Baryza *et al.* (2004) [217] for their binding affinity to a mixture of PKC isozymes isolated from rat brain and the translocation of PKCδ coupled to green-fluorescent protein (PKCδ-GFP) in rat basophilic leukaemia (RBL) cells. The binding affinity to PKC, a thermodynamic parameter, indicates biological interactions, but it is not associated with a certain biological function, a kinetic parameter. For this reason, PKCδ-GFP translocation is considered to be a more appropriate method, than the PKC binding affinity, to assess the biological function of the examined compounds. Furthermore, activated PKCδ is linked to crucial cellular functions such as cell growth control, differentiation, and apoptosis [218, 219]. In the first set of experiments, analogues 59 and 60 exhibited an inhibitory concentration (K_i) of 3.0 nM and 0.25 nM respectively to a mixture of PKC isolated from rat brain in comparison to 1.35 nM for bryostatin 1 (22) [214]. It was reported in a later publication that variation in K_i values was observed depending on the specific batch of PKC [214]. Moreover, the synthesised analogues 59 and 60 possess higher translocation potencies and rates of PKCδ-GFP translocation to the nucleus. Their minimum dose was lower (1 nM and 100 pM respectively) than bryostatin 1 (22) (5 nM). Wender *et al.* (2002) [220] showed that, in the cellular membrane, analogues 59 and 60 form a tertiary complex with PKC. The distinct lipophilicities of the derivatives might alter thermodynamic or kinetic parameters of this interaction and result in more potent PKC-translocation activity. These results correlated with the more potent growth inhibition activity of the "bryologs" in 24 out of 35 cancer cell lines assays compared to bryostatin 1 (22) [220].

Table 1. PKC inhibitory activity of "bryologs" modified at position C9, C13 or C26. The PKC kinase inhibitory concentrations (K_i), obtained from the listed references, are given in nM

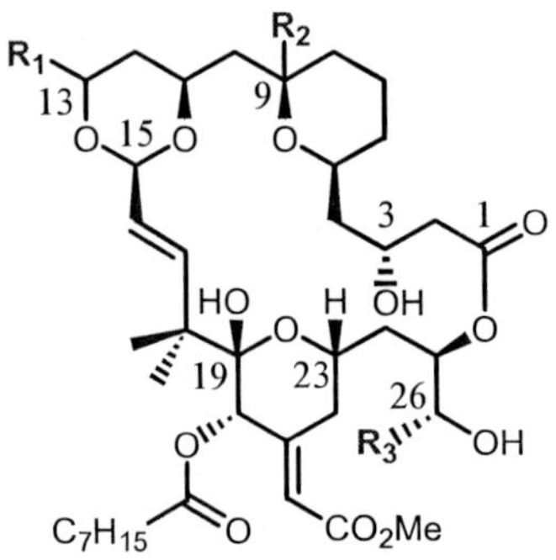

structure	R_1 =	R_2 =	R_3 =	K_i (nM)	Reference
61	H	H	Me	3.4	[220]
62	H	H	H	0.25	[214, 220]
63	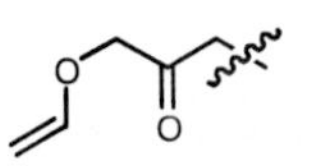	H	H	1.2	[275]
64		H	H	0.67	[275]
65	H	OH	H	2	[276]
66	H	OH	Me	4	[276]

Table 2. PKC inhibitory activity "bryologs", lacking an A-ring, and modifications at C13. The PKC kinase inhibitory concentrations (K_i), obtained from the listed references, are given in nM

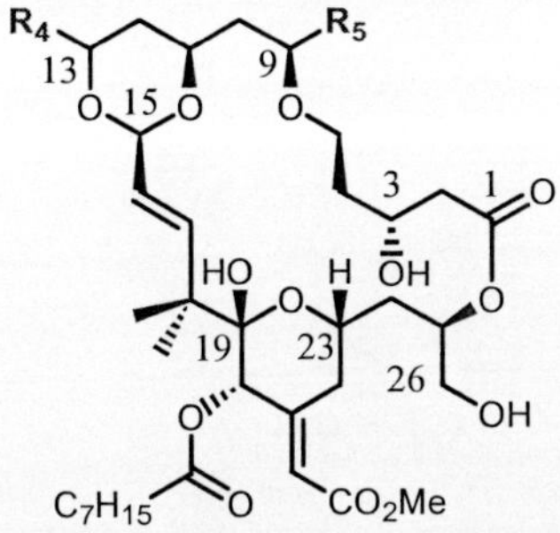

structure	R_4 =	R_5 =	K_i (nM)	Reference
67	H	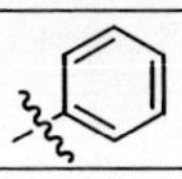	2.3	[226]
68	H	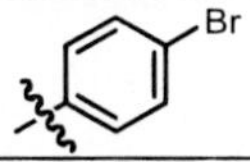	1.6	[226]
69	H	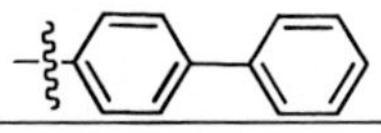	3.1	[226]
70	H	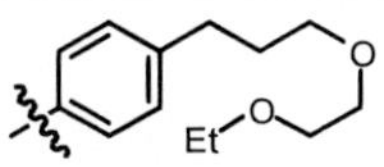	3.8	[226]
71	H	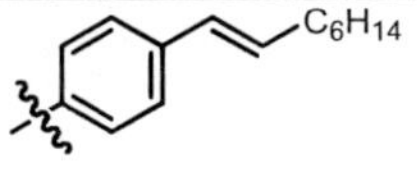	4.6	[226]
72	H	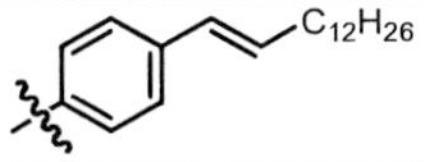	30	[226]
73	H		6.5	[221]
74	H	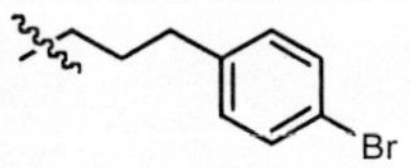	1.9	[221]
75	H	H	47	[210]
76	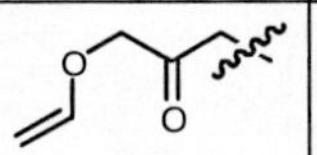	H	3.0	[275]
77		H	2.6	[275]

Table 3. PKC inhibitory activity of "bryologs" with a five-membered B-ring. The PKC kinase inhibitory concentration (K_i), obtained from [226], is given in nM

structure		K_i (nM)	Reference
78	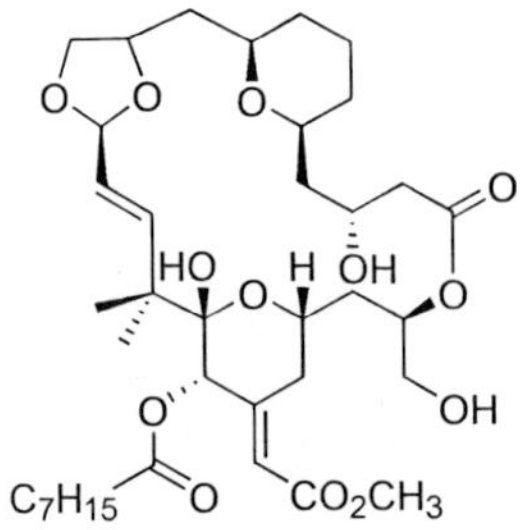	5.4 nM	[215]

Table 4. PKC inhibitory activity of "bryologs" with different side chains at C20. The PKC kinase inhibitory concentrations (K_i), obtained from the listed references, are given in nM

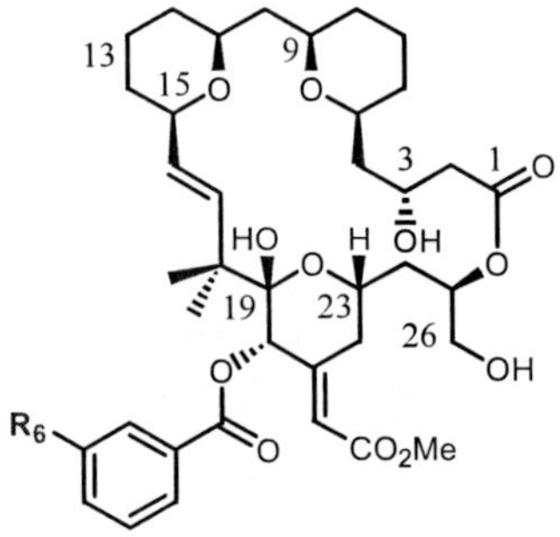

structure	R_6 =	K_i (nM)	Reference
79	NO_2	10	[216]
80	NH_2	60	[216]
81	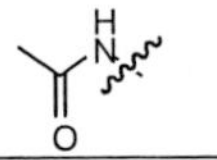	91	[216]
82	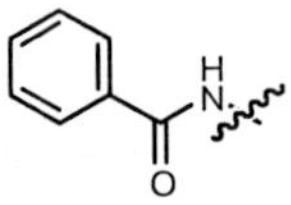	18	[216]
83	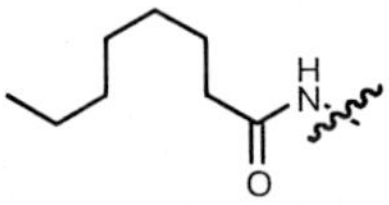	12	[216]
84	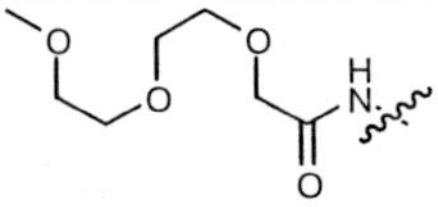	77	[216]
85	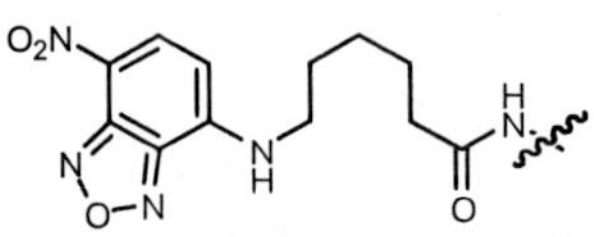	21	[216]

Table 5. PKC inhibitory activity of "bryologs" with a hydropyranyl A-ring. The PKC kinase inhibitory concentrations (K_i), obtained from the listed references, are given in nM

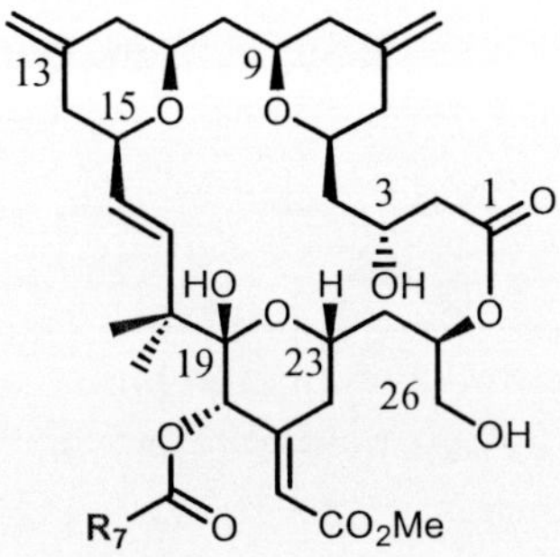

structure	R_7 =	K_i (nM)	Reference
86	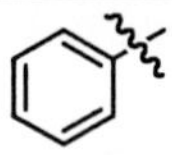	0.70	[227]
87	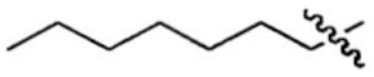	1.05	[227]
88	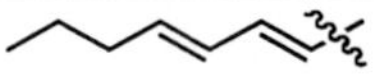	0.70	[227]

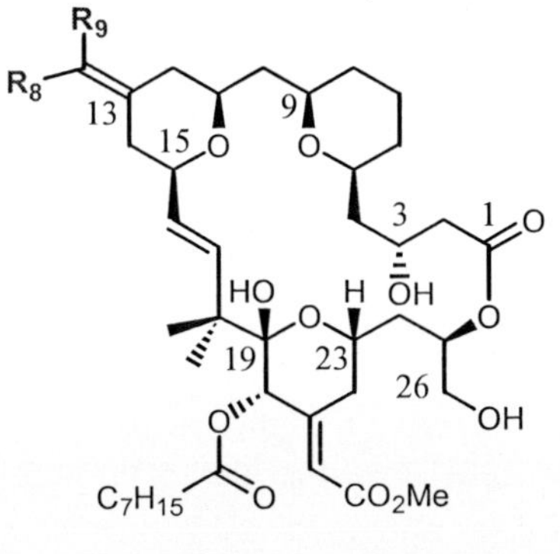

structure	R_8 =	R_9 =	K_i (nM)	Reference
89	H	H	1.6	[214]
90	H	CO_2Me	2.5	[214]
91	CO_2Me	H	0.9	[214]
92	/	/	3.1	[214, 220]

Interestingly, the missing of the A-ring does not alter the binding affinity of bryostatin analogues to PKC, even the sterically less demanding *p*-bromo-phenylpropyl derivative 74 expressed a higher binding affinity than the bulkier *tertio*-butyl substituent 73 [221]. However, a single hydrogen at this position in 75 resulted in a significant reduction of binding affinity [210]. A five-membered B-ring analogue (78) exhibited a binding affinity of

5.4 nM to a mixture of PKC isozymes isolated from rat brain and the translocation from PKCδ was rapid and complete, similar to the lead the bryostatin 1 analogue 60. However, the selectivity for PKCβI was significantly decreased in comparison to bryostatin 1 (22) [222].

Finally, lipophilic side-chains do not significantly affect the binding affinity to an isozyme mixture of PKC, except for the highly lipophilic derivative 74, for which the lower binding affinity is explained by reduced ability to partition to the phospholipid vesicles, as already observed in the case of phorbol esters [215, 223].

New SAR studies on the PKC inhibitory activity of brystatin 1 (22) are currently in progress. But the bryostatin 1 (22) analogues synthesized so far by Wender *et al.* (1998) [224, 225] have already been shown to have highly promising *in vitro* biological activities, warranting a very exciting future for this field of research.

5.3. Chemical Synthesis and NF-κB-related SAR of Hymenialdisine and its Analogues

The bromopyrrole alkaloid hymenialdisine (33) isolated from various marine sponges was first reported as an NF-κB inhibitor in 1997 by Breton *et al.* and by Roshak *et al.*, but no information about the molecular targets of the compound was provided at that time [228, 229]. The first synthesis of hymenialdisine (33), involving 8 steps starting from commercially available pyrrole-2-carboxylic acid, had already been reported two years earlier by Annoura and Tatsuoka [230].

The mode of action of hymenialdisine (33) was investigated in depth in 2000 by Meijer *et al.* [231]. Meijer *et al.* (2000) [231] showed that hymenialdisine targets the kinases glycogen synthase kinase-3b (GSK-3b), cyclin dependent kinase 1 (CDK1/cyclin1) and 2 (CDK5/p25), and casein kinase 1 (CK1). These kinases are not directly involved in the canonical NF-kB activation pathway, but casein kinases have been reported to induce IκBα degradation through a non-canonical pathway in mammalian cells exposed to UV radiation [74, 117]. GSK-3b, CDK1, CDK5, and CK1 are also known to be involved in hyperphosphorylation of tau proteins, resulting in neurofibrillary tangles, one of the pathological symptom of Alzheimer's disease [232-234], and in cell cycle regulation [235]. Hymenialdisine (33) was reported to inhibit the latter kinases at nanomolar concentrations [231]. To date, a wide range of natural and synthetic analogues of hymenialdisine (33) have been reported [236-239], and have been the subject of several SAR studies [240-242] (scheme 11).

Meijer *et al.* (2000) [231] reported that the diacetyl- and diacetyldebromo- derivatives of hymenialdisine 93 and 94, and the structurally related metabolites odiline 95 axinohydantoin 96 were less active than the natural compound. The crystal structure of the CDK2-hymenialdisine complex revealed that hymenialdisine (33) and the kinase interacted through direct hydrogen bonds and through van der Waals interactions at the His^{84}, Phe^{82} and Leu^{134} residues of the enzyme. Because of the nature of these interactions, even small modifications to the core structure of hymenialdisne (33) can lead to dramatic losses of activity, as observed by Meijer *et al.* [243] for the analogues 94-96. The results were consistent with the findings of Tasdemir *et al.* (2002) [242] evaluated the MEK1 inhibition potency and the cytotoxicity of the hymenialdisine derivatives 97-100 isolated from the marine sponge *Stylissa massa*.

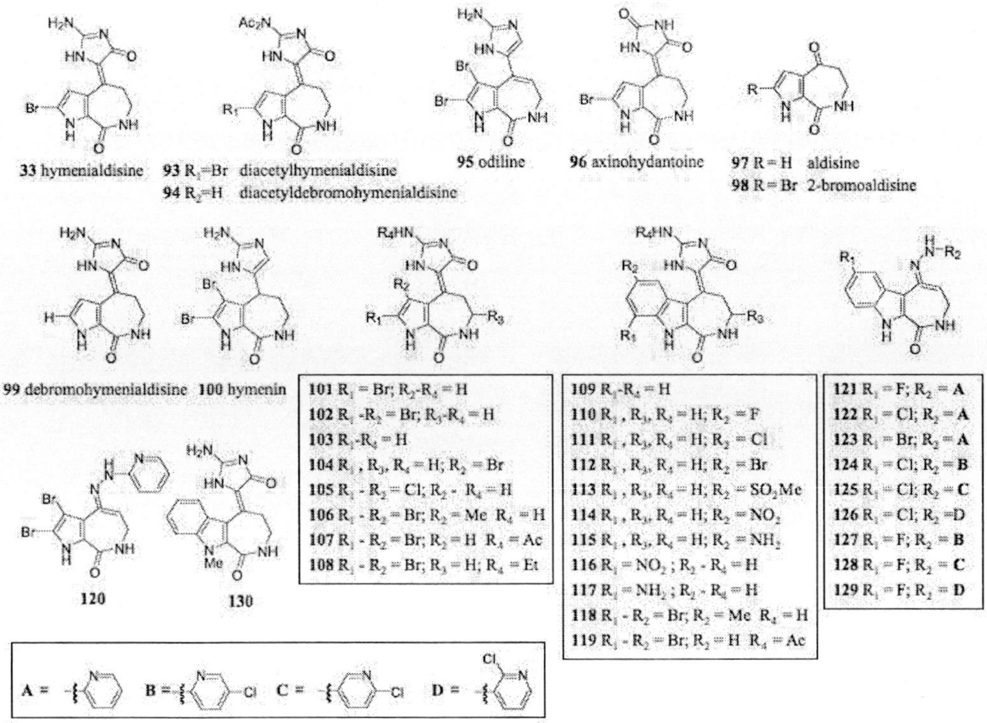

Scheme 11. Natural and synthetic analogues of hymenialdisine (33). Because of the high number of existing hydrazone analogues of hymenialdisine (33), only a selection of them are presented in the scheme. (Ac = acetyl group; Et = ethyl group; Me = methyl group).

Wan *et al.* (2004) [241] carried out an extensive SAR study using synthetic hymenialdisine analogues 101-129. As a first step, they evaluated structural modifications at the pyrroloazepine ring. Halogenation at the C2 and C3 positions in the pyrrole ring did not significantly affect inhibition potency against CDK5/p25 or GSK3β, but led to a 3-4 fold lowered activity against CDK1/cyclin B. CDK5/CDK1 inhibition selectivity was increased in case of a bromine in the C3 position (101). For all three kinases, methylation of the azepine ring (106) resulted in clearly reduced inhibition potential. Acylation or ethylation of the free guanidine amine moiety, which is involved in the hydrogen bonds between hymenialdisine and the kinases was shown to lower the inhibition potential significantly (107 and 108) [241]. The two novel hydrazone-indole analogues 128 and 129 were shown to have a particularly high CDK1/cyclinβ potential. Further, they were shown to selectively inhibit CDK5/p25 over GSK3β. The three analogues 121-124 were shown to significantly inhibit cell proliferation [244], and the three compounds 125-127 (all containing a pyridin-3-yl-hydrazine moiety) induced G2/M cell cycle arrest at a ten-fold lower concentration than hymenialdisine (33).

Sharma *et al.* (2004) [245] synthesized the two indoloazepine analogues 109 and 130 and investigated their effects on IL-2 and TNF-α production and on cell growth. To their surprise, the two compounds showed weaker bioactivity than hymenialdisine (13) [170, 245]. The weakness of the bioactivity of 130 may be explained by the methylation of the free indole-amine, which has been recognized as a major pharmacophore in hymenialdisine (33) since the first SAR studies [170, 245].

The results of the SAR studies performed on hymenialdisine (33) and its analogues are summarized in scheme 12. In brief, with the exception of a few hydrazone-indole analogues, even small modifications of the core unit of hymenialdisine (33) lead to dramatic decreases in bioactivity. None of the analogues of hymenialdisine (33) has entered clinical trials to date, but their potential use as medicinal drugs has been patented (US Patent 7098204).

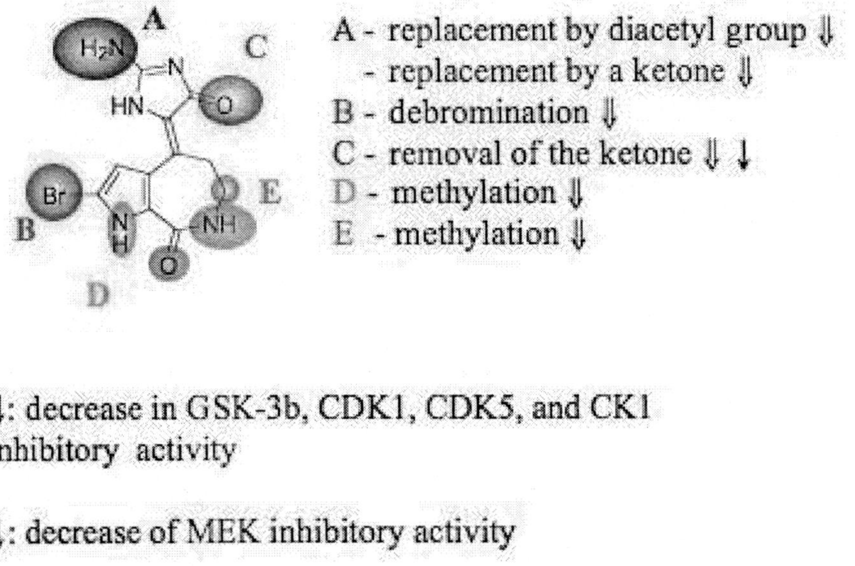

Scheme 12. Summary of the SAR of hymenialdisine (13) and its analogues in the context of the inhibition of the kinases GSK-3b, CDK1, CDK5, CK1, and MEK. The major pharmacophores are highlighted in colour. Some essential elements of the core structure are highlighted in grey.

5.4. Chemical Synthesis and NF-κB-related SAR of Avarol and Its Analogues

The sesquiterpene hydroquinone avarol (32) has been isolated from marine sponges since the mid 1970's, and it has, since then, been thoroughly investigated for anti-viral and anti-cancer activity [246-252]. The anti-viral and anti-cancer properties of avarol (32) have been patented under US Patent 4946869 and US patent 5082865, respectively, and the compound has advanced into clinical trials in HIV-infected patients in Germany [181]. The NF-κB inhibitory activity of avarol (32), which might play a critical role in the observed anti-viral and anti-cancer bioactivity of the compound, has only been described very recently [246]. The precise target of avarol (32) along the NF-κB activation pathway remains to be determined [246].

Avarol (32) is a rather unique marine natural product in the sense that it can be produced in large amounts through mariculture or through the ex situ culture of primary sponge cells [180] [181, 253]. Nevertheless, chemical synthesis, which was achieved for the first time in 1982 by Sarma *et al.* [254], still plays a critical role in the production of avarol and of its analogues [255]. The stereocontrolled synthesis of (±)-avarol by Sarma *et al.* (1982) was realised in an eight-step linear synthesis starting from a ene ketol [254, 256]. The synthetic analogues of avarol (32) are shown in scheme 13, together with some natural analogues of the compound.

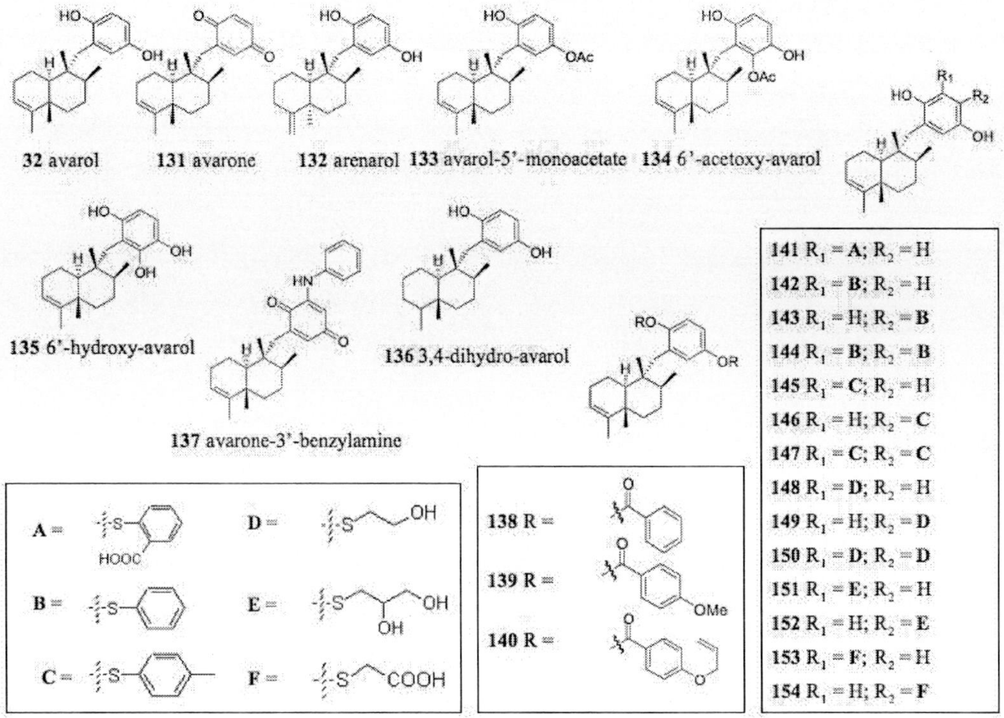

Scheme 13. Synthetic and natural analogues of avarol (32). Ac represents an acetyl group and Me represents a methyl group.

SAR studies on the anti-cancer activity of avarol (32) and its analogues have shown that some analogues, including 133 [257], are much more potent than the lead compound, while others, including the spongean natural product 132, which possess a *cis*-decaline moiety are significantly weaker anti-cancer compounds [258] . The cytotoxiciy of avarol (32) and its analogues is thought to be caused mainly by DNA damage. The damage of DNA is induced by the hydroxyl radicals produced by avarol (32) in the presence of oxygen and of low levels of superoxide dismutase [251]. This hypothesis has been corroborated by the finding that tryptophan, which is known to be a potent radical scavenger, inhibits avarol-induced DNA damage (32) [249]. Thiosalicylate analogues of avarol (32) have been shown to stabilize the produced radical and decreased its cytotoxicity [248, 249]. Belisario *et al.* (1992) reported that the radical scavenger property of avarol (32) is due to the presence of an easily donatable proton [259]. The departure of the proton is hampered by the esterification of the hydroxyl functionalities within the molecule [259].

SAR studies related to the anti-inflammatory activity of avarol (32) and its analogues revealed that the monophenyl-thio and thiosalicylate analogues (141-143 in particular) and the avarone-3'-benzylamine derivative 137 were the strongest inhibitors of UVB-induced NF-κB activation and TNF-α production [260, 261]. These results suggest that a benzylamine or thiosalicylate moiety at the C3' position plays a crucial role in the anti-inflammatory activity of avarol analogues.

In terms of HIV-1 inhibition, the presence of a 6'-hydroxyl-group (135) has been shown to dramatically increase the bioactivity of the lead compound.

The results obtained from the various SAR studies performed on avarol (32) and its analogues are summarized in scheme 14.

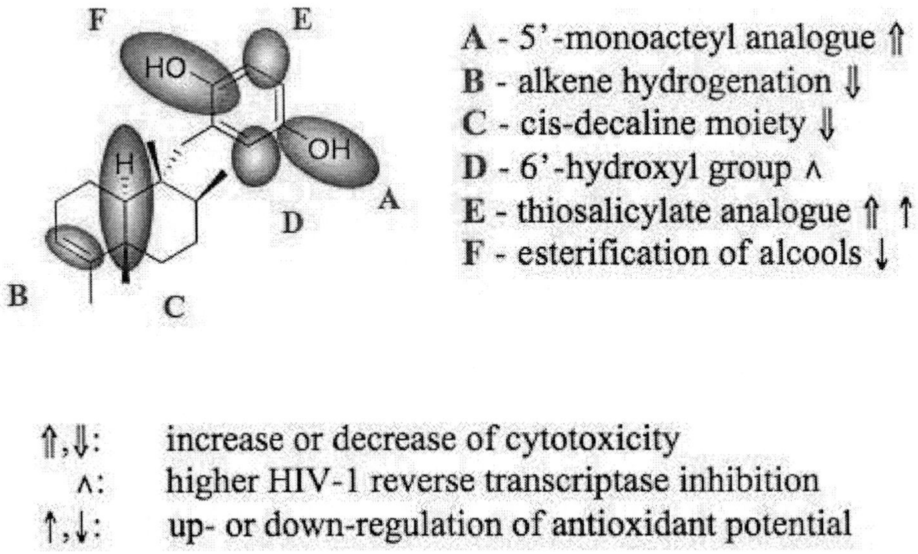

Scheme 14. Summary of the results of the various SAR studies performed on avarol (**32**) and its analogues. The pharmacophores identified through the SAR studies are highlighted in colour.

5.5. Chemical Synthesis and NF-κB-related SAR of Cacospongiolide B and its Analogues

Cacospongionolide B (29), a potent anti-inflammatory natural product isolated from various marine sponges, is closely related to avarol (32). The anti-inflammatory activity of 29 has been associated with its inhibition of the pro-inflammatory secretory phospholipase A_2 (PLA_2) [262-264]. PLA_2 induces the production of arachidonic acid, which is a precursor of leukotrienes and other cytokines implicated in the activation of NF-κB [265], PLA_2 inhibitors can hence be considered as indirect NF-kB inhibitors [266]. The NF-κB inhibitory potential of cacospongiolide B (29) was first reported by D'Acquisto *et al.* [166] and Palanki *et al.* [267] in 2000. Further investigation of the NF-κB inhibitory potential of cacospongiolide B (29) was performed by Posadas *et al.* in 2003 [164]. The synthesis of cacospongiolide B (29) analogues of it has been available since 1998 [268]. The first total synthesis of cacospongionolide B was achieved by Cheung and Snapper in 2002, through Michael addition of an enolate to an enone [269]. This synthetic approach by Cheung *et al.* enabled the access to several cacospongionolide analogues [269, 270]. The synthetic analogues of cacospongiolide B (29) are presented in scheme 15, together with some natural analogues.

The PLA_2 inhibitory activity of cacospongiolide B (29) has been attributed to the masking of an aldehyde group in the enzyme by the γ-hydroxy-butenolide moiety of 29 [271, 272].

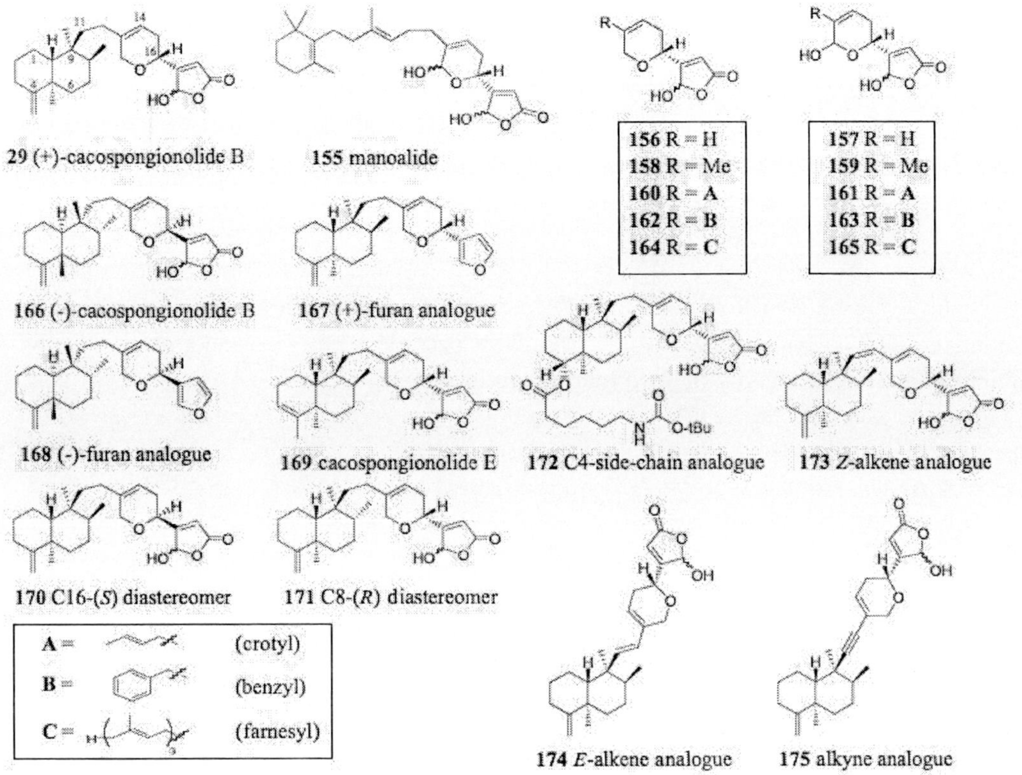

Scheme 15. Synthetic and natural analogues of cacospongiolide B (**29**). Me represents a methyl group.

Like its closely related analogue manoalide (155) isolated from marine sponges, cacospongiolide B (29) has been shown to interfere with the enzymatic activity of PLA_2 through the formation of a Schiff base-like covalent imine bond between the aldehyde generated upon the opening of the γ-hydroxybutenolide of the natural product and a lysine residue at the enzyme-lipid interface of PLA_2 [271, 272]. The hydrophobic region of manoalide (155) and cacospongionolide (29) allows the formation of non-covalent bonds between the enzyme and the natural products, thereby facilitating the formation of the Schiff base-like covalent bond [273].

SAR studies on the PLA_2 inhibitory activity of cacospongiolide B (29) and its analogues have revealed that the presence of a certain level of lipophilicity is a pre-requisite for the inhibition of the enzymatic activity of PLA_2. Cacospongiolide B analogues with short alkyl side-chains, such as methyl-, crotyl- and benzyl- functionalities (160-163), possess a significantly lower inhibition potential than analogues with a farnesyl side-chain (164-165), for which the IC_{50} values are up to five times lower than the IC_{50} of cacospongiolide B (29) [268].

The PLA_2 inhibition-related SAR studies on cacospongiolide B (29) and its analogues have also highlighted a crucial role of the stereochemistry of the compounds in the latter's bioactivity [269, 270]. The PLA_2 inhibitory activity of the synthetic cacospongiolide B (29) enantiomer 166 has been shown to be half the bioactivity of cacospongiolide B (29) [269, 270]. The substitution of the γ-hydroxybutenolide by a furan ring (167-168) led to a slightly lower bioactivity, but did not result in a complete loss of bioactivity, suggesting that the pyranofuranone ring is one, but not the unique key element for PLA_2 inhibition activity [269, 270]. The stereochemical inversion of the C16 and C8 stereocentres has been shown to boost the PLA_2 inhibitory activity in 170 and 171, respectively [269, 270]. An SAR study on the conformationally restricted cacospongionolide B analogues 173-175 recently synthesised by Murelli *et al.* [264] has investigated the impact of the three-dimensional orientation of cacospongiolide B analogues on the latter's PLA_2 inhibitory potential. The study revealed that rigidified three-dimensional structure such as in 173 are the most favourable in terms of PLA_2 inhibitory activity [264, 269, 270].

Finally, the noteworthy structural relatedness of cacospongiolide B (29) to avarol (32) within the *trans*-decaline core could potentially provide some explanations for the NF-κB inhibitory activity shared by the two marine products.

The results of the PLA_2 inhibitory activity-related SAR studies performed on cacospongiolide B (29) and its analogues are summarized in scheme 16. Briefly, the results of the SAR studies indicate that the PLA_2 inhibitory activity of cacospongionolide B (29), which is involves the formation of a Schiff-base-like covalent bond formation between the masked aldehyde generated upon the opening of the and a lysine of the enzyme, is both enantio- and diastereo- selective [264, 269, 270]. The most potent cacospongiolide B analogue (164) has been shown to possess an anti-inflammatory potential equal to the one of indomethacin (Indocin®), which is an approved nonsteroidal anti-inflammatory drug [274]. This finding suggests a highly promising future for cacospongiolide B analogues in the field of anti-inflammatory drug discovery.

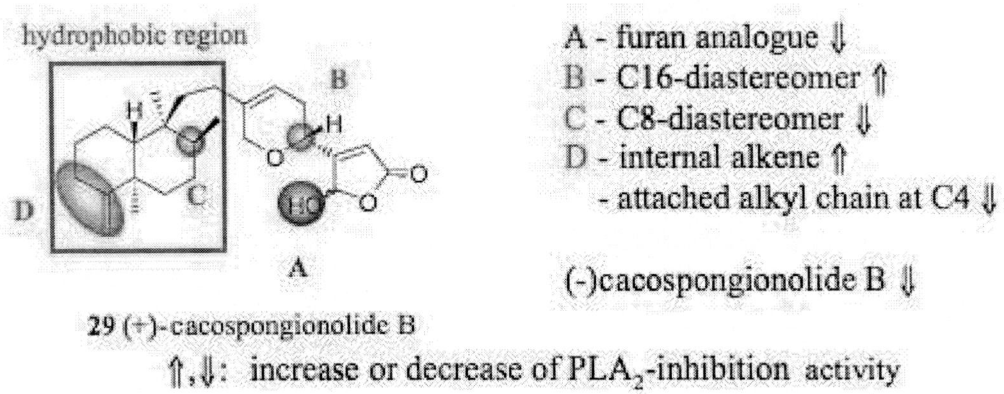

Scheme 16. Summary of the results obtained in the PLA_2 inhibition-related SAR studies performed on cacospongiolide B (29) and its analogues. The pharmacophores identified through the SAR studies are highlighted in colour.

CONCLUSION

There is constantly growing evidence for the important role of NF-κB inhibitors both in marine ecology and in biomedicine. Marine organisms have been shown to be a rich source of highly diverse natural products with very promising NF-κB inhibitory properties. While the detailed mode of action of marine NF-κB inhibitors remains, in many cases, poorly understood, structure activity relationship (SAR) studies on natural and synthetic analogues of lead marine NF-kB inhibitors have provided a tremendous amount of information on the molecular targets of marine NF-κB inhibitors. As described in the present review, the molecular target of marine natural products shown to inhibit NF-κB activation is, in many cases, upstream of the NF-κB activation cascade. NF-κB inhibition-related SAR studies have led to the discovery of several synthetic analogues with significantly higher bioactivities than the corresponding natural products. Some of these recently discovered compounds have reached clinical trials. The chemical synthesis of marine NF-κB inhibitors has played a very important role in marine drug discovery, not just in terms of identifying potent analogues of the bioactive natural products, but also in terms of providing a sustainable source of the bioactive molecules for clinical trials and for the future marketing of the compounds as medicinal drugs.

Acknowledgments

MJ is the recipient of a BBSRC Research Development Fellowship, and MS is a recipient of a postdoctoral research fellowship (BFR06/016) from the Luxembourg government. Research at the Laboratoire de Biologie Moléculaire et Cellulaire du Cancer (LBMCC) is financially supported by "Recherche Cancer et Sang" foundation, by the "Recherches Scientifiques Luxembourg" association, by "Een Häerz fir kriibskrank Kanner" a.s.b.l. (Luxembourg), and by Télévie Luxembourg.

References

[1] Barnes RSK, Hughes R. *An introduction to marine ecology*. Oxford, U.K.: Blackwell Science, 1988.

[2] Simmons TL, Andrianasolo E, McPhail K, Flatt P, Gerwick WH. Marine natural products as anticancer drugs. *Mol. Cancer Ther*. 2005;4:333-42.

[3] Pawlik JR. Marine Invertebrate Chemical Defenses. *Chem. Rev*. 1993;93:1911-22.

[4] Paul VJ, Puglisi MP, Ritson-Williams R. Marine chemical ecology. *Nat. Prod. Rep*. 2006;23:153-80.

[5] Scheuer PJ. Some Marine Ecological Phenomena - Chemical Basis and Biomedical Potential. *Science* 1990;248:173-7.

[6] Garson M. The biosynthesis of marine natural products. *Chem. Rev*. 1993;93:1699-733.

[7] Faulkner DJ. Biomedical Uses for Natural Marine Chemicals. *Oceanus* 1992;35:29-35.

[8] Paul VJ, Ritson-Williams R. Marine chemical ecology. *Nat. Prod. Rep*. 2008;*in press* (DOI: 10.1039/b702742g).

[9] Williams DH, Stone MJ, Hauck PR, Rahman SK. Why are secondary metabolites (natural products) biosynthesized? *J. Nat. Prod*. 1989;52:1189-208.

[10] Hay MF. Marine chemical ecology: What's known and what's next? *Journal of experimenta marine biology and ecology* 1996;200:103-34.

[11] Fenical W. Marine pharmaceuticals - Past, present, future. *Oceanus* 2006;19:111-9.

[12] Jensen PR, Fenical W. Strategies for the discovery of secondary metabolites from marine bacteria: ecological perspectives. *Annu. Rev. Microbiol*. 1994;48:559-84.

[13] Battershill CN, Jaspars M, Long P. Marine biodiscovery: new drugs from the ocean depths. *Biologist* 2005;52:107-14.
[14] Butler M. Natural products to drugs: natural product-derived compounds in clinical trials. *Nat. Prod. Rep.* 2008;25:475-516.
[15] Aggarwal BB, Takada Y, Shishodia S, Gutierrez AM, Oommen OV, Ichikawa H, et al. Nuclear transcription factor NF-kappa B: role in biology and medicine. *Indian J. Exp. Biol.* 2004;42:341-53.
[16] Ichikawa H, Nakamura Y, Kashiwada Y, Aggarwal BB. Anticancer drugs designed by mother nature: ancient drugs but modern targets. *Curr. Pharm. Des.* 2007;13:3400-16.
[17] Aggarwal BB, Ichikawa A, Garodia P, Weerasinghe P, Sethi G, Bhatt I, et al. From traditional Ayurvedic medicine to modern medicine: identification of therapeutic targets for suppression of inflammation and cancer. *Expert Opin. Ther Targets* 2006;10:87-118.
[18] Kingston DG, Newman DJ. Natural products as drug leads: an old process or the new hope for drug discovery? *IDrugs* 2005;8:990-2.
[19] Faulkner DJ. Marine pharmacology. *Antonie Van Leeuwenhoek* 2000;77:135-45.
[20] Latchman DS. Transcription factors: an overview. *Int. J. Biochem. Cell Biol.* 1997;29:1305-12.
[21] Karin M. Too many transcription factors: positive and negative interactions. *New Biol.* 1990;2:126-31.
[22] Sen R, Baltimore D. Multiple nuclear factors interact with the immunoglobulin enhancer sequences. *Cell* 1986;46:705.
[23] Ghosh S, May MJ, Kopp EB. NF-kappa B and Rel proteins: evolutionarily conserved mediators of immune responses. *Annual Review on Immunology* 1998;16:225-60.
[24] Pahl HL. Activators and target genes of Rel/NF-kappaB transcription factors. *Oncogene* 1999;18:6853-66.
[25] Li Q, Verma IM. NF-kappaB regulation in the immune system. *Nature Rev. Immunol.* 2002;2:725-34.
[26] Gerondakis S, Grossmann M, Nakamura Y, Pohl T, Grumont R. Genetic approaches in mice to understand Rel/NF-kappaB and IkappaB function: transgenics and knockouts. *Oncogene* 1999;18:6888-95.
[27] Gerondakis S, Grumont R, Rourke I, Grossmann M. The regulation and roles of Rel/NF-kappa B transcription factors during lymphocyte activation. *Curr. Opin. Immunol.* 1998;10:353-9.
[28] Sen R, Baltimore D. Inducibility of kappa immunoglobulin enhancer-binding protein Nf-kappa B by a posttranslational mechanism. *Cell* 1986;47:921-8.
[29] Jost PJ, Ruland J. Aberrant NF-kappaB signaling in lymphoma: mechanisms, consequences, and therapeutic implications. *Blood* 2007;109:2700-7.
[30] Bargou RC, Leng C, Krappmann D, Emmerich F, Mapara MY, Bommert K, et al. High-level nuclear NF-kappa B and Oct-2 is a common feature of cultured Hodgkin/Reed-Sternberg cells. *Blood* 1996;87:4340-7.
[31] Zhou H, Du MQ, Dixit VM. Constitutive NF-kappaB activation by the t(11;18)(q21;q21) product in MALT lymphoma is linked to deregulated ubiquitin ligase activity. *Cancer Cell* 2005;7:425-31.

[32] Davis RE, Brown KD, Siebenlist U, Staudt LM. Constitutive nuclear factor kappaB activity is required for survival of activated B cell-like diffuse large B cell lymphoma cells. *J. Exp. Med.* 2001;194:1861-74.

[33] Keller SA, Schattner EJ, Cesarman E. Inhibition of NF-kappaB induces apoptosis of KSHV-infected primary effusion lymphoma cells. *Blood* 2000;96:2537-42.

[34] Sun SC, Ballard DW. Persistent activation of NF-kappaB by the tax transforming protein of HTLV-1: hijacking cellular IkappaB kinases. *Oncogene* 1999;18:6948-58.

[35] Xiao G, Cvijic ME, Fong A, Harhaj EW, Uhlik MT, Waterfield M, et al. Retroviral oncoprotein Tax induces processing of NF-kappaB2/p100 in T cells: evidence for the involvement of IKKalpha. *Embo. J.* 2001;20:6805-15.

[36] Rayet B, Gelinas C. Aberrant rel/nfkb genes and activity in human cancer. *Oncogene* 1999;18:6938-47.

[37] Spano JP, Bay JO, Blay JY, Rixe O. Proteasome inhibition: a new approach for the treatment of malignancies. *Bull. Cancer* 2005;92:E61-6, 945-52.

[38] Kumar A, Lnu S, Malya R, Barron D, Moore J, Corry DB, et al. Mechanical stretch activates nuclear factor-kappaB, activator protein-1, and mitogen-activated protein kinases in lung parenchyma: implications in asthma. *Faseb. J.* 2003;17:1800-11.

[39] Hart LA, Krishnan VL, Adcock IM, Barnes PJ, Chung KF. Activation and localization of transcription factor, nuclear factor-kappaB, in asthma. *Am. J. Respir. Crit. Care Med.* 1998;158:1585-92.

[40] Marok R, Winyard PG, Coumbe A, Kus ML, Gaffney K, Blades S, et al. Activation of the transcription factor nuclear factor-kappaB in human inflamed synovial tissue. *Arthritis Rheum.* 1996;39:583-91.

[41] Gilston V, Jones HW, Soo CC, Coumbe A, Blades S, Kaltschmidt C, et al. NF-kappa B activation in human knee-joint synovial tissue during the early stage of joint inflammation. *Biochem. Soc. Trans* 1997;25:518S.

[42] Miyazawa K, Mori A, Yamamoto K, Okudaira H. Constitutive transcription of the human interleukin-6 gene by rheumatoid synoviocytes: spontaneous activation of NF-kappaB and CBF1. *Am. J. Pathol.* 1998;152:793-803.

[43] Ellis RD, Goodlad JR, Limb GA, Powell JJ, Thompson RP, Punchard NA. Activation of nuclear factor kappa B in Crohn's disease. *Inflamm Res.* 1998;47:440-5.

[44] Bonetti B, Stegagno C, Cannella B, Rizzuto N, Moretto G, Raine CS. Activation of NF-kappaB and c-jun transcription factors in multiple sclerosis lesions. Implications for oligodendrocyte pathology. *Am. J. Pathol.* 1999;155:1433-8.

[45] Gveric D, Kaltschmidt C, Cuzner ML, Newcombe J. Transcription factor NF-kappaB and inhibitor I kappaBalpha are localized in macrophages in active multiple sclerosis lesions. *J. Neuropathol. Exp. Neurol.* 1998;57:168-78.

[46] Strait K, Li Y, Dillehay DL, Weitzmann MN. Suppression of NF-kappaB activation blocks osteoclastic bone resorption during estrogen deficiency. *Int. J. Mol. Med.* 2008;21:521-5.

[47] Aggarwal BB, Takada Y, Oommen OV. From chemoprevention to chemotherapy: common targets and common goals. *Expert Opin. Invest Drugs* 2004;13:1327-38.

[48] Feinman R, Koury J, Thames M, Barlogie B, Epstein J, Siegel DS. Role of NF-kappaB in the rescue of multiple myeloma cells from glucocorticoid-induced apoptosis by bcl-2. *Blood* 1999;93:3044-52.

[49] Ahn KS, Sethi G, Aggarwal BB. Reversal of chemoresistance and enhancement of apoptosis by statins through down-regulation of the NF-kappaB pathway. *Biochem. Pharmacol*. 2008;75:907-13.
[50] Perona R, Sanchez-Perez I. Signalling pathways involved in clinical responses to chemotherapy. *Clin. Transl. Oncol.* 2007;9:625-33.
[51] Wu M, Lee H, Bellas RE, Schauer SL, Arsura M, Katz D, et al. Inhibition of NF-kappaB/Rel induces apoptosis of murine B cells. *Embo J.* 1996;15:4682-90.
[52] Wang CY, Mayo MW, Baldwin AS, Jr. TNF- and cancer therapy-induced apoptosis: potentiation by inhibition of NF-kappaB. *Science* 1996;274:784-7.
[53] Winkler JD, Eris T, Sung CM, Chabot-Fletcher M, Mayer RJ, Surette ME, et al. Inhibitors of coenzyme A-independent transacylase induce apoptosis in human HL-60 cells. *J. Pharmacol. Exp. Ther.* 1996;279:956-66.
[54] Coussens LM, Werb Z. Inflammation and cancer. *Nature* 2002;420:860-7.
[55] Courtois G, Gilmore TD. Mutations in the NF-kappaB signaling pathway: implications for human disease. *Oncogene* 2006;25:6831-43.
[56] Kumar A, Takada Y, Boriek AM, Aggarwal BB. Nuclear factor-kappa B: its role in health and disease. *J. Mol. Med.* 2004;82:434-48.
[57] Karin M, Greten FR. NF-kappaB: linking inflammation and immunity to cancer development and progression. *Nature Rev. Immunol.* 2005;5:749-59.
[58] Folmer F, Jaspars M, Diederich M. Marine natural products as targeted modulators of the transcription factor NF-κB. *Biochem. Pharmacol.* 2008;75:603-17.
[59] May MJ, Ghosh S. Signal transduction through NF-kappa B. *Immunol. Today* 1998;19:80-8.
[60] Keutgens A, Robert I, Viatour P, Chariot A. Deregulated NF-κB activity in haemological malignancies. *Biochem. Pharmacol.* 2006;72:1069-80.
[61] Jacobs MD, Harrison SC. Structure of an IkappaBalpha/NF-kappaB complex. *Cell* 1998;95:749-58.
[62] Ghosh S. Handbook of Transcription Factor NF-kappaB. Boca Raton, FL (U.S.A.): CRC Press, 2006.
[63] Baldwin AS, Jr. The NF-kappa B and I kappa B proteins: new discoveries and insights. *Annual Review on Immunology* 1996;14:649-83.
[64] Chen FE, Huang DB, Chen YQ, Ghosh G. Crystal structure of p50/p65 heterodimer of transcription factor NF-kappaB bound to DNA. *Nature* 1998;391:410-3.
[65] Bergqvist S, Croy C, Kjaergaard M, Huxford T, Ghosh G, Komives E. Thermodynamics reveal that helix four in the NLS of NF-κB p65 anchors IκBα, forming a very stable complex. *J. Mol. Biol.* 2006;360:421-34.
[66] Huxford T, Huang DB, Malek S, Ghosh G. The crystal structure of the IkappaBalpha/NF-kappaB complex reveals mechanisms of NF-kappaB inactivation. *Cell* 1998;95:759-70.
[67] Koradi R, Billeter M, Wüthrich K. MOLMOL: a program for display and analysis of macromolecular structures. *Journal of Molecular Graphics* 1996;14:51-5.
[68] Bours V, Bonizzi G, Bentires-Alj M, Bureau F, Piette J, Lekeux P, et al. NF-kappaB activation in response to toxical and therapeutical agents: role in inflammation and cancer treatment. *Toxicology* 2000;153:27-38.
[69] Garg AK, Aggarwal BB. Reactive oxygen intermediates in TNF signaling. *Mol. Immunol.* 2002;39:509-17.

[70] Karin M, Takahashi T, Kapahi P, Delhase M, Chen Y, Makris C, et al. Oxidative stress and gene expression: the AP-1 and NF-kappaB connections. *Biofactors* 2001;15:87-9.
[71] Ichikawa H, Aggarwal BB. Guggulsterone inhibits osteoclastogenesis induced by receptor activator of nuclear factor-kappaB ligand and by tumor cells by suppressing nuclear factor-kappaB activation. *Clin. Cancer Res.* 2006;12:662-8.
[72] Kim JM, Min SK, Kim H, Kang HK, Jung SY, Lee SH, et al. Vacuolar-type H+-ATPase-mediated acidosis promotes in vitro osteoclastogenesis via modulation of cell migration. *Int. J. Mol. Med.* 2007;19:393-400.
[73] Karin M. How NF-kappaB is activated: the role of the IkappaB kinase (IKK) complex. *Oncogene* 1999;18:6867-74.
[74] Karin M. Nuclear factor-kB in cancer development and progression. *Nature* 2006;441:431-6.
[75] Bowie A, O'Neill LA. Oxidative stress and nuclear factor-kappaB activation: a reassessment of the evidence in the light of recent discoveries. *Biochem. Pharmacol.* 2000;59:13-23.
[76] Hochstrasser M. Lingering mysteries of ubiquitin-chain assembly. Cell 2006;124:27-34.
[77] Mukhopadhyay D, Riezman H. Proteasome-independent functions of ubiquitin in endocytosis and signaling. *Science* 2007;315:201-5.
[78] Evans P. Regulation of pro-inflammatory signalling networks by ubiquitin: identification of novel targets for anti-inflammatory drugs. *Expert Reviews in Molecular Medicine* 2005;7:1-19.
[79] Haefner B. NF-kappa B: arresting a major culprit in cancer. *Drug Disc. Today* 2002;7:653-63.
[80] Greten FR, Karin M. The IKK/NF-kappaB activation pathway-a target for prevention and treatment of cancer. *Cancer Lett.* 2004;206:193-9.
[81] Bremner P, Heinrich M. Natural products as targeted modulators of the nuclear factor-kappa B pathway. *J. Pharm. Pharmacol.* 2002;54:453-72.
[82] Rungeler P, Castro V, Mora G, Goren N, Vichnewski W, Pahl HL, et al. Inhibition of transcription factor NF-kappa B by sesquiterpene lactones: a proposed molecular mechanism of action. *Bioorg. Med. Chem.* 1999;7:2343-52.
[83] Drahl C, Cravatt BF, Sorensen EJ. Protein-reactive natural products. *Angew. Chem.* 2005;44:5788-809.
[84] Garcia-Pineres AJ, Castro V, Mora G, Schmidt TJ, Strunck E, Pahl HL, et al. Cysteine 38 in p65/NF-kappa B plays a crucial role in DNA binding inhibition by sesquiterpene lactones. *J. Biol. Chem.* 2001;276:39713-20.
[85] Lyss G, Schmidt TJ, Merfort I, Pahl HL. Helenalin, an anti-inflammatory sesquiterpene lactone from *Arnica*, selectively inhibits transcription factor NF-κB. *Biol. Chem.* 1997;378:951-61.
[86] Adams J. The proteasome as a novel target for the treatment of breast cancer. *Breast Disease* 2002;15:61-70.
[87] Kisselev AF, Goldberg AL. Proteasome inhibitors: from research tools to drug candidates. *Chem. Biol.* 2001;8:739-58.
[88] Delcros JG, Baudy Floc'h M, Prigent C, Arlot-Bonnemains Y. Proteasome inhibitors as therapeutic agents: Current and future strategies. *Curr. Med. Chem.* 2003;10:479-503.
[89] Ahn KS, Sethi G, Chao TH, Neuteboom ST, Chaturvedi MM, Palladino MA, et al. Salinosporamide A (NPI-0052) potentiates apoptosis, suppresses osteoclastogenesis,

and inhibits invasion through down-modulation of NF-kappaB regulated gene products. *Blood* 2007;110:2286-95.

[90] Karin M, Yamamoto Y, Wang QM. The IKK NF-kappa B system: A treasure trove for drug development. *Nature Rev. Drug Disc.* 2004;3:17-26.

[91] Ogino S, Tsuruma K, Uehara T, Nomura Y. Herbimycin A abrogates nuclear factor-kappaB activation by interacting preferentially with the IkappaB kinase beta subunit. *Mol. Pharmacol.* 2004;65:1344-51.

[92] Sasaki K, Rinehart KL, Jr., Slomp G, Grostic MF, Olson EC. Geldanamycin. I. Structure assignment. *J. Am. Chem. Soc.* 1970;92:7591-3.

[93] Shrivastava A, Aggarwal BB. Antioxidants differentially regulate activation of Nuclear Factor-κB, Activator Protein-1, c-Jun Amino-Terminal Kinases, and apopoptosis induced by Tumor Necrosis Factor: Evidence that JNK and NF-κB activation are not linked to apoptosis. *Antiox Redox Signalling* 1999;1:181-91.

[94] Palladino MA, Bahjat FR, Theodorakis EA, Moldawer LL. Anti-TNF-alpha therapies: The next generation. *Nature Rev. Drug Disc.* 2003;2:736-46.

[95] Takamatsu S, Hodges TW, Rajbhandari I, Gerwick WH, Hamann MT, Nagle DG. Marine natural products as novel antioxidant prototypes. *J. Nat. Prod.* 2003;66:605-8.

[96] Strivastava A, Aggarwal BB. Antioxidants differentially regulate activation of NFKB, APA, CJNK and apoptosis induced by TNF. *Antiox Redox Signalling* 1999;1:181-91.

[97] Wang XW, Tan NS, Ho B, Ding JL. Evidence for the ancient origin of the NF-{kappa}B/I{kappa}B cascade: Its archaic role in pathogen infection and immunity. *Proc. Natl. Acad. Sci. USA* 2006;103:4204-9.

[98] Friedman R, Hughes AL. Molecular evolution of the NF-kappaB signaling system. *Immunogenetics* 2002;53:964-74.

[99] Povelones M, Tran K, Thanos D, Ambron RT. An NF-kappaB-like transcription factor in axoplasm is rapidly inactivated after nerve injury in *Aplysia. J. Neurosci.* 1997;17:4915-20.

[100] Pancer Z, Rast JP, Davidson EH. Origins of immunity: transcription factors and homologues of effector genes of the vertebrate immune system expressed in sea urchin coelomocytes. *Immunogenetics* 1999;49:773-86.

[101] Mitta G, Vandenbulcke F, Roch P. Original involvement of antimicrobial peptides in mussel innate immunity. *FEBS Lett* 2000;486:185-90.

[102] Merlo E, Freudenthal R, Romano A. The I kappa B kinase inhibitor sulfasalazine impairs long-term memory in the crab *Chasmagnathus. Neuroscience* 2002;112:161-72.

[103] Shimada M, Satoh N, Yokosawa H. Involvement of Rel/NF-kappaB in regulation of ascidian notochord formation. Dev, Growth, *Differentiation* 2001;43:145-54.

[104] Gueguen Y, Cadoret JP, Flament D, Barreau-Roumiguiere C, Girardot AL, Garnier J, et al. Immune gene discovery by expressed sequence tags generated from hemocytes of the bacteria-challenged oyster, *Crassostrea gigas. Gene* 2003;303:139-45.

[105] Bettencourt R, Roch P, Stefanni S, Rosa D, Colaço A, Santos R. Deep sea immunity: Unveiling immune constituents from the hydrothermal vent mussel *Bathymodiolus azoricus. Marine Environmental Research* 2007;64:108-27.

[106] Waterhouse RM, Kriventseva EV, Meister S, Xi Z, Alvarez KS, Bartholomay LC, et al. Evolutionary dynamics of immune-related genes and pathways in disease-vector mosquitoes. *Science* 2007;316:1738-43.

[107] Gauthier MJ, Degnan BM. The transcription factor NF-kappaB in the demosponge *Amphimedon queenslandica*: insights on the evolutionary origin of the Rel homology domain. *Development Genes and Evolution* 2008;doi 10.1007/s00427-007-0197-5.

[108] Escoubas JM, Briant L, Montagnani C, Hez S, Devaux C, Roch P. Oyster IKK-like protein shares structural and functional properties with its mammalian homologues. *FEBS Lett.* 1999;453:293-8.

[109] Xiong X, Feng Q, Chen L, Xie L, Zhang R. Cloning and characterization of an IKK homologue from pearl oyster, *Pinctada fucata. Developmental and Comparative Immunology* 2008;32:15-25.

[110] Goodson MS, Kojadinovic M, Troll JV, Scheetz TE, Casavant TL, Soares MB, et al. Identifying components of the NF-κB pathway in the beneficial *Euprymna scolopes-Vibrio fischeri* light organ symbiosis. *Applied and Environmental Microbiology* 2005;71:6934-46.

[111] Schlezinger JJ, Blickarz CE, Mann KK, Doerre S, Stegeman JJ. Identification of NF-kappaB in the marine fish *Stenotomus chrysops* and examination of its activation by aryl hydrocarbon receptor agonists. *Chem. Biol. Interactions* 2000;126:137-57.

[112] Muller WE, Gamulin V, Rinkevich B, Spreitzer I, Weinblum D, Schroder HC. Ubiquitin and ubiquitination in cells from the marine sponge *Geodia cydonium. Bio Chem. Hoppe-Seyler* 1994;375:53-60.

[113] Muller WE, Koziol C, Muller IM, Wiens M. Towards an understanding of the molecular basis of immune responses in sponges: the marine demosponge *Geodia cydonium* as a model. *Microscopy Res. Tech.* 1999;44:219-36.

[114] Muller WE, Wiens M, Adell T, Gamulin V, Schroder HC, Muller IM. Bauplan of urmetazoa: basis for genetic complexity of metazoa. *Int. Rev. Cytol.* 2004;235:53-92.

[115] Muller WE, Wiens M, Muller IM, Schroder HC. The chemokine networks in sponges: potential roles in morphogenesis, immunity and stem cell formation. *Prog. Mol. Subcell Biol.* 2004;34:103-43.

[116] Li N, Karin M. Ionizing radiation and short wavelength UV activate NF-κB through two distinct mechanisms. *Proc. Natl. Acad. Sci. USA* 1998;95:13012-7.

[117] Kato T, Delhase M, Hoffmann A, Karin M. CK2 is a C-terminal IκB kinase responsible for NF-κB activation during the UV response. *Mol. Cell* 2003;12:829-39.

[118] Bender K, Gottlicher S, Whiteside H, Rahmsdorf H, Herrlich P. Sequential DNA damage-independent and -dependent activation of NF-κB by UV. *EMBO Journal* 1998;17:5170-81.

[119] Karentz D. Chemical defenses of marine organisms against solar radiation exposure: UV-absorbing mycosporine-like amino acids and scytonemin. *in*: Marine Chemical Ecology. Boca Raton: CRC Press, 2001.

[120] Patel A, Mishra S, Ghosh PK. Antioxidant potential of C-phycocyanin isolated from cyanobacterial species Lyngbya, Phormidium and Spirulina spp. *Indian Journal of Biochemistry and Biophysics* 2006;43:25-31.

[121] Pannunzio T, Storey K. Antioxidant defenses and lipid peroxidation during anoxia stress and aerobic recovery in the marine gastropod *Littorina littorea. J. Exp. Mar. Biol. Ecol.* 1997;221:277-92.

[122] Michiels C, Minet E, Mottet D, Raes M. Regulation of gene expression by oxygen: NF-kappaB and HIF-1, two extremes. *Free Radicals in Biology and Medicine* 2002;33:1231-42.

[123] Rius J, Guma M, Schachtrup C, Akassoglou K, Zinkernagel A, Nizet V, et al. NF-κB links innate immunity to the hypoxic response through transcriptional regulation of HIF-1α. *Nature* 2008;453:807-11.

[124] van Uden P, Kenneth NS, Rocha S. Regulation of hypoxia inducible factor-1 alpha by NF-kappaB. *Biochem J.* 2008;412:477-84.

[125] Dai J, Liu Y, Zhou YD, Nagle DG. Hypoxia-selective anti-tumor agents: norsesterterpene peroxides from the marine sponge *Diacarnus levii* preferentially suppresses the growth of tumor cells under hypoxic conditions. *J. Nat. Prod.* 2007;70:130-3.

[126] Seo SB, Jeong D, Chung HS, Lee JD, You YO, Kajiuchi T, et al. Inhibitory effect of high molecular weight water-soluble chitosan an hypoxia-induced inflammatory cytokine production. *Biol. Pharm. Bull* 2003;26:717-21.

[127] Mohammed KA, Hossain CF, Zhang L, Bruick RK, Zhou YD, Nagle DG. Laurenditerpenol, a new diterpene from the tropical marine alga *Laurencia intricata* that potently inhibits HIF-1 mediated hypoxic signaling in breast tumor cells. *J. Nat. Prod.* 2004;67:2002-7.

[128] Dai J, Fishback JA, Zhou YD, Nagle DG. Sodwanone and yardenone triterpenes from a South African species of the marine sponge *Axinella* inhibit hypoxia-inducible factor-1 (HIF-1) activation in both breast and prostate tumor cells. *J. Nat. Prod.* 2006;69:1715-20.

[129] Dai J, Liu Y, Zhou YD, Nagle DG. Benzochromenones from the marine crinoid *Comantheria rotula* inhibit hypoxia-inducible factor-1 (HIF-1) in cell-based reporter assays and differentially suppress the growth of certian tumor cell lines. *J. Nat. Prod.* 2007;70:1462-6.

[130] Liu R, Liu Y, Zhou YD, Nagle DG. Molecular-targeted antitumor agents. 15. Neolamellarins from the marine sponge Dendrilla nigra inhibit hypoxia-inducible factor-1 activation and secreted vascular endothelial growth factor production in breast tumor cells. *J. Nat. Prod.* 2007;70:1741-5.

[131] Mohammed KA, Jaduico RC, Bugni TS, Harper MK, Sturdy M, Ireland C. Strongylophorines: natural product inhibitors of hypoxia-inducible factor-1 transcriptional pathway. *J. Med. Chem.* 2008;51:1402-5.

[132] El Sayed KA, Khanfar MA, Shallal HM, Muralidharan A, Awate B, Youssef DT, et al. Latrunculin A and Its C-17-O-carbamates Inhibit prostate tumor cell invasion and HIF-1 activation in breast tumor cells. *J. Nat. Prod.* 2008;71:396-402.

[133] Tsai JR, Lin HC. V-type H^+-ATPase and Na^+,K^+-ATPase in the gills of 13 euryhaline crabs during salinity acclimation. *J. Exp. Biol.* 2007;210:620-7.

[134] Beutler JA, McKee TC. Novel marine and microbial natural product inhibitors of vacuolar ATPase. *Curr. Med. Chem.* 2003;10:787-96.

[135] Xie XS, Padron D, Liao X, Wang J, Roth MG, De Brabander JK. Salicylihalamide A Inhibits the V_0 Sector of the V-ATPase through a Mechanism Distinct from Bafilomycin A_1. *J. Biol. Chem.* 2004;279:19755-63.

[136] James ER, Green DR. Manipulation of apoptosis in the host-parasite interaction. *Trends Parasitol.* 2004;20:280-7.

[137] Lisowska K, Witkowski JM. Viral strategies in modulation of NF-kappaB activity. *Arch. Immunol. Ther. Exp.* 2003;51:367-75.

[138] Neish AS. Bacterial inhibition of eukaryotic pro-inflammatory pathways. *Immunol. Res.* 2004;29:175-86.

[139] Hiscott J, Kwon H, Genin P. Hostile takeovers: viral appropriation of the NF-kappa B pathway. *J. Clin. Invest* 2001;107:143-51.

[140] Lane N. Marine microbiology: Origins of Death. *Nature* 2008;453:583-5.

[141] Jayakumar A, Donovan MJ, Tripathi V, Ramalho-Ortigao M, McDowell MA. *Leishmania major* infection activates NF-κB and interferon regulatory factors 1 and 8 in human dendritic cells. *Infection and immunity* 2008;76:2138-48.

[142] Heussler VT, Rottenberg S, Schwab R, Kuenzi P, Fernandez PC, McKellar S, et al. Hijacking of host cell IKK signalosomes by the transforming parasite *Theileria. Science* 2002;298:1033-6.

[143] Umezawa K, Chaicharoenpong C. Molecular design and biological activities of NF-kappaB inhibitors. *Molecules and Cells* 2002;14:163-7.

[144] Le Negrate G, Faustin B, Welsh K, Loeffler M, Krajewska M, Hasegawa P, et al. Salmonella secreted factor L deubiquitinase of *Salmonella typhimurium* inhibits NF-κB, supresses IκBα ubiquitination and modulates innate immune responses. *Journal of* I*nmmunology* 2008;180:5045-56.

[145] Merlo E, Freudenthal R, Maldonado H, Romano A. Activation of the transcription factor NF-kappaB by retrieval is required for long-term memory reconsolidation. *Learning and Memory* 2005;12:23-9.

[146] Folmer F, Jaspars M, Dicato M, Diederich M. Marine natural products as targeted modulators of the transcription factor NF-kappaB. *Biochem. Pharmacol.* 2008;75:603-17.

[147] Terracciano S, Aquino M, Rodriquez M, Monti MC, Casapullo A, Riccio R, et al. Chemistry and biology of anti-inflammatory marine natural products: molecules interfering with cyclooxygenase, NF-kappaB and other unidentified targets. *Curr. Med. Chem.* 2006;13:1947-69.

[148] Groll M, Huber R, Potts BC. Crystal structures of Salinosporamide A (NPI-0052) and B (NPI-0047) in complex with the 20S proteasome reveal important consequences of beta-lactone ring opening and a mechanism for irreversible binding. *J. Am. Chem. Soc.* 2006;128:5136-41.

[149] Tsukamoto S. The search for inhibitors of the ubiquitin–proteasome system from natural resources for drug development. *Journal of Natural Medicines* 2006;60:273-8.

[150] Sharp KH, Davidson SK, Haygood MG. Localization of '*Candidatus Endobugula sertula*' and the bryostatins throughout the life cycle of the bryozoan *Bugula neritina. The ISME Journal* 2007;1:693-702.

[151] Banerjee S, Wang Z, Mohammad M, Sarkar F, Mohammad R. Efficacy of selected natural products as therapeutic agents against cancer. *J. Nat. Prod.* 2008;71:492-6.

[152] Wender P, Cribbs C, Koehler M, Sharkey N, Herald CL, Kamano Y, et al. Modeling of the bryostatins to the phorbol ester pharmacophore on protein kinase C. *Proc. Natl. Acad. Sci. USA* 1988;85:7197-201.

[153] Stone RM, Sariban E, Pettit GR, Kufe DW. Bryostatin 1 activates protein kinase C and induces monocytic differentiation of HL-60 cells. *Blood* 1988;72:208-13.

[154] La Porta CA, Comolli R. PKC-dependent modulation of IκBα-NFκB pathway in low metastatic B16F1 murine melanoma cells and in highly metastatic *BL6 cells. Anticancer Res.* 1998;18:2591-7.

[155] Mortenson MM, Galante JG, Gilad O, Schlieman MG, Virudachalam S, Kung HJ, et al. BCL-2 functions as an activator of the AKT signaling pathway in pancreatic cancer. *J. Cell Biochem.* 2007;102:1171-9.

[156] Silberman DM, Zorrilla-Zubilete M, Cremaschi GA, Genaro AM. Protein kinase C-dependent NF-κB activation is altered in T cells by chronic stress. *Cell Mol. Life Sci.* 2005;62:1744-54.

[157] Folmer F, Harrison WT, Tabudravu JN, Jaspars M, Aalbersberg W, Feussner K, et al. NF-kappaB-inhibiting naphthopyrones from the Fijian echinoderm Comanthus parvicirrus. *J. Nat. Prod.* 2008;71:106-11.

[158] Lee SJ, Bai SK, Lee KS, Namkoong S, Na HJ, Ha KS, et al. Astaxanthin inhibits nitric oxide production and inflammatory gene expression by suppressing I kappa B kinase-dependent NF-kappa B activation. *Molecules and Cells* 2003;16:97-105.

[159] Stevenson CS, Capper EA, Roshak AK, Marquez B, Eichman C, Jackson JR, et al. The identification and characterization of the marine natural product scytonemin as a novel antiproliferative pharmacophore. *J. Pharmacol. Exp. Ther.* 2002;303:858-66.

[160] Terracciano S, Rodriquez M, Aquino M, Monti MC, Casapullo A, Riccio R, et al. Chemistry and biology of anti-inflammatory marine natural products: molecules interfering with cyclo-oxygenase, NF-κB and other unidentified targets. *Curr. Med. Chem.* 2006;13:1947-69.

[161] Kamata K, Okamoto S, Oka S, Kamata H, Yagisawa H, Hirata H. Cycloprodigiosin hydrocloride suppresses tumor necrosis factor (TNF) alpha-induced transcriptional activation by NF-kappa B. *FEBS Lett* 2000;507:74-80.

[162] Choi IK, Shin HJ, Lee HS, Kwon HJ. Streptochlorin, a marine natural product, inhibits NF-kappaB activation and supresses angiogenesis in vitro. *Journal of Microbiology and Biotechnology* 2007;17:1338-43.

[163] Oda T, Namikoshi K, Akamo K, Honma Y, Kasahara T. Verrucarin A inhibition of MAP kinase activation in a PMA-stimulated promyelocytic leukemia cell line. *Mar. Drugs* 2005;3:64-73.

[164] Posadas I, De Rosa S, Terencio MC, Paya M, Alcaraz MJ. Cacospongionolide B suppresses the expression of inflammatory enzymes and tumour necrosis factor-alpha by inhibiting nuclear factor-kappa B activation. *Br. J. Pharmacol.* 2003;138:1571-9.

[165] Posadas I, Terencio MC, Randazzo A, Gomez-Paloma L, Paya M, Alcaraz MJ. Inhibition of the NF-kappa B signaling pathway mediates the anti-inflammatory effects of petrosaspongiolide M. *Biochem. Pharmacol.* 2003;65:887-95.

[166] D'Acquisto F, Lanzotti V, Carnuccio R. Cyclolinteinone, a sesterterpene from sponge *Cacospongia linteiformis*, prevents inducible nitric oxide synthase and inducible cyclo-oxygenase protein expression by blocking nuclear factor-kappa B activation in J774 macrophages. *Biochem. J.* 2000;346:793-8.

[167] Amigo M, Paya M, Braza-Boïls A, De Rosa M, Terencio MC. Avarol inhibits TNF-alpha generation of NF-kappaB activation in human cells and in animal models. *Life Sci.* 2008;82:256-64.

[168] Breton JJ, ChabotFletcher MC. The natural product hymenialdisine inhibits interleukin-8 production in U937 cells by inhibition of nuclear factor-kappa B. *J. Pharmacol. Exp. Ther.* 1997;282:459-66.

[169] Roshak A, Jackson JR, Chabot Fletcher M, Marshall LA. Inhibition of NF kappa B-mediated interleukin-1 beta-stimulated prostaglandin E-2 formation by the marine natural product hymenialdisine. *J. Pharmacol. Exp. Ther.* 1997;283:955-61.

[170] Sharma V, Lansdell TA, Jin G, Tepe JJ. Inhibition of cytokine production by hymenialdisine derivatives. *J. Med. Chem.* 2004;47:3700-3.

[171] Lu PH, Chuch SC, Kung FL, Pan SL, Shen YC, Guh JH. Ilimaquinone, a marine metabolite, displays anticancer activity via GADD153-mediated pathway. *Eur. J. Pharmacol.* 2007;556:45-54.

[172] Lucas R, Casapullo A, Ciasullo L, Gomez-Paloma L, Paya M. Cycloamphilectenes, a new type of potent marine diterpenes: inhibition of nitric oxide production in murine macrophages. *Life Sci.* 2003;72:2543-52.

[173] Rawat DS, Joshi MC, Atheaya H. Marine peptides and related compounds in clinical trial. *Anticancer Agents and Medicinal Chemistry* 2006;6:33-40.

[174] D'Incalci M, Simone M, Tavecchio M, Damia G, Garbi A, Erba E. New drugs from the sea. *J. Chemotherapy* 2004;16 Suppl 4:86-9.

[175] Newman DJ, Cragg GM. Advanced preclinical and clinical trials of natural products and related compounds from marine sources. *Curr. Med. Chem.* 2004;11:1693-713.

[176] Schwartsmann G, Da Rocha AB, Mattei J, Lopes R. Marine-derived anticancer agents in clinical trials. *Expert Opin. Invest Drugs* 2003;12:1367-83.

[177] Jimeno J, Faircloth G, Fernandez Sousa-Faro JM, Scheuer PJ, Rinehart KL. New marine derived anticancer therapeutics - a journey from the sea to clinical trials. *Mar. Drugs* 2004;2:14-29.

[178] Haefner B. Drugs from the deep: marine natural products as drug candidates. *Drug Disc. Today* 2003;8:536-44.

[179] Jaspars M. Pharmacy of the Deep - Marine organisms as sources of anticancer agents. In: Harvey A, editor. *Advances in Drug Discovery Techniques*. London, UK: John Wiley and Sons Ltd., 1998. p. 65-84.

[180] Sipkema D, Osinga R, Schatton W, ., Mendola D, Tramper J, Wijffels RH. Large-scale production of pharmaceuticals by marine sponges: sea, cell, or synthesis? *Biotechnol. Bioeng.* 2005;90:201-22.

[181] De Rosa S, De Caro S, Iodice C, Tommonaro C, Stefanov K, Popov S. Development in primary cell culture of demosponges. *J. Biotechnol.* 2003;100:119-25.

[182] Jensen PR, Mincer TJ, Williams PG, Fenical W. Marine actinomycete diversity and natural product discovery. *Antonie Van Leeuwenhoek* 2005;87:43-8.

[183] Newman D, Hill R. New drugs from marine microbes: The tide is turning. *J. Ind. Microbiol. Biotechnol.* 2006;33:539-44.

[184] Mutter R, Wills M. Chemistry and clinical biology of the bryostatins. *Bioorg. Med. Chem.* 2000;8:1841-60.

[185] Neidle S, Thurston DE. Chemical approaches to the discovery and development of cancer therapies. *Nat. Rev. Cancer* 2005;5:285-96.

[186] Macherla VR, Mitchell SS, Manam RR, Reed KA, Chao TH, Nicholson B, et al. Structure-activity relationship studies of salinosporamide A (NPI-0052), a novel marine derived proteasome inhibitor. *J. Med. Chem.* 2005;48:3684-7.

[187] Reed KA, Manam RR, Mitchell S, Xu J, Teisan S, Chao T, et al. Salinosporamides D-J from the marine actinomycete *Salinispora tropica*, bromosalinosporamide, and

thioester derivatives are potent Inhibitors of the 20S proteasome. *J. Nat. Prod.* 2007;70:269-76.

[188] Ahn KS, Sethi G, Chao TH, Neuteboom ST, Chaturvedi MM, Palladino MA, et al. Salinosporamide A (NPI-0052) potentiates apoptosis, suppresses osteoclastogenesis, and inhibits invasion through down-modulation of NF-κB-regulated gene products. *Blood* 2007;110:2286-95.

[189] Feling RH, Buchanan GO, Mincer TJ, Kauffman CA, Jensen PR, Fenical W. Salinosporamide A: a highly cytotoxic proteasome inhibitor from a novel microbial source, a marine bacterium of the new genus *Salinospora. Angew Chem.* 2003;42:355-7.

[190] Newman D, Cragg G. Recent progress of natural marine products as anticancer drugs. *International Oncology Updates: Marine anticancer compounds in the Era of targetet therapies* 2008.

[191] Reddy RL, Saravan P, Corey EJ. A simple stereocontrolled synthesis of salinosporamide A. *J. Am. Chem. Soc.* 2004;126:6230-1.

[192] Hogan P, Corey EJ. Proteasome Inhibition by a Totally Synthetic β-Lactam Related to Salinosporamide A and Omuralide. *J. Am. Chem. Soc.* 2005;127:15386-7.

[193] Endo A, Danishefsky SJ. Total synthesis of salinosporamide A. *J. Am. Chem. Soc.* 2005;127:8298-9.

[194] Reddy LR, Fournier JF, Reddy BV, Corey EJ. New synthetic route for the enantioselective total synthesis of salinosporamide A and biologically active analogues. *Org. Lett 2005*;7:2699-701.

[195] Takahashi K, Midori M, Kawano K, Ishihara J, Hatakeyama S. Entry to heterocycles based on indium-catalyzed conia-ene reactions: asymmetric synthesis of (-)-salinosporamide A. *Angew Chem.* 2008;*in press* (PMID 18600821).

[196] Udwary DW, Zeigler L, Asolkar RN, Singan V, Lapidus A, Fenical W, et al. Genome sequencing reveals complex secondary metabolome in the marine actinomycete *Salinospora tropica*. *Proc. Natl. Acad. Sci. USA* 2007;104:10376-81.

[197] Eustaquio AS, Pojer F, Noel JP, Moore BS. Discovery and characterization of a marine bacterial SAM-dependent chlorinase. *Nature Chem. Biol.* 2008;4:69-74.

[198] Aikawa S, Matsuzawa F, Satoh Y, Kadota Y, Doi H, Itoh K. Prediction of the mechanism of action of omuralide (clasto-lactacystin beta-lactone) on human cathepsin A based on a structural model of the yeast proteasome beta5/PRE2-subunit/omuralide complex. *Biochim. Biophys Acta* 2006;1764:1372-80.

[199] Wender P, Hinkle K, Koehler M, Lippa M. The rational design of potential chemotherapeutic agents: Synthesis of bryostatin analogues. *Med. Res. Rev.* 1999;19:388-407.

[200] Wall NR, Mohammad RM, Reddy KB, Al-Katib AM. Bryostatin 1 induces ubiquitination and proteasome degradation of Bcl-2 in the human acute lymphoblastic leukemia cell line, Reh. *Int. J. Mol. Med.* 2000;5:165-71.

[201] Evans DAC, P. H.; Carreira, E. M.; Charette, A. B.;, Prunet JAL, M. Total Synthesis of Bryostatin 2. *J. Am. Chem. Soc.* 1999;121:7540-52.

[202] Kageyama MT, T.; Nantz, M. H.; Roberts, J. C.; Somfai, P.; Whritenour DCM, S. Synthesis of bryostatin 7. *J. Am. Chem. Soc.* 1990;112:7407-8.

[203] Manaviazar S, Frigerio M, Bhatia GS, Hummersone MG, Aliev AE, Hale KJ. Enantioselective formal total synthesis of the antitumor macrolide bryostatin 7. *Org. Lett* 2006;8:4477-80.
[204] Ohmori KO, Y.; Obitsu, T.; Ishikawa, Y.; Nishiyama, S.;, Yamamura S. Total Synthesis of Bryostatin 3. *Angew Chem,* Int Ed 2000;39:2290-4.
[205] Mendola D. Aquaculture of three phyla of marine invertebrates to yield bioactive metabolites: process developments and economics. *Biomol. Eng.* 2003;20:441-58.
[206] Piel J. Bacterial symbionts: prospects for the sustainable production of invertebrate-derived pharmaceuticals. *Curr. Med. Chem.* 2006;13:39-50.
[207] Hildebrand M, Waggoner LE, Liu H, Sudek S, Allen S, Anderson C, et al. bryA: an unusual modular polyketide synthase gene from the uncultivated bacterial symbiont of the marine bryozoan Bugula neritina. *Chem. Biol.* 2004;11:1543-52.
[208] Bogi K, Lorenzo PS, Szallasi Z, Acs P, Wagner GS, Blumberg PM. Differential selectivity of ligands for the C1a and C1b phorbol ester binding domains of protein kinase Cdelta: possible correlation with tumor-promoting activity. *Cancer Res.* 1998;58:1423-8.
[209] Statsuk AV, Bai R, Baryza JL, Verma VA, Hamel E, Wender PA, et al. Actin is the primary cellular receptor of bistramide A. *Nature Chem. Biol.* 2005;1:383-8.
[210] Wender P, DeBrabander J, Harran P, Jimenez J, Koehler M, Lippa B, et al. The design, computer modeling, solution structure, and biological evaluation of synthetic analogs of bryostatin 1. *PNAS* 1998;95:6624-9.
[211] Wender PA, Koehler KF, Sharkey NA, Dell'Aquila ML, Blumberg PM. Analysis of the phorbol ester pharmacophore on protein kinase C as a guide to the rational design of new classes of analogs. *Proc. Natl. Acad. Sci. USA* 1986;83:4214-8.
[212] Wender PA, Cribbs CM, Koehler KF, Sharkey NA, Herald CL, Kamano Y, et al. Modeling of the bryostatins to the phorbol ester pharmacophore on protein kinase C. *Proc. Natl. Acad. Sci. USA* 1988;85:7197-201.
[213] Wender PA, Mayweg AV, VanDeusen CL. A concise, selective synthesis of the polyketide spacer domain of a potent bryostatin analogue. *Org. Lett.* 2003;5:277-9.
[214] Wender PA, Dechristopher BA, Schrier AJ. Efficient synthetic access to a new family of highly potent bryostatin analogues via a Prins-driven macrocyclization strategy. *J. Am. Chem. Soc.* 2008;130:6658-9.
[215] Wender PA, Verma VA. Design, synthesis, and biological evaluation of a potent, PKC selective, B-ring analog of bryostatin. *Org. Lett.* 2006;8:1893-6.
[216] Wender PA, Baryza JL. Identification of a tunable site in bryostatin analogs: C20 Bryologs through late stage diversification. *Org. Lett.* 2005;7:1177-80.
[217] Baryza JL, Brenner SE, Craske ML, Meyer T, Wender P. Simplified analogs of bryostatin with anticancer activity display greater potency for translocation of PKCδ-GFP. *Chemistry and Biology* 2004;11:1261-7.
[218] Brodie C, Blumberg PM. Regulation of cell apoptosis by protein kinase c delta. *Apoptosis* 2003;8:19-27.
[219] Kikkawa U, Matsuzaki H, Yamamoto T. Protein kinase C delta (PKC delta): activation mechanisms and functions. *J. Biochem.* 2002;132:831-9.
[220] Wender PA, Baryza JL, Bennett CE, Bi FC, Brenner SE, Clarke MO, et al. The practical synthesis of a novel and highly potent analogue of bryostatin. *J. Am. Chem. Soc.* 2002;124:13648-9.

[221] Wender PA, Clarke MO, Horan JC. Role of the A-ring of bryostatin analogues in PKC binding: synthesis and initial biological evaluation of new A-ring-modified bryologs. *Org. Lett.* 2005;7:1995-8.

[222] Lallemend F, Hadjab S, Hans G, Moonen G, Lefebvre PP, Malgrange B. Activation of protein kinase CbetaI constitutes a new neurotrophic pathway for deafferented spiral ganglion neurons. *J. Cell Sci.* 2005;118:4511-25.

[223] Wang QJ, Fang TW, Fenick D, Garfield S, Bienfait B, Marquez VE, et al. The lipophilicity of phorbol esters as a critical factor in determining the pattern of translocation of protein kinase C delta fused to green fluorescent protein. *J. Biol. Chem.* 2000;275:12136-46.

[224] Wender PA, DeBrabander J, Harran PG, Jimenez JM, Koehler MF, Lippa B, et al. The design, computer modeling, solution structure, and biological evaluation of synthetic analogs of bryostatin 1. *Proc. Natl. Acad. Sci. USA* 1998;95:6624-9.

[225] Wender PA, DeBrabander J, Harran PG, I-Iinke KW, Lippa B, Pettit GR. Synthesis and Biological Evaluation of Fully Synthetic Bryostatin Analogues. *Tetrahedron Lett* 1998;39:8625-8.

[226] Wender PA, Horan JC. Synthesis and PKC binding of a new class of a-ring diversifiable bryostatin analogues utilizing a double asymmetric hydrogenation and cross-coupling strategy. *Org. Lett.* 2006;8:4581-4.

[227] Keck GE, Kraft MB, Truong AP, Li W, Sanchez CC, Kedei N, et al. Convergent assembly of highly potent analogues of bryostatin 1 via pyran annulation: bryostatin look-alikes that mimic phorbol ester function. *J. Am. Chem. Soc.* 2008;130:6660-1.

[228] Breton JJ, Chabot-Fletcher MC. The natural product hymenialdisine inhibits interleukin-8 production in U937 cells by inhibition of nuclear factor-kappaB. *J. Pharmacol. Exp. Ther.* 1997;282:459-66.

[229] Roshak A, Jackson JR, ChabotFletcher M, Marshall LA. Inhibition of NF kappa B-mediated interleukin-1 beta-stimulated prostaglandin E-2 formation by the marine natural product hymenialdisine. J. *Pharmacol. Exp. Ther.* 1997;283:955-61.

[230] Annoura H, Tatsuoka T. Total Syntheses of Hymenialdisine and Debromohymenialdisine: Stereospecific Construction of the 2-Amino-4-Oxo-2-Imidazolin-5(Z)-Disubstituted Ylidene Ring System. *Tetrahedron Letters* 1995;36:413-6.

[231] Meijer L, Thunnissen AM, White AW, Garnier M, Nikolic M, Tsai LH, et al. Inhibition of cyclin-dependent kinases, GSK-3beta and CK1 by hymenialdisine, a marine sponge constituent. *Chem. Biol.* 2000;7:51-63.

[232] Imahori K, Uchida T. Physiology and pathology of tau protein kinases in relation to Alzheimer's disease. J. *Biochem.* 1997;121:179-88.

[233] Mandelkow EM, Mandelkow E. Tau in Alzheimer's disease. *Trends Cell Biol.* 1998;8:425-7.

[234] Singh TJ, Grundke-Iqbal I, Iqbal K. Phosphorylation of tau protein by casein kinase-1 converts it to an abnormal Alzheimer-like state. *J. Neurochem.* 1995;64:1420-3.

[235] Sielecki TM, Boylan JF, Benfield PA, Trainor GL. Cyclin-dependent kinase inhibitors: useful targets in cell cycle regulation. *J. Med. Chem.* 2000;43:1-18.

[236] Chacun-Lefèvre L, Joseph B, Mérour J-Y. Synthesis and reactivity of azepino[3,4-b]indol-5-yl trifluoromethanesulfonate. *Tetrahedron* 2000;56:4491-9.

[237] Kaiser HM, Zenz I, Lo WF, Spannenberg A, Schroder K, Jiao H, et al. Preparation of novel unsymmetrical bisindoles under solvent-free conditions: synthesis, crystal structures, and mechanistic aspects. *J. Org. Chem.* 2007;72:8847-58.
[238] Papeo G, Posteri H, Borghi D, Varasi M. A new glycociamidine ring precursor: syntheses of (Z)-hymenialdisine, (Z)-2-debromohymenialdisine, and (+/-)-endo-2-debromohymenialdisine. *Org. Lett.* 2005;7:5641-4.
[239] Xu Yz Y, Yakushijin K, Horne DA. Synthesis of C(11)N(5) Marine Sponge Alkaloids: (+/-)-Hymenin, Stevensine, Hymenialdisine, and Debromohymenialdisine. *J. Org. Chem.* 1997;62:456-64.
[240] Meijer L, Raymond E. Roscovitine and other purines as kinase inhibitors. From starfish oocytes to clinical trials. *Acc. Chem. Res.* 2003;36:417-25.
[241] Wan Y, Hur W, Cho CY, Liu Y, Adrian FJ, Lozach O, et al. Synthesis and target identification of hymenialdisine analogs. *Chem. Biol.* 2004;11:247-59.
[242] Tasdemir D, Mallon R, Greenstein M, Feldberg LR, Kim SC, Collins K, et al. Aldisine alkaloids from the Philippine sponge Stylissa massa are potent inhibitors of mitogen-activated protein kinase kinase-1 (MEK-1). *J. Med. Chem.* 2002;45:529-32.
[243] Meijer L, Thunnissen AM, White AW, Garnier M, Nikolic M, Tsai LH, et al. Inhibition of cyclin-dependent kinases, GSK-3beta and CK1 by hymenialdisine, a marine sponge constituent. *Chem. Biol.* 2000;7:51-63.
[244] Wan Y, Hur W, Cho CY, Liu Y, Adrian FJ, Lozach O, et al. Synthesis and target identification of hymenialdisine analogs. *Chem. Biol.* 2004;11:247-59.
[245] Sharma V, Tepe JJ. Potent inhibition of checkpoint kinase activity by a hymenialdisine-derived indoloazepine. *Bioorg. Med. Chem. Lett.* 2004;14:4319-21.
[246] Amigo M, Paya M, Braza-Boils A, De Rosa S, Terencio MC. Avarol inhibits TNF-alpha generation and NF-kappaB activation in human cells and in animal models. *Life Sci.* 2008;82:256-64.
[247] Amigo M, Paya M, De Rosa S, Terencio MC. Antipsoriatic effects of avarol-3'-thiosalicylate are mediated by inhibition of TNF-alpha generation and NF-kappaB activation in mouse skin. *Br. J. Pharmacol.* 2007;152:353-65.
[248] Amigo M, Terencio MC, Mitova M, Iodice C, Paya M, De Rosa S. Potential antipsoriatic avarol derivatives as antioxidants and inhibitors of PGE(2) generation and proliferation in the HaCaT cell line. *J. Nat. Prod.* 2004;67:1459-63.
[249] Schroder HC, Wenger R, Gerner H, Reuter P, Kuchino Y, Sladic D, et al. Suppression of the modulatory effects of the antileukemic and anti-human immunodeficiency virus compound avarol on gene expression by tryptophan. *Cancer Res.* 1989;49:2069-76.
[250] Loya S, Hizi A. The inhibition of human immunodeficiency virus type 1 reverse transcriptase by avarol and avarone derivatives. *FEBS Lett.* 1990;269:131-4.
[251] Muller WE, Sladic D, Zahn RK, Bassler KH, Dogovic N, Gerner H, et al. Avarol-induced DNA strand breakage in vitro and in Friend erythroleukemia cells. *Cancer Res.* 1987;47:6565-71.
[252] Muller WE, Maidhof A, Zahn RK, Schroder HC, Gasic MJ, Heidemann D, et al. Potent antileukemic activity of the novel cytostatic agent avarone and its analogues in vitro and in vivo. *Cancer Res.* 1985;45:4822-6.
[253] Müller WEG, Böhm M, Batel R, De Rosa S, Tommonaro G, Müller IM, et al. Application of cell culture for the production of bioactive compounds from sponges: synthesis of avarol by primmorphs from *Dysidea avara*. *J. Nat. Prod.* 2000;63:1077-81.

[254] Sarma AS, Chattopadhyay P. Synthetic studies of trans-clerodane diterponoids and congeners: stereocontrolled total synthesis of (±)-avarol. *J. Org. Chem.* 1982;47:1727-31.
[255] Sipkema D, Osinga R, Schatton W, Mendola D, Tramper J, Wijffels RH. Large-scale production of pharmaceuticals by marine sponges: sea, cell, or synthesis? *Biotechnol. Bioeng.* 2005;90:201-22.
[256] Dutcher JS, Macmillan JG, Heathcock CH. Pentacyclic Triterpene Synthesis. 5. Synthesis of Optically Pure Ring AB Precursors. *J. Org. Chem.* 1976;41:2663-9.
[257] Crispino A, de Giulio A, de Rosa S, Strazzullo G. A new bioactive derivative of avarol from the marine sponge Dysidea avara. *J. Nat. Prod.* 1989;52:646-8.
[258] Schmitz FJ, Lakshmi V, Powell DR, Van der Helm D. Arenarol and arenarone: sesquiterpenoids with rearranged drimane skeletons from the marine sponge Dysidea arenaria. *J. Org. Chem.* 1984;49:241-4.
[259] Belisario MA, Maturo M, Pecce R, De Rosa S, Villani GR. Effect of avarol and avarone on in vitro-induced microsomal lipid peroxidation. *Toxicology* 1992;72:221-33.
[260] Amigo M, Terencio MC, Paya M, Iodice C, De Rosa S. Synthesis and evaluation of diverse thio avarol derivatives as potential UVB photoprotective candidates. *Bioorg Med. Chem. Lett.* 2007;17:2561-5.
[261] Amigo M, Schalkwijk J, Olthuis D, De Rosa S, Paya M, Terencio MC, et al. Identification of avarol derivatives as potential antipsoriatic drugs using an in vitro model for keratinocyte growth and differentiation. *Life Sci.* 2006;79:2395-404.
[262] De Rosa S, Crispino A, De Giulio A, Iodice C, Benrezzouk R, Terencio MC, et al. A new cacospongionolide inhibitor of human secretory phospholipase A2 from the Tyrrhenian sponge *Fasciospongia cavernosa* and absolute configuration of cacospongionolides. *J. Nat. Prod.* 1998;61:931-5.
[263] Garcia Pastor P, De Rosa S, De Giulio A, Paya M, Alcaraz MJ. Modulation of acute and chronic inflammatory processes by cacospongionolide B, a novel inhibitor of human synovial phospholipase A2. *Br. J. Pharmacol.* 1999;126:301-11.
[264] Murelli RP, Cheung AK, Snapper ML. Conformationally restricted (+)-cacospongionolide B analogues. Influence on secretory phospholipase A2 inhibition. *J. Org. Chem.* 2007;72:1545-52.
[265] Anthonsen MW, Solhaug A, Johansen B. Functional coupling between secretory and cytosolic phospholipase A2 modulates tumor necrosis factor-alpha- and interleukin-1beta-induced NF-kappa B activation. *J. Biol. Chem.* 2001;276:30527-36.
[266] Potts BC, Faulkner DJ, Jacobs RS. Phospholipase A2 inhibitors from marine organisms. *J. Nat. Prod.* 1992;55:1701-17.
[267] Palanki MS, Erdman PE, Gayo-Fung LM, Shevlin GI, Sullivan RW, Goldman ME, et al. Inhibitors of NF-kappaB and AP-1 gene expression: SAR studies on the pyrimidine portion of 2-chloro-4-trifluoromethylpyrimidine-5-[N-(3', 5'-bis(trifluoromethyl) phenyl)carboxamide]. *J. Med. Chem.* 2000;43:3995-4004.
[268] De Rosa M, Giordano S, Scettri A, Sodano G, Soriente A, Pastor PG, et al. Synthesis and comparison of the antiinflammatory activity of manoalide and cacospongionolide B analogues. *J. Med. Chem.* 1998;41:3232-8.
[269] Cheung AK, Snapper ML. Total syntheses of (+)- and (-)-cacospongionolide B: new insight into structural requirements for phospholipase A(2) inhibition. *J. Am. Chem. Soc.* 2002;124:11584-5.

[270] Cheung AK, Murelli R, Snapper ML. Total syntheses of (+)- and (-)-cacospongionolide B, cacospongionolide e, and related analogues. Preliminary study of structural features required for phospholipase a2 inhibition. *J. Org. Chem.* 2004;69:5712-9.
[271] Potts BC, Faulkner DJ, Jacobs RS. Phospholipase A2 inhibitors from marine organisms. *J. Nat. Prod.* 1992;55:1701-17.
[272] Potts BC, Faulkner DJ, De Carvalho MS, Jacobs RS. Chemical mechanism of inactivation of bee venom phospholipase A2 by the marine natural products manoalide, luffariellolide, and scalaradial. *J. Am. Chem. Soc.* 1992;114:5093-100.
[273] Glaser KB, de Carvalho MS, Jacobs RS, Kernan MR, Faulkner DJ. Manoalide: structure-activity studies and definition of the pharmacophore for phospholipase A2 inactivation. *Mol. Pharmacol.* 1989;36:782-8.
[274] De Rosa M, Giordano S, Scettri A, Sodano G, Soriente A, Pastor PG, et al. Synthesis and comparison of the antiinflammatory activity of manoalide and cacospongionolide B analogues. *J. Med. Chem.* 1998;41:3232-8.
[275] Wender PA, Horan JC, Verma VA. Total Synthesis and Initial Biological Evaluation of New B-Ring-Modified Bryostatin Analogs. *Org. Lett.* 2006; 8: 5299-302.
[276] Wender PA, Hilinski MK, Mayweg AV. Late-Stage Intermolecular CH Activation for Lead Diversification: A Highly Chemoselective Oxyfunctionalization of the C-9 Position of Potent Bryostatin Analogues. *Org. Lett.* 2005; 7:79-82.

In: New Developments in Medicinal Chemistry
Editors: Marta P. Ortega and Irene C. Gil

ISBN 978-1-60456-810-3

Chapter 8

LAYERED DOUBLE HYDROXIDES AND THEIR INTERCALATION COMPOUNDS IN PHOTO-CHEMISTRY AND IN MEDICINAL CHEMISTRY*

Umberto Costantino*† and Morena Nocchetti
Dipartimento di Chimica, Università degli Studi di Perugia, Via Elce di Sotto, 8, I-06123 Perugia, Italy

INTRODUCTION

Many efforts have been made in the past few years to build up, into the interlayer region of layered solids, supramolecular assemblies with special functionalities in the field of photochemistry, electrochemistry, molecular recognition, chiral recognition and catalysis.[1-5] Furthermore, the interlayer region of layered solids is starting to be used as a privileged reaction vessel to perform chemical reactions between the guest themselves (polymerisation reactions) or between the guests and the host (topotactic and grafting reactions).[6] In addition, the interlayer region of a layered solid may be considered a micro-container where guest species are stored, protected from oxidation or photolysis, and withdrawn for use by a chemical signal, i. e., by a deintercalation process.[4] Other interesting reactions performed with layered solids are "exfoliation reactions", that consist of separating the sheets of a layered compound into individual lamellae. This goal is reached with the aid of specific intercalation or de-intercalation reactions and leads to colloidal dispersion of lamellae. These dispersions can be used to obtain materials with a very high specific surface area useful in catalysis or films and thin layers with applications ranging from optical coating to microelectronics.[7]

As a consequence of these findings, research in intercalation chemistry has shifted from fundamental aspects related to the determination of insertion mechanisms and of the disposition of guest species in the interlayer region, to the preparation of new materials, not

---

* A version of this chapter was also published in *Layered Double Hydroxides: Present and Future*, edited by Vicente Rives published by Nova Science Publishers, Inc. It was submitted for appropriate modifications in an effort to encourage wider dissemination of research.

† Tel.**39.075.585.5565; Fax **39.075.585.5566; E-mail: ucost@unipg.it

obtainable through other synthetic procedures. Relatively "poor" materials such as natural clays, synthetic layered phosphates and oxychlorides of transition metal ions, acquire a great added value after intercalation of substrates with special functionalities. For example, insertion of dyes and chromophores into lamellar hosts has produced materials for use as tunable lasers, fluorescent, and non linear optics devices.[8]

Layered Double Hydroxides (hereafter LDHs) compare favourably with other lamellar hosts in the production of High-Tech materials for several reasons. First of all, they are practically the only class of lamellar hosts consisting of positively charged layers balanced by exchangeable anions (other are hydroxyacetates). Thus, functional organic or organo-metallic species with anionic groups (carboxylic, phosphonic, sulphonic) which, in general, are much more numerous than those with cationic groups, can be intercalated, via anion exchange processes, only in this class of layered hosts. Furthermore, LDHs are a versatile class of compounds in so far as it is possible to modulate their properties by changing the nature of the divalent(s) and/ or trivalent cations in the brucite-like sheet and it is possible to alter the charge density of the layers by altering the molar ratio of divalent to trivalent cation.[9-13] These considerations have induced many research groups to use LDHs as hosts for the intercalation of species with magnetic, photochemical, and optical proprerties with the aim of immobilizing these species on a solid support and/or of modulating their properties by confining them in a constrained medium. It is also relevant to recall that hydrotalcites Mg-Al or Mg-Fe are used in medicine as antiacids and against sideremia,[9] respectively, and that they can be used as drug-carriers, once the drug contains an anionic function and it has been inserted into the interlayer region. In this chapter we shall try to rationalize the work done for the application of LDHs in the field of photophysics and photochemistry and in medicinal chemistry, taking into account important reviews published recently.[8,12,13] For sake of clarity, we will initially recall some fundamental concepts of photochemistry.

Photophysical and Photochemical Processes in a Constrained Media

A full understanding of photo-processes requires the knowledge of the nature and properties of electronic excited states. A molecule with all electrons spin-paired possesses a total spin quantum number equal to zero and a spin multiplicity equal to one. Such an electronic ground state is denoted by the symbol S_0. When this molecule is suitably irradiated, electrons are promoted to the LUMO frontier orbitals. The excitation does not involve a change in the electron spin, the total spin quantum number is still zero and the excited state is a singlet state denoted S_1 , S_2 and so on. If there exists some means of changing the spin of the excited electron, an excited state is generated with a spin multiplicity of three, denoted as a triplet state and abbreviated by the symbol T_1, T_2, etc. For the most part, photochemistry of chomophores deals with the first excited singlet and triplet states.

A species that does not react with another substrate after excitation cannot persist in the excited state over time and de-excitation processes occur, the excess of energy being released as thermal (non radiative transitions) or radiation (radiative transitions) energy. Radiative transition is called fluorescence if it involves a de-excitation process $S_1 \rightarrow S_0$ and phosphorescence if it originates in the de-excitation of an excited state of spin multiplicity different from that of the ground state (i.e., $T_1 \rightarrow S_0$). Radiationless transitions involve processes that are called Internal Conversion (IC), if they occur between states of the same spin multiplicity, and Intersystem Crossing (ISC), if they occur between states of different spin multiplicity. Figure 1 reports a Jablonsky diagram that is a useful schematic portrayal of the possible transitions that may occur between the different energy levels of an excited species.[14]

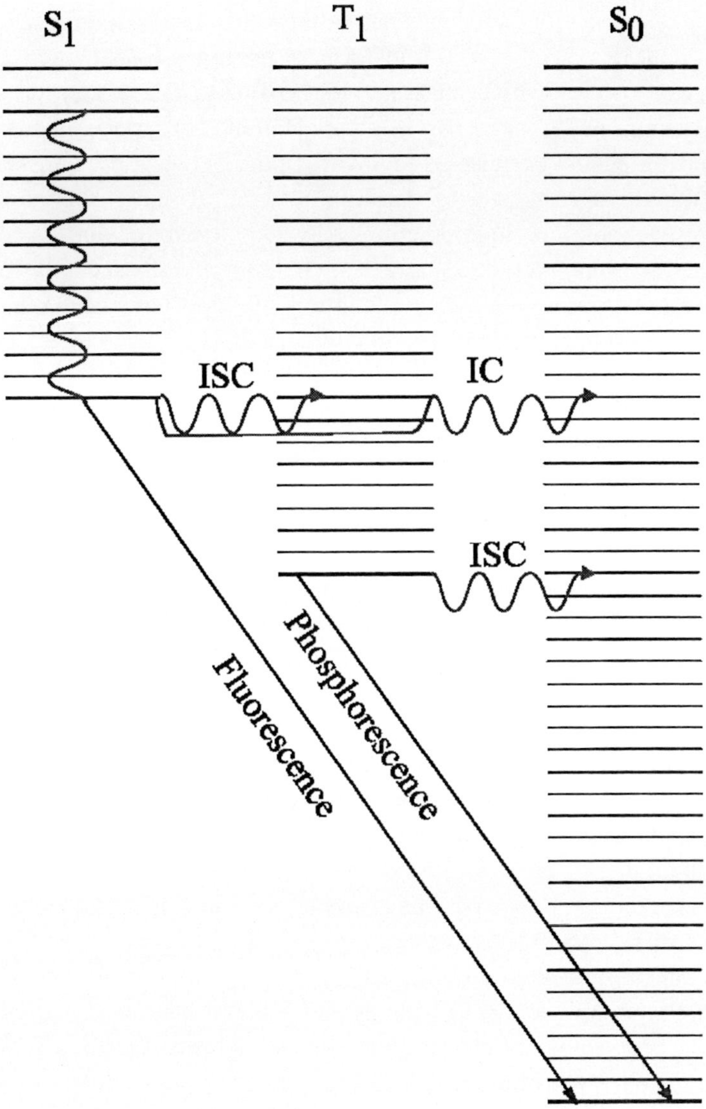

Figure 1. Jablonsky diagram showing the relative positions of the electronic energy levels of a molecole. (IC, internal conversion; ISC, intersystem crossing).

The processes above described are essentially photophysical processes. In many instances the light excitation of a species involves a chemical transformation with structural changes and even with formation of new products. Light excitation may induce cis-trans isomerisation, cycloaddition, electrocyclization, dissociation or formation of a tautomeric equilibrium. The species that, as a consequence of these structural rearrangements, change colour (or in general absorb light in a different spectral region) are called photochromic. Thus, photochromism may be represented by the simple schemes A + $h\nu_1 \rightarrow$ B; B +$\Delta(h\nu_2) \rightarrow$ A. The A species absorbs light of suitable frequency, undergoes a structural change and is converted into the B species. Photochromism is generally reversible in so far as the B species transforms into the A species because of light ($h\nu_2$) or heat (Δ) absorption.[15]

Another de-excitation path of a light-excited species is energy transfer to a reactive molecule, usually denoted as a quencher, that extinguishes the luminescence emission. In the condensed state, intermolecular energy transfer, at a relatively long range (up to 10 nm), can take place between a species, called sensitiser, that is able to absorb the excitation light, reach a singlet or triplet status and transfer part of the excitation energy to another species that emits light. Energy transfer occurs because of the overlapping between the emission spectra of the the sensitiser (the donor) and the absorption spectra of the acceptor. Energy transfer mechanisms involve dipole-dipole interactions between the excited donor and the non excited acceptor. In some cases the interaction between the donor and the acceptor allows an electron transfer with the formation of an ionic couple. Many attempts have been made to stabilise the charge separation, since the system obtained could be used for the storage and conversion of sunlight.

Finally, mention should be made of non linear optic effects shown by non centrosymmetric species when interacting with light. These materials produce a new light field that is different in wavelength or phase. For example, it has been found that many chromophores with a large molecular hyperpolarisability present the phenomenon of second harmonic generation (SHG), that is, they are able to convert a light signal into a ligth emission with frequency twice that of the absorbed light. This effect is of great importance in the development of optical systems for transmission and recording of data.[16]

All the above mentioned photophysical and photochemical effects of interaction of light with matter are highly sensitive to the microenvironment of the excited chromophore, so that the emission maximum of an excited species depends on its aggregation state. The fluorescence maximum of a dye in the microcrystalline state is red-shifted in comparison to that of the same dye in diluted solution. The energy difference between the fluorescence maxima has been attributed to the change of the emitting state energy in the crystal (compared to the single molecule) caused by coulombic or electron exchange type interactions of the excited species with the neighboring unexcited molecules. Hence, one can control fluorescence, photochromism, quenching, SHG and other properties of photoreactive species by organising them into matrices with appropriate geometry and chemical environment. With the development of inclusion chemistry and of supramolecular chemistry much research has been done on the preparation of inclusion compounds or composites in which the chromophores are immobilised in organic [17] (urea and thiourea, deoxycholic acid, cyclodextrins, micelles, Langmuir-Blodgett films and liquid crystals) and inorganic [1] (zeolites and related materials, lamellar solids and glasses) systems, in order to organise them into ordered structures, with a given orientation and chemical environment. Among the various systems, lamellar solids have a special role. At least two levels of organisation of guest species in lamellar solids may be envisaged: the first refers to the organisation on the surface of the microcrystals, the second to the arrangement in the interlayer region. Both depend on the nature and density of the active sites and on the shape and size of the guest. Figure 2 reports a pictorial representation of the possible arrangement of given species on the surface (see Fig. 2a and 2b) and in the interlayer region of LDHs with different charge densities. Fluorescence emission of a chromophore depends on its orientation and on the packing density in the interlayer region. These systems are candidates for solid state tunable lasers. It is possible to impose a given orientation on the molecular anions if the intercalation process is performed under an electric field and obtain materials with SHG proprerties are obtained (Fig.2g). It is also possible to cointercalate two different species, a donor and an acceptor, to obtain energy (or charge) transfer in the interlayer to produce systems for energy storage and conversion (Fig.2h).

Finally, it is evident that these systems offer guest sites of relatively restricted dimensions and therefore the photophysical and photochemical properties of the chromophore can be modified for a variety of reasons including restricted molecular motion, limited access of potential reactants or hydrophilic and hydrophobic effects. In turn, changes in the photophysical properties may provide important information on the microenvironment of the intercalated chromophore and thus on the microstructure of the host-guest system. Taking into account the above mentioned concepts, a review of the synthetic procedures and the characterisation of LDH-chromophore complexes recently published together with their potential applications will be made.

Synthetic Strategies to Obtain Intercalation Compound and Composite LDH-Chromophores

The insertion of a guest species into the interlayer region of a layered host is a fascinating phenomenon. The layers maintain their integrity but move apart in a direction perpendicular to the layer plane to give room to the incoming guests. It is not conceivable that the layered crystals open at once to accomodate the guest species. It is very likely that the layers bend and curl at the edge of the micro-crystals and the process goes on towards the core of the microcrystals with an advanced phase boundary mechanism. The rigidity of the layers will thus play an important role in the intercalation mechanism and in the energetics of the process. Layered crystals with charged layers contain exchangeable counterions in the interlayer region and the intercalation of a given guest will require an ion exchange reaction between the two species. Layered double hydroxides possess relatively rigid layers (the rigidity parameters has been evaluated to be 4.8 [18]), positively charged as a consequence of the isomorphous substitution of a divalent cation by a trivalent cation, and contain exchangeable anions.

The first way to obtain LDH-chromofore intercalated compounds is therefore to perform an anion exchange reaction. In designing the reaction, attention should be given to the nature of the counterion originally present in the LDH. Since chromophores are much bulkier anions than common counterions, it is important to choose a counterion that is weakly held and that determines a large gallery height to facilitate the diffusional processes. For the most common counterions, the selectivity scale [19] follows the order $CO_3^{2-} > SO_4^{2-} >> OH^- > F^- > Cl^- > Br^- > NO_3^- > ClO_4^-$. LDHs containing nitrate or perchlorate anions are, therefore, the most suitable precursors for the uptake of anionic dyes. They have also a relatively higher interlayer distance. LDHs containing chloride anions are also frequently used. LDHs in the carbonate form are excellent precursors for investigating the surface uptake of the chromophores, since the interlayer carbonates are not involved in the exchange process, whereas the surface carbonates or hydrogencarbonates are easily replaced by the dye.

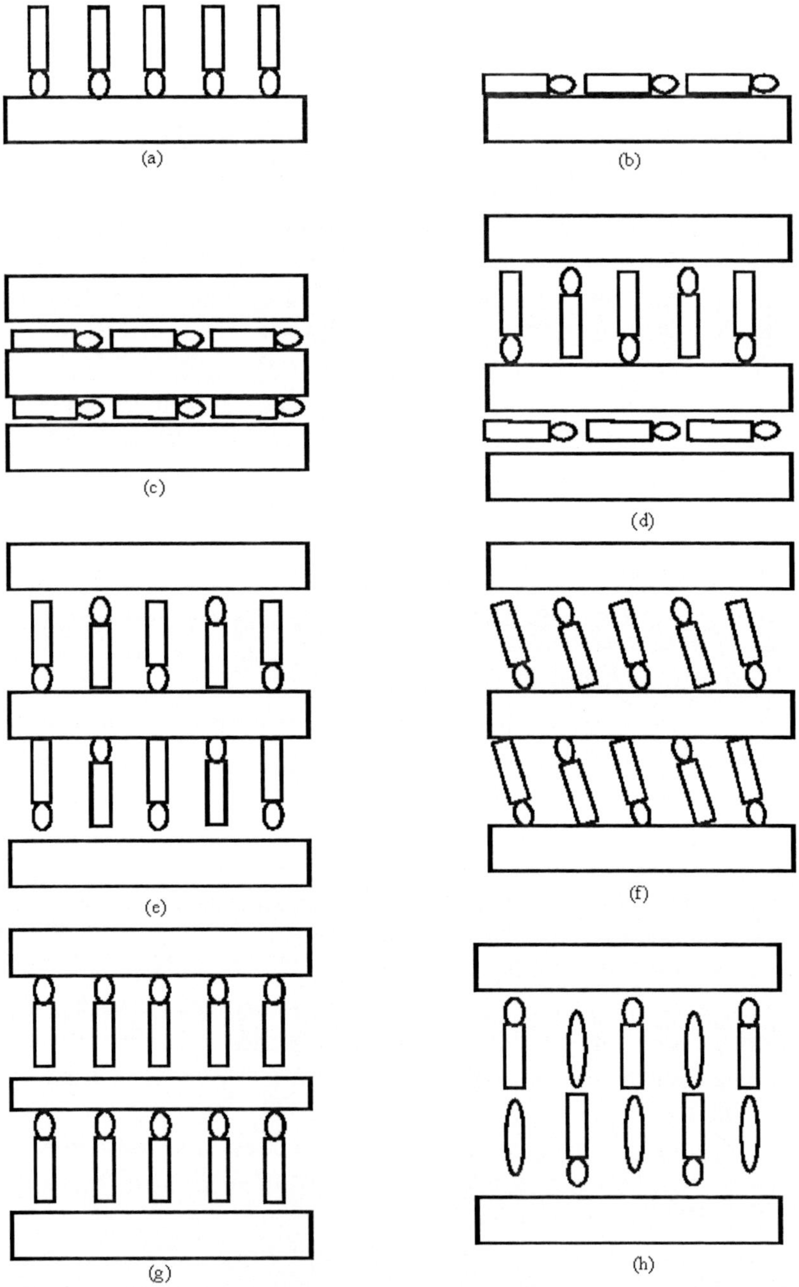

Figure 2. Possible arrangement of given species on the surface (a)and (b) and in the interlayer region of LDHs. Case (g) refers to an oriented disposition of guests; case (h) to the cointercalation of two different guests (e.g. donor-accepter).

To study the intercalation mechanism it is worth constructing the anion exchange isotherm,[20] in order to obtain the relative selectivity coefficient and to follow the structural changes occurring at different degrees of exchange, by recording the X-ray diffraction powder patterns of samples with different loadings. As an example, Figures 3a and 3b show the uptake curve of methylorange anions in Zn-Al-Cl LDH and the PXRD patterns of samples at different exchange percentages. It may be seen that the dye is exchanged with high selectivity and that the ion exchange process occurs with a first order phase transition from the Cl-phase (interlayer distance 7.74 Å) to the methylorange-phase (interlayer distance 24.2 Å).[21]

If high steric hindrance prevents direct exchange, two other procedures may be used to obtain the intercalation compound. The first procedure involves direct synthesis by coprecipitation, the second procedure is concerned with the so called "memory effect" of hydrotalcites.

The direct synthesis procedure requires the precipitation of the LDHs in the presence of the anionic form of the chromophore. To succeed in this, M(II) and M(III) cations must be introduced as salts of anions with low affinity for the host (nitrate, chloride, acetate) and the presence of strongly held anions such as carbonates or sulphates must be carefully avoided. Coprecipitation is performed at controlled pH (between 9 and 10) with the addition of NaOH solution to the solution containing the metal cations and the chromophores. The M(III)/M(II) molar ratio and hence the charge density of the obtained LDH must be chosen according to the size and charge of the chromophore. The cross section of the guest should not exceed the free area around each charge on the layer plane in order to avoid the "covering effect" and to obtain the fully converted LDH. Well crystallised samples are formed when the guests have a high self-assembly tendency. In other cases a hydrothermal treatment of the intercalates formed may improve the crystallinity of the products.[22]

The second procedure, typical of Mg-Al-CO_3 LDH, concerns the possibility of reconstructing the layered structure when conctacting the material previously calcinated at 300-500°C in water or in a solution of given anions.[23] Recent results indicate that at these temperatures the original hydrotalcite is converted into a mixture of magnesium and aluminum oxides and the oxides have "memory" of the original structure.[24] In fact, in the presence of water they regenerate the double hydroxides in the form of brucite-like sheets and the positive charges are balanced by OH^- ions. If the reconstruction is obtained in a solution containing anions, these could be incorporated to produce intercalation compound of LDH with this anion. An advantage of this method in comparison with the direct synthesis procedure is that the incorporation of competing inorganic counterions is avoided and that the chromophore may be present in solution as a free acid. Possible contamination by carbonates is still a problem and the whole procedure must be carried out in N_2 atmosphere. The procedure is not general. Among LDHs only Mg-Al and in some instances Zn-Al show the "memory effect" in a manner so clean as to obtain the intercalation compound as a single phase.

Table 1 summarises the literature data on the immobilation of different chromophores in LDHs. The table shows the type of host, the type of guest, the synthetic procedure used and, where available, the composition and the main photophysical and photochemical behaviour. It can be seen that the above mentioned synthetic strategies have been employed to obtain the incorporation into LDH of different types of chromophores. It is also interesting to note that, in spite of the large variety of LDHs presently available, the choice is limited to Mg-Al and Zn-Al. Both these LDHs are easily obtained with a good crystallinity degree, show the "memory effect" and are white. Some attention has also been paid to $[LiAl_2(OH)_6]^-$ which consists of sheets of aluminium octahedra with vacancies filled by lithium ions.[9]

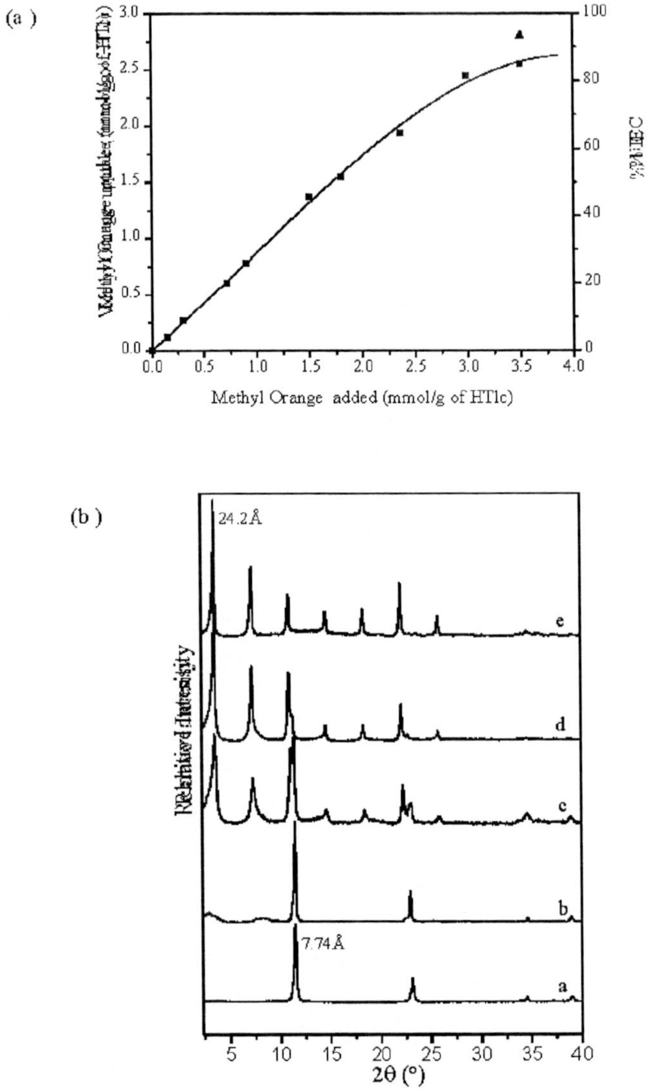

Figure 3. (a) Methylorange uptake of [Zn0.67Al0.33(OH)2]Cl0.33·0.6H2O as a function of the amount of methylorange offered in solution. (b) X-ray powder diffraction patterns of Zn-Al-LDH samples having increasing methylorange exchange percentage: (a) 4%, (b) 9%, (c) 46%, (d) 70%, (e) 84%. Reprinted from U. Costantino, N. Coletti, M. Nocchetti, G. G. Aloisi, F. Elisei, Anion exchange of methyl orange into Zn-Al synthetic hydrotalcite and photophysical characterization of the intercalates obtained, Langmuir 15 (1999) 4454, © 1999, with permission from The American Chemical Society.

In the following part, with reference to Table 1, the properties of the intercalation compounds or of the composites prepared will be discussed in relationship to the nature of the guest and hence of the photoprocesses investigated.

LDH - Anionic Dye Intercalation Compounds and their Fluorescence Properties

Relatively few papers deal with the preparation of composites containing anionic dyes. For the reader convenience, figure 4 shows the structural formula of the dyes considered. Miyata determined the anion exchange isotherm Cl^-/Naphthol YellowS^{2-} in a Mg-Al LDH. He found that the dye is exchanged with high selectivity and that it is accommodated in the interlayer region as a divalent anion with the main axis almost perpendicular to the layer plane. He suggested the use of LDH for the removal of this acid dye from waste solutions.[19]

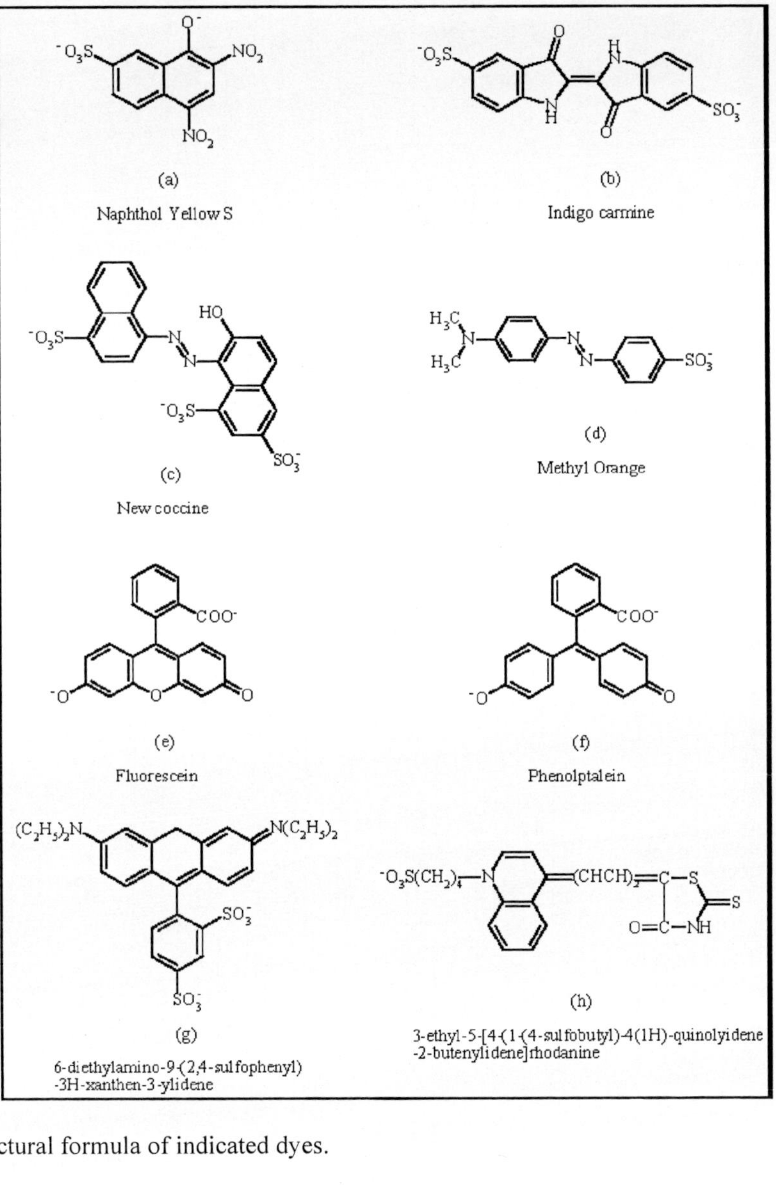

Figure 4. Structural formula of indicated dyes.

Table 1. Layered Double Hydroxides-chromophores intercalation compounds and composites. The synthetic procedure, composition and photofunctions of the obtained systems are also reported. References in round brackets.

LDH HOST	GUEST(S)	SYNTHETIC PROCEDURES	COMPOSITION	PHOTOFUNCTIONS (REFS)
Mg/Al	Naphthol Yellow S	Anion exchange	100% of AEC	(19)
Mg/Al	Indigo carmine, New coccine	Direct Synthesis and Anion exchange	75% of AEC	(25)
Zn/Al	Methyl Orange	Anion exchange	94% of AEC	Fluorescence (21)
Zn/Al	Fluorescein	Anion exchange	67% of AEC	Fluorescence (26)
Zn/Al	Phenolptalein	Anion exchange	54% of AEC	Fluorescence
Mg/Al and Zn/Al	Substituted Rhodanine and Xanthene dyes	Reconstruction	48-100% of AEC	Fluorescent Probe (27)
Mg/Al	Ruthenium polypyridine complex	Direct Synthesis	27% of AEC	Luminescent Probe (28)
Mg/Al	Dioxorhenium (V) ions	Direct Synthesis	15% of AEC	Excited-state properties (29)
Mg/Al	Derivative Porphyrin	Anion exchange	81% of AEC	(30)
Mg/Al	Derivative Porphyrin	Anion exchange	≈ 100% of AEC	Hole burning technology (31)
Mg/Al	Co(II) phtolocyanine	Reconstruction		Photocatalysis (33,34)
Mg/Al	Co(II), Cu and Zn(II) phthalocyanine	Reconstruction	40-99.5 μmol/g	Photooxydation of sulfur-containing compounds (35)
Mg/Al	Co(II) and Zn(II) phthalocyanine	Reconstruction	40 μmol/g	Photooxydation of sulfide and thiosulfate (photocalysis) (36)
Mg/Al	Mn(II) phtolocyanine	Reconstruction		Photocatalysis (37)
Li/Al/Myristate	Porphyrin-TiO_x	In situ hydrolysis of Ti butoxide		Photochemical Assemblies (38)
Li/Al/Myristate	Pyrene	Absorption		Photoluminescence (39,40)
Mg/Al	Na - reduced Fullerene	Anion exchange	1% of AEC	Photoluminescence (42)
Mg/Al-Dodecylsulfate	Fullerene	Absorption		Photoluminescence (41)
Li/Al	2-nitrohyppuric acid			Nonlinear optical properties (43)
Mg/Al	CdS-ZnS	Anion exchange	Precipitation "in situ"	Photocatalysts (44)
Mg/Al	Cinnamate	Anion exchange		Photodimerization and Photoisomerization (45)
Mg/Al	Arylacrylate, p-phenilcinnamate, 1-naphthylacrylate, Stilbenecarboxylate	Anion exchange	≈ 40% of AEC	Photochemical cycloaddition (46,47)
Mg/Al	Aromatic Ketocarboxylates	Anion exchange	68-92% of AEC	Photolysis (Norrish type reactions) (48)
Mg/Al	4-benzoylbenzoate, 4-(2-phenylethenyl) benzoate	Anion exchange	≈ 100% of AEC	Photochemical cycloaddition (49,50)
Mg/Al	4-vinylbenzoate, m- and p-phenylenediacrylates	Anion exchange	32-71% of AEC	Photopolymerization (51)
Mg/Al/p-toluensulfonate Zn/Al/p-toluensulfonate	Indolinespirobenzopyran	Reconstruction	1.3-7% (w/w)	(52)

Indigo Carmine (i.c.) and New coccine (n.c.), two dyes which are used as colour additives for food, have been intercalated in Mg-Al LDH by anion exchange and by the direct synthesis procedure. The PXRD patterns of the samples obtained following both procedures are very similar. The anionic dyes were easily exchanged, starting from the nitrate (or chloride) form of the LDH, probably because of the high tendency of these two dyes to be aggregated as a monofilm of oriented species in the interlayer region. The values of the interlayer distance (22.4 Å for i.c. and 16.5 Å for n.c.) support this hypothesis. The U.V. and visible reflectance spectra of the intercalation compounds are similar to those of the pure dyes in the solid state, suggesting that the guests are packed in the interlayer region in a manner similar to that of the solids.[25]

The azoic dye methylorange has been intercalated in Zn-Al-Cl LDH by anion exchange procedure and the ion exchange mechanism has been studied in detail (see figure 3a and 3b). A knowledge of the structure of the host and of the composition and interlayer distance of the intercalation compound has allowed the calculation of the packing coefficient of the dye in the interlayer region, expressed as the ratio between the volume occupied by the methylorange anions and the volume available to them in the interlayer space. The packing coefficient, 0.48, is very close to that of methylorange in the solid state (0.55). In addition, a computer-generated model of the intercalation compound has been proposed (see figure 5). The study of fluorescence emission spectra of samples containing the chromophore on the surface of the microcrystals or intercalated (in the presence or in the absence of co-intercalated water molecules) has shown a considerable red shift of the λ_{max} of fluorescence emission when compared with the λ_{max} of the free dye (see figure 6). If the fluorescence of methylorange microcrystals is considered, a wide range of wavelengths (400-850 nm) is covered. These characteristics are of interest for the design of materials for tunable dye lasers. Fluorescence response is highly sensitive to the microenvironment of the dye. It can be seen from figure 7 that the fluorescence maximum shifted to the red as a consequence of the dehydration of the sample and hence of the increase in the packing coefficient of the dye.[21]

Fluorescein and phenolphtalein, two xantenic dyes, are very different in shape and size from methylorange. In addition, they behave as divalent anions at pH values of 8-9. Because of their large size, intercalation by anion exchange required the use of precursors with a high interlayer distance. The perchlorate form, $[Zn_{0.67}Al_{0.33}(OH)_2]$ $[ClO_4]_{0.33}$ 0.6 H_2O, with an interlayer distance of 11.0 Å has been used to intercalate fluoresceinate anions. The anion exchange reaction occurs with a first-order phase transition from the ClO_4^- phase to a phase containing 0.11 mol of fluoresceinate per mol of the host and an interlayer distance of 16.5 Å. The benzoate form of Zn-AL LDH, with an interlayer distance of 15.5 Å, has been used to exchange phenolphtalein anions. The intercalation compound obtained contains 0.09 mol of guest per mol of host and has an interlayer distance of 19.4 Å. The surface uptake was investigated by equilibrating $[Zn_{0.67}Al_{0.33}(OH)_2]$ $[CO_3]_{0.165}$ $0.4H_2O$ with diluted solutions of the dyes. Only carbonates present on the surface of microcrystals were exchanged. Computer-generated models, based on the structure of the host, composition and interlayer distance of the intercalation compounds and van der Waals dimensions of guests, showed that both fluorescein and phenolphtalein anions are incorporated in the interlayer region as a bi-layer of species with the main axis parallel to the layer plane.[26]

The photophysical characterisation was performed by determining the absorption, excitation and fluorescence spectra and fluorescence lifetimes of the dyes under different experimental conditions. The emission properties of the guests, adsorbed or intercalated in the inorganic host, have been compared to those of dyes as pure solids or dissolved in solution. Again, the emission fluorescence spectra of intercalated dyes are shifted toward that of dyes in the solid state, while those of dyes adsorbed on the surface of the microcrystals are closer to those of free dyes in solution. There is thus the tendency of the dyes to be arranged in the interlayer region in a manner similar to that present in the crystals of the pure dyes. However, the intermolecular interactions responsible for the large red shift observed when the fluorescence maximum of the dye in solution is compared with that of the dye in the solid state, are lower in the intercalated dyes. Diffuse reflectance laser flash photolysis experiments performed with the fluorescinate systems has allowed the determination of absorption spectra and decay lifetimes of the triplet state of the dye bound to the inorganic matrix.[26]

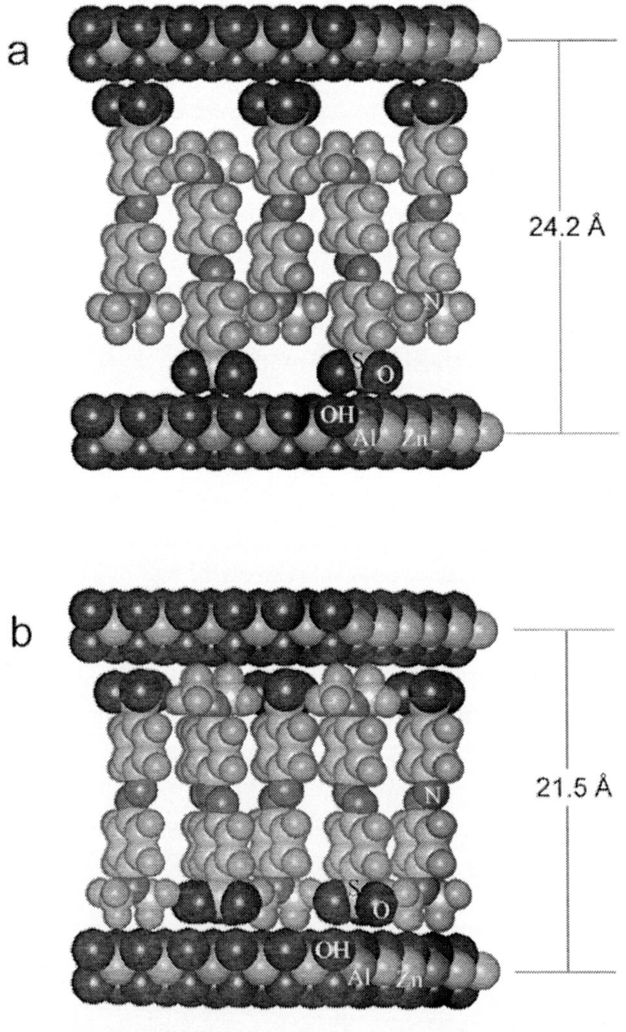

Figure 5. Computer-generated models showing the most probable arrangement of methylorange anions between the LDH layers. (a) Hydrated sample; (b) anhydrous sample. Reprinted from U. Costantino, N. Coletti, M. Nocchetti, G. G. Aloisi, F. Elisei, Anion exchange of methyl orange into Zn-Al synthetic hydrotalcite and photophysical characterization of the intercalates obtained, Langmuir 15 (1999) 4454, © 1999, with permission from The American Chemical Society.

It should be finally mentioned the intercalation of a substituted Xanthene and Rhodamine dyes (see Fig. 4g and 4h respectively) into Zn-Al-LDH by the reconstruction procedure. Photophysical studies on these composites have not been reported.[27]

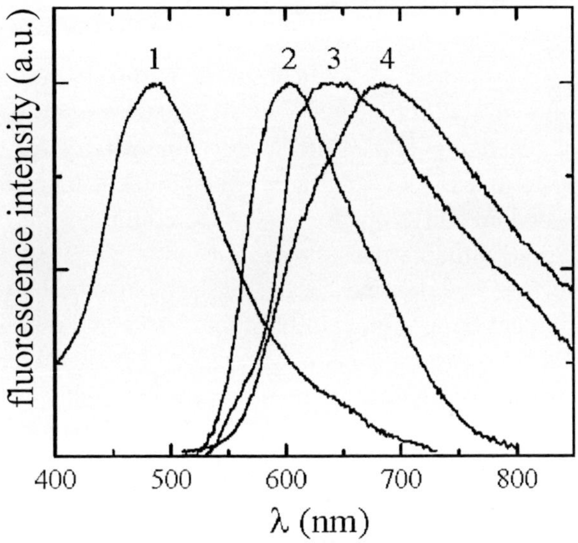

Figure 6. Normalised emission spectra of methyl orange (1) in ethanol solution, (2) absorbed on Zn-Al-LDH (sample a), (3) intercalated in Zn-Al-LDH (sample e) and (4) pure solid (λexc = 334 nm). Reprinted from U. Costantino, N. Coletti, M. Nocchetti, G. G. Aloisi, F. Elisei, Anion exchange of methyl orange into Zn-Al synthetic hydrotalcite and photophysical characterization of the intercalates obtained, Langmuir 15 (1999) 4454, © 1999, with permission from The American Chemical Society.

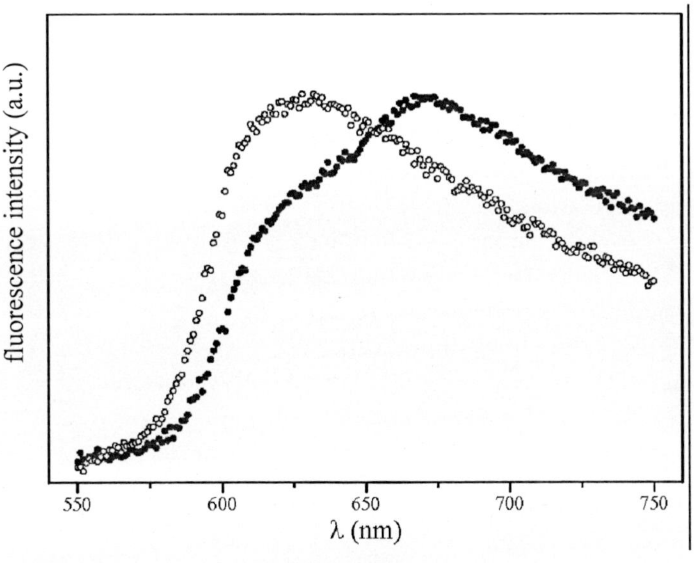

Figure 7. Normalised fluorescence spectra of methyl orange intercalated in Zn-Al-LDH (94%); anhydrous (full circles) and hydrated (open circles) samples (λexc = 434 nm). Reprinted from U. Costantino, N. Coletti, M. Nocchetti, G. G. Aloisi, F. Elisei, Anion exchange of methyl orange into Zn-Al synthetic hydrotalcite and photophysical characterization of the intercalates obtained, Langmuir 15 (1999) 4454, © 1999, with permission from The American Chemical Society.

LDH Intercalates Containing Luminescent Inorganic Complexes and their Photophysical Properties

The photochemistry of many complexes of transition metal ions with organic ligands has been thoroughly investigated to produce supramolecular metallocatalysts and supramolecular photochemical devices. Among these complexes, $[Ru(bpy)_3]^{2+}$ plays a special role. The complex is thermally stable and its photochemical excitation involves the promotion of an electron from a molecular orbital of essentially metal character to one of an essentially ligand character, after which its oxidation is possible. Many composite systems have been developed in which the ruthenium complex has been used for the photochemical decomposition of water for solar energy conversion and storage. Furthermore, the excited state of the Ru-bpy complex has a relative long lifetime that allows its complete photochemical characterisation (steady-state and time resolved luminescence studies) and a study of the effect of the chemical micro-environment on the fluorescent properties. The luminescence of $[Ru(bpy)_3]^{2+}$ incorporated in ion exchange resins, micelles, glass, zeolites and layered materials has been investigated and information at nanolevel scale has been obtained.[1]

The intercalation in LDHs requires the introduction of ligands with anionic functions. Figure 8a shows the structural formula of a Ru(II) complex with phenanthroline ligand functionalised with sulphonic groups. The $[Ru(BPS)_3]^{4-}$luminescent complex has been partially incorporated into Mg-Al-LDH by the co-precipitation procedure and subsequent hydrothermal treatment. The product obtained has the formula $[Mg_{2.98}Al_{1.04}(OH)_8]Cl_{0.72}$ $(Ru(BPS)_3)_{0.07}$ and an interlayer distance of 22 Å. This value can be predicted by the presence, in the interlayer region, of a monolayer of $[Ru(BPS)_3]^{4-}$ units with their C_3 axes normal to the layer plane. However, the Ru-complex moieties are not randomly distributed among the chlorides anions, but seem to be segregated at the edges of the crystallites. In fact, the emission decay profiles of the luminescent intercalate display multiexponential form due to excited-state self-quenching processes. Co-intercalation with the analogous Zn(BPS) complex causes a dilution of the emitting Ru complex with a diminution in the self quenching reaction rate.[28]

High-valent transition metal-oxo compounds such as trans-dioxorhenium(V) complexes have a highly emissive excited state. The Re(V) complex, shown in Figure 8b, has been intercalated by the co-precipitation procedure into Mg-Al-LDH. The interlayer distance of the intercalate (9.5Å) is consistent with the effective C_3 axis of the pseudo-octahedral $[ReO_2(CN)_4]^{3-}$ complex perpendicular to the layers. The complex does not show luminescent properties, in contrast to the strong fluorescence observed when a positive charge Re-complex is intercalated in hectorite. Probably, fast non radiative transitions take place in the LDH intercalate because the emitting ReO_2^+ core interacts, via hydrogen bonds, with the hydroxyl of the layers.[29] This is a significant example of the role of host-guest interactions on the excited state properties of an intercalated chromophore.

PORPHYRIN AND PHTALOCYANINE-LDH INTERCALATION COMPOUNDS AND THEIR PHOTO REACTIVITY

Porphyrins, phtalocyanines and their metallo complexes are of extraordinary importance in many fields from biology to material science. Metalloporphyrins are responsible in cytochromes for electron transport around the cell and in myoglobins and haemoglobins for molecular oxygen transport around the whole organisms. Phtalocyanines and their metallo complexes are used as dyes and efficient catalysts. They find application in material science for non linear optical devices, optical data storage, rectifying devices and as electrochromic substances. In addition, they are used in various processes involving visible light such as photo oxidation reactions in solution and photosensitizers. The organization of porphyrins and phtalocyanines in a constrain media, their proximity and relative orientation, may play an important role in modulating their attractive properties and much work has been done to intercalate (or include) these species in a variety of organic and inorganic hosts matrices.

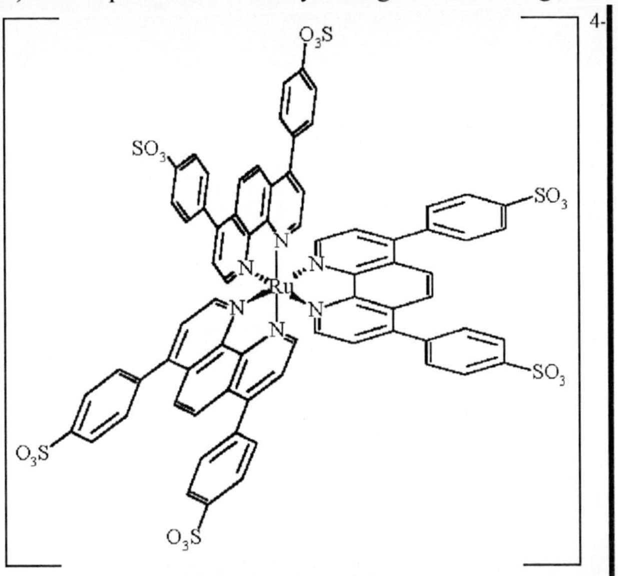

(a)

$Ru(BPS)_3^{4-}$

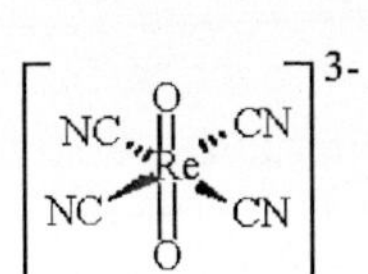

(b)

$ReO_2(CN)_4^{3-}$

Figure 8. Structural formula of some luminescent complexes.

Incorporation into LDHs requires the functionalisation of the molecules with anionic groups. The structural formulae of the derivative porphyrin and phtalocyanines employed are reported in Figure 9. Anionic (sulfonic and carboxylic) tetraphenylporphyrines have been incorporated with the three mentioned procedures into the interlayer region of Zn-Al-LDHs having different charge density. The interlayer distance of the products obtained ranges from 18.54 to 22.91 Å, indicating that porphyne anions can arrange in the interlayer region with different orientations according to the layer charge density of the particular LDH.[30]

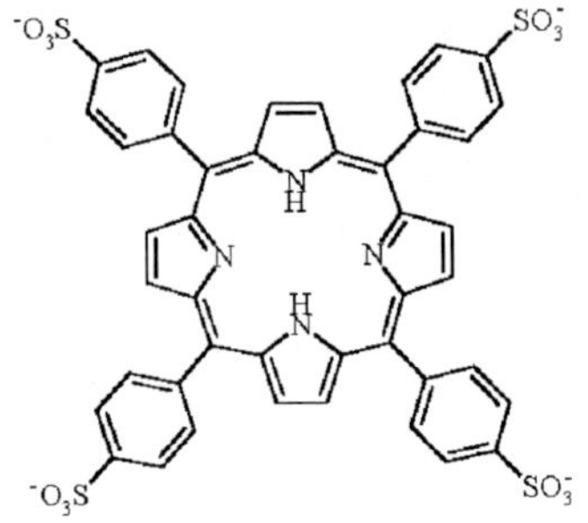

(a)

Tetra(4-sulfonatophenyl)porphyrin

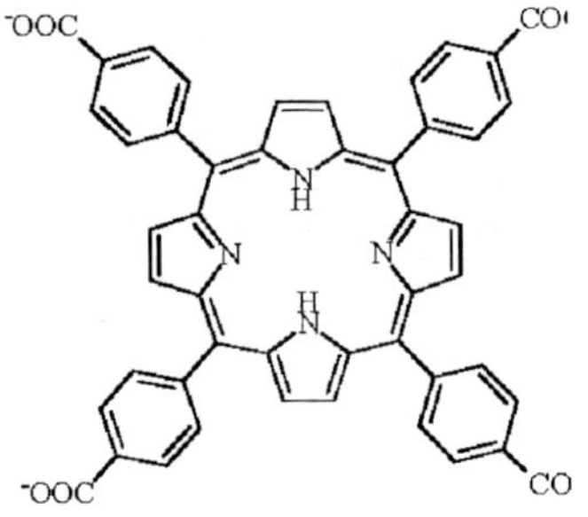

(b)

Tetra(4-carboxyphenyl)porphyrin

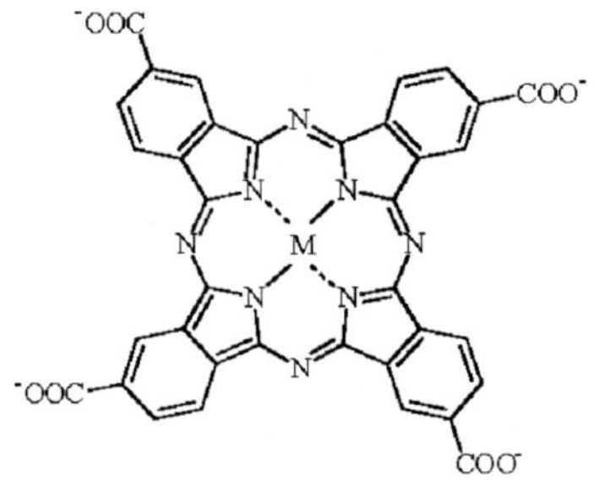

(c)

Metallo tetracarboxy- phthalocyanine (M=Co, Zn, Cu)

Figure 9. Structural formula of indicated porphyrin and phtalocyanines.

The anion of tetrasodium tetra(4-sulfonatophenyl) porphyne has been intercalated into a Mg-Al-Cl LDH by the anion exchange procedure. Almost complete replacement of the Cl^- counter-ions takes place and the interlayer distance of porphyrin-LDH intercalation compound was found to be 22.4 Å. This value is consistent with a model where the intercalated porphyrin anions are arranged in the interlayer with their molecular planes perpendicular to the hydroxide layers. The authors suggested the use of this intercalation compound in the photochemical hole burning technology.[31] The analogous tetra(4-carboxyphenyl) porphyrin has been intercalated into Mg-Al and Zn-Al LDHs by the reconstruction procedure. The compounds obtained were contaminated by the presence of a small amount of the carbonate phase, but the interlayer distance of the porphyrin phase was similar to that of the corresponding sulfonate derivative.[27] It is interesting to note that the intercalation by anion exchange of Ni(II) phtalocyaninetetrasulfonate ions into [Li

$Al_2(OH)_6]Cl$ produces an increase of the interlayer distance of 5.81Å, indicating that the phthalocyanines anions lie with their main plane parallel to the aluminate layer.[32]

Cu(II) and Co(II) phtalocyanine tetrasulfonates have been intercalated into Mg-Al LDH by the reconstruction procedure. The interlayer distance of the intercalation compounds is about 23 Å, suggesting again that guests are arranged with the molecular plane perpendicular to the sheet. ESR spectra of the Co-derivative seem to confirm the orientation deduced from the PXRD patterns.[33]

Catalytic studies indicate that these systems are efficient biomimetic catalysts. The Co phtalocyanine-LDH complex has been found to be an excellent heterogeneous catalyst for selective autoxidation of 2,6-di-*tert*-butylphenol in the presence of activated oxygen. The catalyst was still active after 3200 catalytic turnovers.[34] Zn and Co phtalocyanines intercalated into Mg-Al LDH by the reconstruction procedure have been used for the oxidation and photooxidation of tiols, sulfides and thiosulfates.[35,36] Mn(III) tetrakis(4-sulfonatophenyl)porphyrin intercalated by the reconstruction procedure into Mg-Al-LDH was found to be a good catalyst for cyclooctene epoxidation.[37]

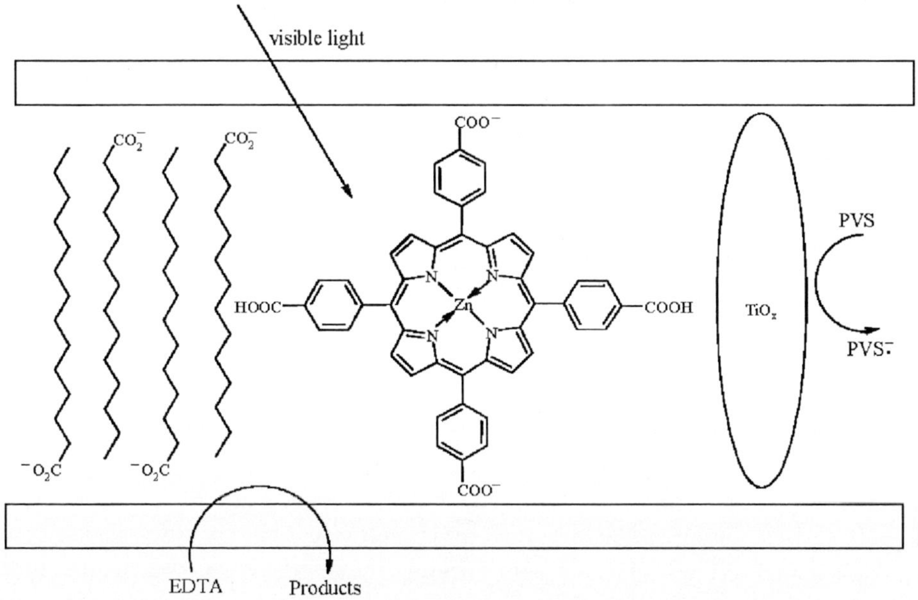

Figure 10. Scheme of the LDH photochemical assembly. Adapted from Ref. 38.

Li-Al-LDH, intercalated with myristate anions, $[Li\ Al_2(OH)_6](OOC(CH_2)_{12}CH_3)$, has been used to support a photochemical assembly for the photolysis of pollutants. The assembly, schematically depicted in Figure 10, consists of Zn-tetrakis(4-carboxyphenyl) porphyne and semiconducting TiO_x particles dispersed in myristate anions. Titanium(IV) butoxide was intercalated into the hydrophobic interlayer of the Li-Al-myristate. The interlayer distance increases from 21 to about 25 Å. Particles of TiO_x were grown "in situ" by exposing the myristate -Ti(IV) butoxide composite to a controlled hydrated environment in order to slowly hydrolyze the butoxide. The metallo-porphyrin was then introduced into the interlayer region of the Li-Al-LDH by exchanging some of the myristate anions with Zn-tetrakis(4-carboxyphenyl)porphyne anions. At the end of this procedure the composite contained 2% w/w of Zn and 6% w/w of Ti. The authors were not able to distinguish between TiO_x particles grown in the interlayer and those present on the surface of the LDH crystallites. However, they found that porphyrin is intercalated with the plane perpendicular to the layers. Furthermore, the emission spectra of the composite were very similar to those of the Zn-porphyrin in diluted solution, where no aggregation of porphyrin species occurs. This is a clear evidence that the Zn-porphyrin anions are randomly dispersed in the Li-Al-LDH-myristate phase. The efficiency of the photochemical assembly was tested with the photoreduction of viologens (a cationic viologen, heptylviologen, and a zwitterionic viologen, propylviologen sulphonate) in the presence of EDTA as sacrifical electron donor. The detection of viologen radical in solution was mainly attributed to the sensitised action of porphyrin on semiconducting titanium oxide particles.[38]

OTHER LDH-COMPOSITES WITH PHOTOLUMINESCENCE AND NON LINEAR OPTIC PROPERTIES

Interesting substrates with photoluminescence, non linear optics or semiconducting properties have been intercalated or adsorbed in LDHs and the photophysical behaviour of the nanocomposites obtained have been investigated.

The Li-Al-LDH exchanged with myristate anions (or exchanged with hexanoate anions) has the ability to partition pyrene from a methanol / water solutions containing this policyclic aromatic compound. No absorption of pyrene was observed in LDH exchanged with the dicarboxylate succinate anions. This observation suggests that the uptake of pyrene requires a hydrophobic environment having some degree of flexibility. In these conditions also the pyrene molecules retain enough mobility to form excimers, detected from the study of the emission spectra.[39,40]

Fullerene molecules have great interest in the field of chemistry, physics and material science. Several studies deal with the immobilation of these molecules, as such or suitably functionalized, in different host matrices, with the aim to study the effects of the microenvironment on the optoelectronic and chemical properties of the molecules and to prepare materials for application in optical and electronic devices. Attention of researchers has been also addressed to LDHs, and two different routes have been followed to incorporate C_{60} molecules into Mg-Al-LDH. In the first route, C_{60} molecules have been incorporated by dissolving the molecule into the hydrophobic interlayer region of LDH intercalated with dodecilsulfate. After heating under vacuum the resultant material to decompose the dodecyl sulfate, C_{60} molecules were sandwiched in between the hydroxide layers.[41] In the second route, fullerene was reduced with sodium to form anionic sites on the molecule and the reduced species were exchanged in hydrotalcite. The presence of a low intensity reflection at d =16.2 Å in the PXRD pattern of the material obtained seems to confirm that a small amount (about 1%) of anionic fullerene has been intercalated. Strong photoluminescence of C_{60} incorporated between LDH layers has been observed. This phenomenon has been attributed to the coulombic interaction of reduced fullerene and the positive charges of the matrix. This interaction alters the photophysical properties of the guest, reduces the inhibition of the electron transition and enhance the fluorescence emission.[42]

Non linear optical (NLO) materials provide the basis for an array of active optical devices under development in high-tech industries. A NLO chromophore should possess a high molecular hyperpolarizability and should be aligned in a non-centrosymmetric assembly, in order to present non linear phenomena such as second harmonic generation. Intercalation compounds in which the non centrosymmetric guests give rise to ordered arrangements of dipoles shall present the SHG phenomena. 2-nitrohyppuric acid, $NO_2C_6H_4\,CONHCH_2COOH$, does not present SHG in the solid state because of packing effects. It has been however found that the intercalation compound of Mg-Al with nitrohyppuric acid exhibits second harmonic generation: 532 nm radiation from incident 1064 nm radiation.[43] Very probably, nitrohyppuric moieties are accommodated in the interlayer region as a monolayer of oriented species to give rise a bulk dipole moment. Intercalation compounds showing SHG effects could be obtained if an external electric field is applied to the system during the intercalation procedure of a non-centrosymmetric chromophore. The guest species will be, in fact, aligned in the interlayer region with the orientation imposed by the electrical field.

The interlayer region of hydrotalcite has also been used to incorporate photo conducting particles of CdS and or of mixtures of CdS and ZnS particles. The co-intercalation of the $Cd(EDTA)^{2-}$ and S^{2-} anions allowed the "in situ" formation of the CdS (or CdS together with ZnS) particles. The composites, irradiated with visible light in the presence of sodium thiosulfate as sacrificial donor, were capable of efficient hydrogen evolution.[44]

Organic Photochemistry Inside the LDH Interlayer Region

Light was recognized early as powerful reagent for the development of organic synthesis. As already described, the excitation of a substrate by irradiation can produce photolytic breakage of bonds, with the formation of radicals and induces photoionization, isomerization or cyclization. It also greatly enhances the rate of reactions even at low temperatures. A key problem of organic photochemists is to perform photochemical reactions with high selectivity. The use of monochromatic radiations, or better, a laser source, allows the excitation of specific states with high precision and thus the direction of the reaction path. Another way to reach specificity and stereoselectivity is to perform photoreactions in organized media.[17] Numerous papers describe photoreactions in different host systems (urea and thiourea, cyclodextrins, micelles, zeolites, layered solids). The interlayer region of LDHs provides a particular microenvironment for reaction of photoactive species and classical photodimerization, photoisomerization and photocycloaddition reactions in LDHs have been reported.

The reactions of *trans*-cinnamic acid in the crystalline state are well known examples of [2+2] photodimerization and such reactions are strictly controlled by the packing arrangement of the molecules in the crystal. It was thus interesting to study the photodimerization of cinnamate anions when accommodated in the interlayer region of LDH, to compare the results with those obtained from solid sodium cinnamate and also to have information concerning the arrangement of cinnamate anions inside the hydroxides. The anions have been intercalated into a Mg-Al-Cl LDH by the anion exchange procedure and it has been found that both isomerization and dimerization of the cinnamate anions occurs between the layers.

The reaction path and the stereoselectivity suggest an arrangement of intercalated anions similar to that in the solid state.[45] [2+2] photocycloaddition of sodium arylacrylates may give rise to a mixture of syn head-to-head, anti head-to-head and syn head-to-tail cyclodimers (see Scheme 1). Various arylacrylates have been intercalated into Mg-Al-LDH by the ion exchange procedure and intercalation compounds have been irradiated with a Hg-lamp. It has been found that p-phenylcinnamate yielded almost exclusively head-to-head dimers, while phenylethenylbenzoates gave significant amounts of head-to-tail dimers. The product selectivity was shown to be controlled by the packing of the anions in the interlayer.[46] By using Mg-Al-LDHs with different charge density, it has been possible to change the packing density of the anions in the interlayer region and to change the selectivity. Low Al-content (x = 0.15-0.20) determines a relatively high distance between the positive charges and a low packing density which favours the formation of anti head-to-head dimers, while syn head-to-head dimers are mainly formed in LDH with high Al-content.[47]

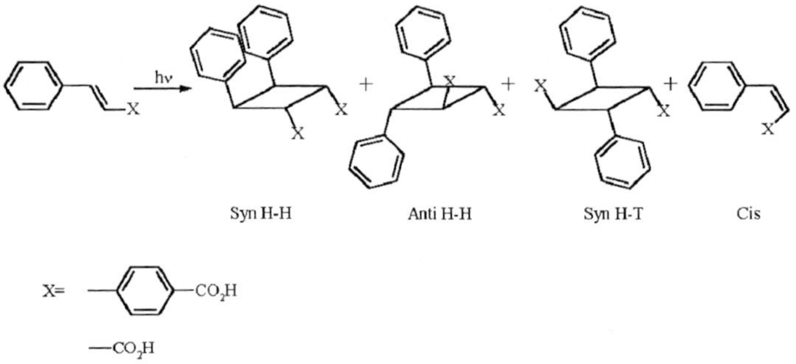

Scheme 1

Other photoreaction widely investigated is the unimolecular rearrangement of aromatic ketocarboxylates (Norrish type II reaction) reported in Scheme 2. In type II reactions, the 1,4-biradical obtained from the excited triplet state leads to cyclization to cyclobutanol or elimination to olefin and ketone *via* intramolecular γ-hydrogen abstraction. The ratio between cyclization and elimination products is significantly sensitive to the environment in which the ketone molecules are accommodated.

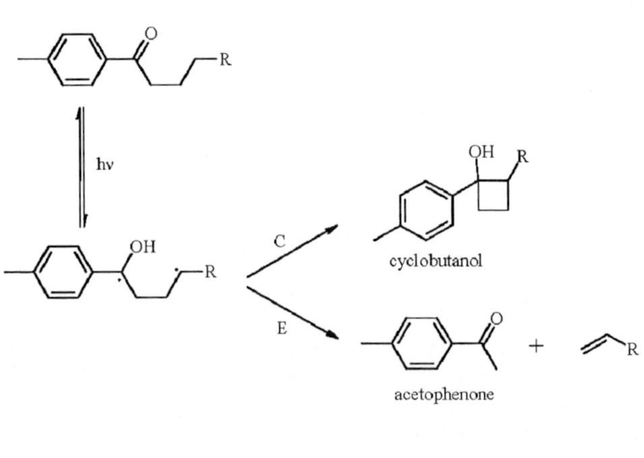

Scheme 2

Aromatic ketocarboxylates ($p\text{-}CH_3C_6H_4CO(CH_2)_nCO_2^-$, n = 4-10) were intercalated both by ion exchange and reconstruction procedures in Mg-Al-LDH. The degree of exchange increases with the increase of *n* and the interlayer distance of the intercalation compounds ranged from ca. 26 to 31 Å. It is likely that the ketocarboxylates are accommodated in the interlayer region as a monofilm of interdigitated species (see Figure 11). U.V. irradiation of the intercalation compounds produces p-methylacetophenone and alkenecarboxylates (elimination pathway) and cyclobutanols (cyclization pathway) in different ratios according to alkyl chain length. The formation of cyclized products was substantial for n=4 and n=5 and negligible for ketocarboxylates with longer chain length. Type II elimination products are constantly formed as major products. These results are different from those obtained in aqueous media, where the ratio of cyclization and elimination products is almost constant whichever of the chain length. These different results imply that the stereochemistry of 1,4-biradicals is significantly governed by the layered constrained media.[48]

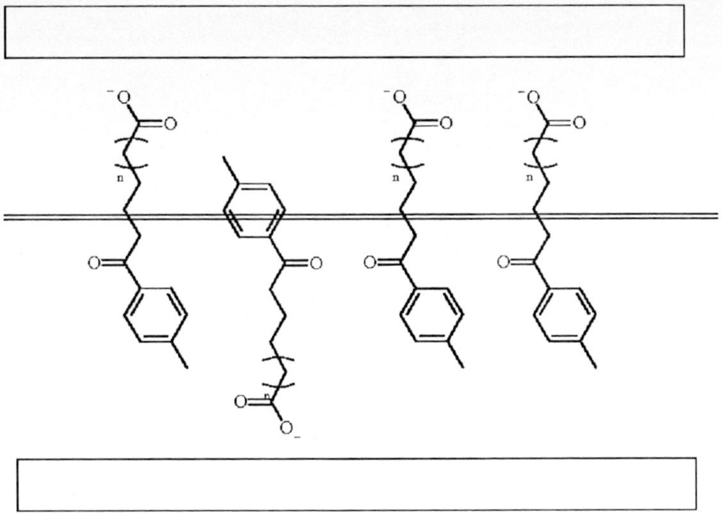

Figure 11. Intercalation of ketocarboxylate in the interlayer of LDH. Adapted from Ref. 48.

Other examples of photoreactions performed in the interlayer region of Mg-Al-LDH concern the photocycloaddition of 4-benzoylbenzoate with unsaturated carboxylates and the photochemical hydrogen abstraction of aliphatic carboxylates by benzoylbenzoate.[49] In both cases the reagents were co-intercalated in LDH by ion exchange procedures. Photoirradiation of a mixture of 4-benzoylbenzoate and 4-(2-phenylethenyl)benzoate yielded regioselectively oxetanes, while hydrogen abstraction from $C_nH_{2n-1}COO^-$ co-intercalated with 4-benzoylbenzoate gave rise to a series of 1:1 adducts with high selectivity.[50]

Well-ordered polymers may be obtained in the interlayer region of a layered solid if monomer guest molecules are regularly juxtaposed and if the polymerisation reaction is suitably promoted. 4-vinylbenzoate and *m*- and *p*- phenylenediacrylates and 4-benzoylbenzoate have been intercalated in Mg-Al-LDH in the chloride form, via anion exchange procedure. U.V.-light irradiation of the intercalation compound containing vinylbenzoate produces the formation of isomeric dimers, predominantly the *syn*-head-to-head cyclodimer. However, polymerisation of vinylbenzoate takes place in the presence of co-intercalated benzoylbenzoate. This latter guest generates 1,4-biradical when irradiated; probably the polymerisation is induced by the presence of this radical. The photolysis of *p*-

phenylenediacrylate intercalation compound produces consecutive [2+2] cycloaddition and oligomers have been obtained having a zig-zag structure [51] similar to that reported in Figure 12.

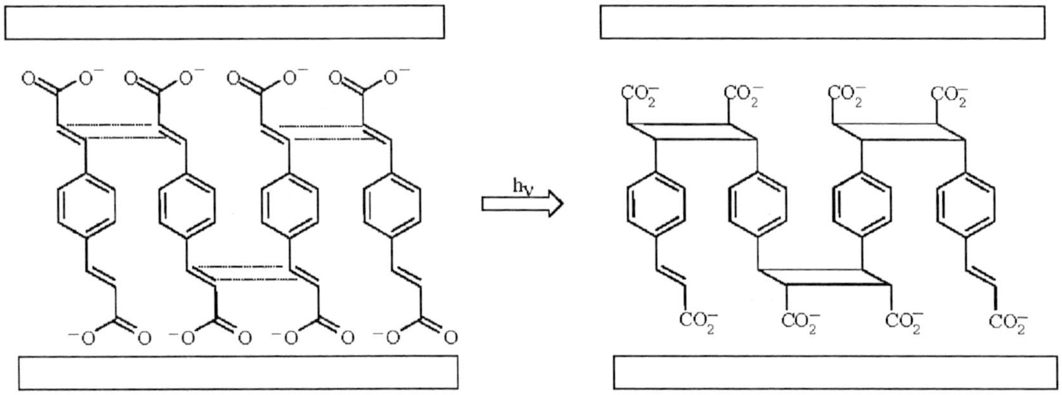

Figure 12. Scheme of the cycloadditive polymerization of the p-phenylenediacrylate anions in hydrotalcite interlayer. Adapted from Ref. 51.

The molecule of indolinespyrobenzopyran is constituted of two heterocyclic parts bounded with a central tetrahedral carbon atom. The two heterocyclic parts lie in two orthogonal planes. As a consequence of u.v.-light irradiation the spyrane C-O bond breaks down with formation of merocyanine, that has a planar structure with delocalizated π electrons. The merocyanine absorbs in the visible region, and thermally reconverts in indolinespyrobenzopyran (see Scheme 3). This system is a typical example of photochromic reaction which has been studied in different media. Attempts have been done to intercalate sulfonated indolinespyrobenzopyran into Mg-Al-LDH by the reconstruction procedure. The interlayer distance of the material obtained was about 7.8 Å. The low expansion of the interlayer region indicates that the sulfonated spyran was mainly adsorbed on the surface. This material showed photoisomerization to merocyanine; the process is irreversible, probably because merocyanine is stabilised in the polar environment of LDH. Co-intercalation of sulfonated indolinespyrobenzopyran and toluene p-sulfonic acid by the reconstruction procedure in slightly acidic medium, produced a material with an interlayer distance of about 25 Å. This composite contains a small amount of sulfonated indolinespyrobenzopyran dispersed into the toluene sulfonate anions present on the surface of the lamellae. In this apolar medium the composite system exhibited reversible chromotropic effect. The yellow powder consisting of spyran cointercalated in LDH-$CH_3C_6H_4SO_3^-$ turns to red, due to the formation of merocyanine, when irradiated with U.V.-light. The red colour changes to yellow for successive visible light irradiation. This cycle has been repeated several times.[52]

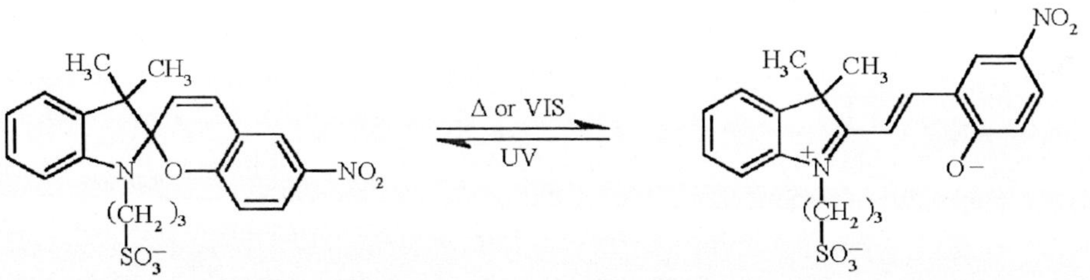

Scheme 3

LDHs in Medicinal Chemistry and in Pharmaceutical Formulations

Many inclusion compounds have found application in pharmaceutical formulations to modify the bioavailability of drugs. For example, β-cyclodextrins are excellent biocompatible hosts for the inclusion of a variety of guests with biological and pharmaceutical activity.[53] The inclusion in biocompatible hosts modifies the physical properties of the drugs with some advantages: non pleasant smells or flavors are masked, the solubility is increased, and the volatility reduced. Furthermore, the drugs can be protected from the action of light, oxygen or water vapor and can be selectively delivered to a given target and released in a modified way. It is also interesting to note that layered materials such as clays (kaolin, bentonite, montmorillonites) are listed in the pharmacopoeia of different countries for their use as excipients or adsorbents. Zirconium phosphate cartridges have been used in portable dialysis machines for the uptake of ammonia obtained from hydrolysis of urea present in blood.[54] In this context, natural and synthetic Mg-Al hydrotalcite finds larger application for its biocompatibility, chemical composition and ability to intercalate anionic drugs. Hydrotalcite and mixtures of similar composition are used in medicine as antacid and antipepsinic agents, and in many ointments and poultices for the protection of damaged skin. The most promising aspect, however, is the use of intercalation compounds with anionic drugs to obtain sustained release formulations. In the following we shall deal with these two aspects of hydrotalcite medical applications, separately.

Antacid and Anti-Pepsin Activity of Hydrotalcite Formulations

To digest food, stomach produces hydrochloric acid and proteolytic enzymes, notably pepsin. Ulcer is a sort of lesion or sore that forms in the lining of the stomach or duodenum where acid and pepsin are present. Both duodenal or stomach ulcers are extremely common diseases, and each year over 4 million people in US develop at least one type of ulcer and thousands of them have surgery because of severe complications. While for many years it was believed that inappropriate diet and stress lifestyle were cause of ulcer, it is now recognized that an imbalance between digestive fluids (HCl and pepsin, in particular) and the ability of the stomach to defend itself from them is the major cause of ulcer. Today, most ulcers are recognized to result from the infection by the *Helicobacter pylori* bacterium. It should be stressed, however, that the *H. pylori* infection produces a pattern of mucosal lesions that are compatible with a marked inability of the stomach to protect itself from acids and proteolytic enzymes.[55] Thus, most of the current therapies for stomach and duodenal ulcer are based on substances able to restore the physiological acidic and proteolytic environment of the stomach, preserving tissues from expanding lesions.[56] Classical prescriptions for ulcer are: (i) anti-H2 (ranitidine, cimetidine), whose effect is to block histamine, a powerful acid stimulant; (ii) proton-pump inhibitors, omeprazole, whose effect is to raise the pH of the stomach by completely inhibiting the final step of acid-production by the stomach. In association with antibiotics is the therapy of choice for *H. pylori*-induced ulcers. (iii) mucosal protective medicaments, which shield the stomach's mucous from damage of acids. (iv) antacids, which often produce substantial pain relief by neutralizing the stomach acidity but also by protecting the mucosal from lesions.

Available without medical prescription, after decades from their introduction, antacids are seeing a renewed interest as first-choice treatment in some ulcer diseases. Among classical antacids, aluminium-magnesium containing compounds (hydrotalcites) are those that display the best pharmacological profiles. For instance, it is known that hydrotalcites have the capability of not only neutralizing acids, but also to inhibit pepsin action at pH values where the pepsin activity is still high.[57-59] More recently, when evaluated in some acid-neutralizing and pH-buffering tests, hydrotalcite demonstrated to be endowed with the highest neutralizing ability and steadier buffering with respect carbonate compounds or magnesium oxide.[60] Furthermore, hydrotalcite has been shown to activate, in the gastric mucosa, genes encoding for the epidermal growth factor (EGF) and its receptor, thus providing the physiological basis for the mechanism of its ulcer healing action.[61] On this basis, many commercial antacid preparations (e.g. Talcid® Bemolan®, Almax®, Maalox®) contain hydrotalcite as the pharmacological active component. Moreover, researches have been carried out on the buffering capacity and efficacy with time of hydrotalcites, when compared with other antiacid preparations,[60,62-64] also taking into account the particle-size of the powders[65] and the presence of added pepsin.[66] Hydrotalcite has prolonged buffering action in an optimum pH range; moreover, the aluminium gastrointestinal adsorption is very low and its constipant action is balanced by the presence of magnesium ions. Many patents have been issued on antacid formulations containing hydrotalcite,[67-71] claiming a better stability because of the presence of eugenol, thymol and oil of cinnamon,[72] or a prolonged gastric residence time for the presence of hydrophobic species,[73] or improved antacid properties in the presence of dihydroxyaluminum aminoacetate,[74] or of polymeric surfactants.[75] Other studies deal with the interaction of drugs with antacids based on hydrotalcite. It has been found that treatment with antacid decreases the rate of penetration of the antibiotic clarithromycin through the gastric mucus layer.[76] Indomethacin, a non-steroidal anti-inflammatory drug, is adsorbed onto Riopan and Rioplus (U.S. Pharmacopoeia), two antacid formulations containing magaldrate, $Al_5Mg_{10}(OH)_{31}(SO_4)_2 2H_2O$. The adsorption mechanism has not been investigated; it has however been found that the bioavailablity of the drug is significantly modified for the contemporary administration of the two medicines.[77]

Many preparations contain Aspirin® and antacids in the same tablet. The stability of these formulations has been tested for long period at different temperatures and relative humidities. It has been found that preparations containing magaldrate have reduced shelf-life in comparison with preparations containing other antacids.[78]

Other Pharmaceutical Applications of Hydrotalcites

Apart from the above-mentioned applications, hydrotalcites are used in many preparations for their property to adsorb anions, some unpleasant odorous or fat materials, and for their alkaline character. In addition, hydrotalcite is often used as excipient in many formulations.

The great affinity of hydrotalcite, converted into the chloride form, for the uptake of phosphate anions[79,80] has suggested its use for the removal of these anions from the gastrointestinal fluid in the prevention of hyperphosphatemia, induced during hemodialysis.[81] Since hydrotalcites are soluble in the gastric fluid they should be used as drug for enteric coating. The removal of about 600 mg of phosphorus per day requires the administration of 3 g of hydrotalcite, three times a day.[81]

Hydrotalcite like compounds containing divalent or trivalent iron in the brucite sheets are considered drugs against sideremia.[82] Patents claim the use of hydrotalcites in dentifrices anti tooth-decay,[83] or as coating material of particles, e.g., nylon spherules, in topical cosmetic preparations,[84] or as a component of skin-moisturing patches containing also seaweed extracts for use in cosmetics and skin pharmaceuticals.[85] Moreover, hydrotalcites have been proposed as excipient to improve the compressibility of crude drugs,[86] in antifungal preparations[87,88] and in adhesives for transdermal delivery of oil soluble drugs.[89]

LDHs in Pharmaceutical Formulations and Drug Release

It is well known that the appropriate choice of a pharmaceutical formulation improves the physico-chemical properties of the active molecule and its bioavailability. The current trend of pharmaceutical technology requires formulations that are able to maintain pharmacologically active drug levels for long periods, avoiding repeated administrations and/or to localize the drug release in its pharmaceutical target. Different controlled drug delivery systems which provide a kinetic control or a spatial control of drug release have been designed and realized.[90] Most of these systems are based on "reservoir devices" constituted by a microporous membrane surrounding a core containing the drug or on "micro-reservoir devices" constituted by microdispersions of the drug encapsulated into a lipophilic polymer. Supramolecular assembly (vesicles, liposomes) with included drugs and β-cyclodextrins inclusion compounds are also employed.

Intercalation compounds of biocompatible layered hosts with pharmaceutically active species could provide other materials to design systems for modified drug release. The interlayer region of a layered host may be, in fact, considered a microvessel in which the drug is stored in an ordered way, while mantaining its individuality, and eventually protected from the action of light and oxygen. After administration of the intercalation compound, the drug may be released *via* a de-intercalation process, occurring because ion exchange or displacement reactions. The release rate is essentially dependent on the rate of the de-intercalation process.

Mg-Al LDHs are the most suitable hosts to prove this idea, because their biocompatibility, their ability to intercalate active molecules containing anionic functions and their collateral antiacid properties. Hydrotalcite has already been used, but only as an excipient, in sustained-release formulations containing nifedipine, an antihypertensive drug[91] or dextromethorphan, a psychoactive drug.[92] A more systematic work has been done and is still in progress on the intercalation of non steroidal active species possessing antiinflammatory action (NSAID).[93-96] Name, abbreviation and structural formula of active molecules taken into account are shown in Figure 13. It may be seen that each of them contains a carboxylic group that can be easily transformed into carboxylate, allowing an exchange process between the hydrotalcite conterions and the active molecules in the anionic form.

Ion exchange procedures have, in fact, been used to obtain the intercalation compounds of $[Mg_{0.67}Al_{0.33}(OH)_2]\ Cl_{0.67}0.6H_2O$ with the most of NSAID molecules. It has been found that hydrotalcite shows a marked preference for these species, probably because of their high tendency to aggregate as a compact monofilm in the interlayer region. Figure 14 shows the ion exchange isotherms of DIK^- and IBU^- anions that replace Cl^- from hydrotalcite. It can be seen that the selectivity towards DIK is very high, so that hydrotalcite could be employed for the recovery of the drug from diluted solutions. Table 2 reports composition, interlayer distance and the drug loading (w/w%) of the various intercalation compounds. It may be observed that Aspirin® readily transforms into salicilate (SA) anions as a consequence of intercalation, even if the process has been carried out in anhydrous acetone. It may be also observed that the values of the interlayer distance of the different intercalation compounds increase with the increasing of the size of the guests. Moreover, these values are consistent with the presence in the interlayer region of a monolayer of NSAID anions, partially interdigitated, with their principal axis almost perpendicular to the layer plane. As an example, Figure 15 shows the probable arrangement of DIK anions in the interlayer region of the intercalation compound. The computer-aided model was obtained taking into account the experimental data of interlayer distance and composition of the intercalate.[93]

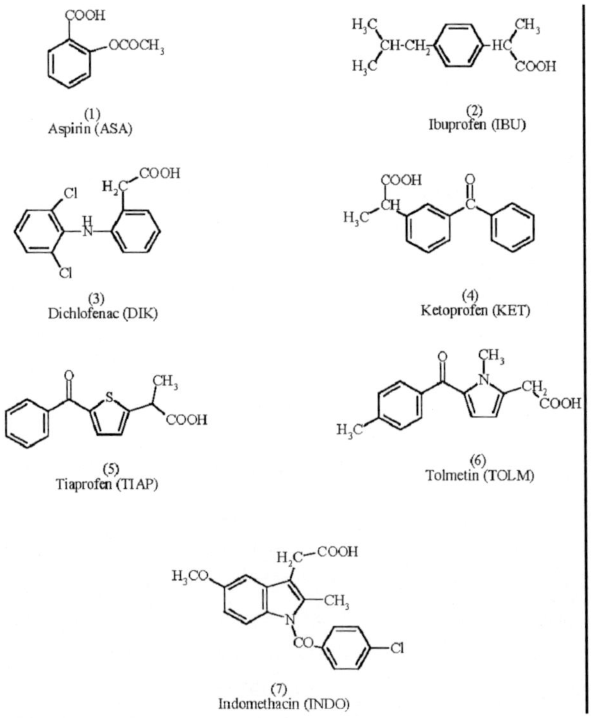

Figure 13. Structural formula of indicated non-steroidal anti-inflammatory drugs (NSAID).

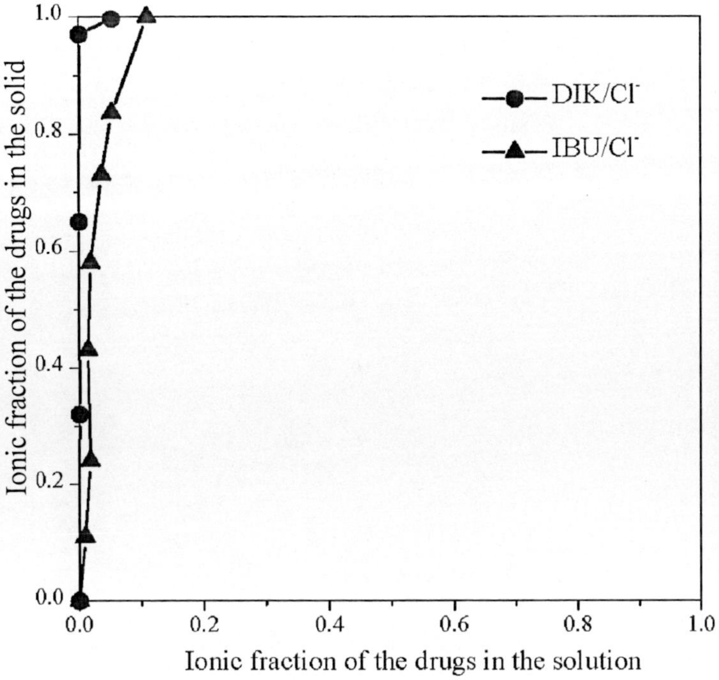

Figure 14. Anionic exchange isotherms of Mg-Al-Cl-LDH toward dichlofenac (DIK) (–□–) and ibuprofen (IBU) (–▲–) anions. Experimental conditions: Concentration 0.1M, Temperature 25°C, reaction time 3 days.

Table 2. Use of Mg-Al hydrotalcites in medicinal chemistry and in pharmaceutical formulation. References in round brackets.

Antacid and anti-pepsinic formulations (55-78)
Adsorbent of intestinal phosphates (prevention of hyperphosphatemia) (79-81)
Drugs for treatment of iron deficiency (sideremia) (82)
Dentifrices, Topical cosmetics and deodorants (83-85)
Excipients (86-89)
Drug delivery systems (91-96)

The intercalation compounds were encapsulated with an enteric coating and *in-vitro* drug release was performed with the paddle type dissolution apparatus,[96] U.S.P.XX ed., in a simulated intestinal fluid at pH = 7.5. The release profile of the Mg-Al-DIK is shown in Figure 16. The same figure reports, for sake of comparison, the release profile of a physical mixture of Mg-Al-Cl and dichlorofenac in the sodium form and that of a commercial controlled release formulation (Dealgic 100®). The dissolution test evidences that the release of the drug from the intercalation compound occurs in a modified way. It has also been found that the de-intercalation process is driven by DIK/phosphates ion exchange reaction.[94]

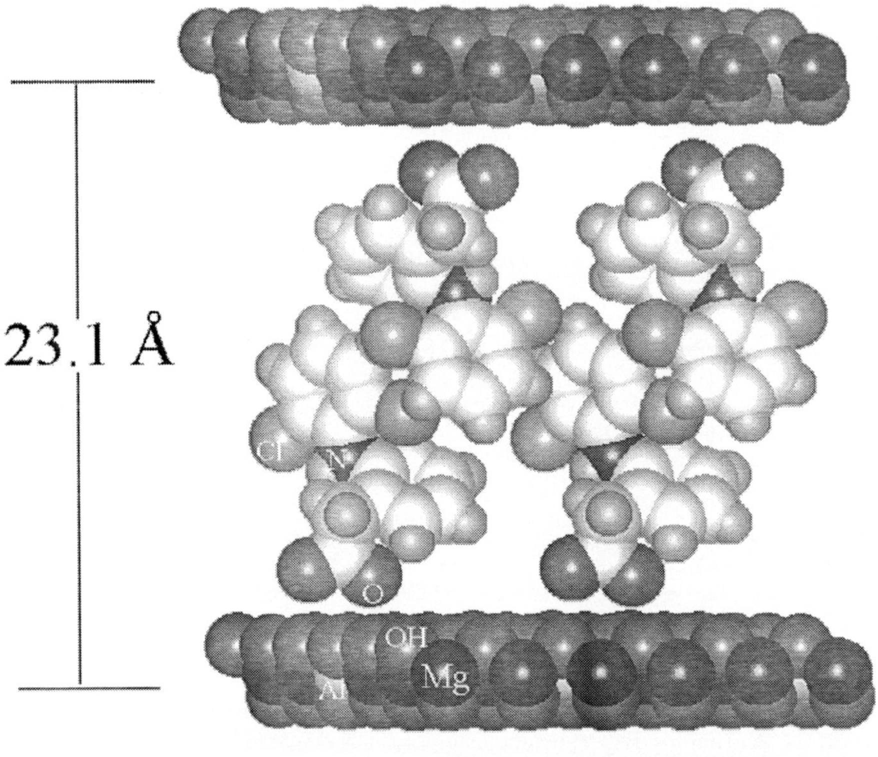

Figure 15. Computer-generated models showing the most probable arrangement of dichlofenac anions between the LDH layers.

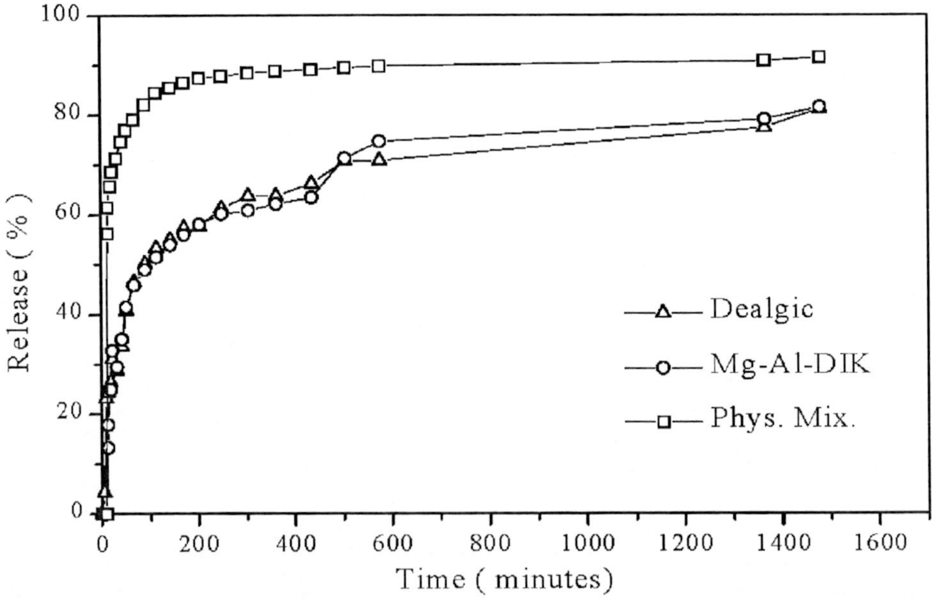

Figure 16. Release profile of dichlofenac from Dealgic 100® (–Δ–), Mg-Al-DIK intercalation compound (–□–) and physical mixture of DIK and LDH-Cl (–□–).

Table 3. Composition, interlayer distance and drug loading of intercalation compounds of Mg-Al LDH with some non sterolidal anti-inflammatory drugs.

Formula	d (Å)	Drug loading (% w/w)
$[Mg_{0.67}Al_{0.33}(OH)_2]SA_{0.167}\ OH_{0.163}\cdot 1H_2O$	15.3	22
$[Mg_{0.67}Al_{0.33}(OH)_2]IBU_{0.33}\cdot 0.25H_2O$	21.8	50
$[Mg_{0.67}Al_{0.33}(OH)_2]DIK_{0.33}\cdot 1.1H_2O$	23.1	55
$[Mg_{0.67}Al_{0.33}(OH)_2]KET_{0.21}\ Cl_{0.12}\cdot 0.36H_2O$	23.7	43
$[Mg_{0.67}Al_{0.33}(OH)_2]TIAP_{0.22}\ Cl_{0.11}\cdot 0.8H_2O$	23.2	42
$[Mg_{0.67}Al_{0.33}(OH)_2]TOLM_{0.28}\ Cl_{0.05}\cdot 0.7H_2O$	24.7	43
$[Mg_{0.67}Al_{0.33}(OH)_2]IND_{0.22}\ Cl_{0.11}\cdot 1.1H_2O$	25.2	48

CONCLUSION

Layered Double Hydroxides offer a large number of possibilities to organize photoactive species on the interlayer and/or on the surface of the microcrystals. Materials for non-linear optics and for energy storage and conversion have been envisaged and numerous other applications can be given. The development of real devices on the basis of modified LDHs, however, still requires additional research effort and the involvement of researchers with different competences.

Application of hydrotalcites in medicinal chemistry has already reached the commercial area and many patents have been issued; however, the development of drug release systems based on intercalation compounds requires much more fundamental and applied research.

On the whole, the work done has a good scientific value and contains the seeds for future development and for a rich harvest of new discoveries.

REFERENCES

[1] G. Alberti and T. Bein (Vol. Eds.) in *Solid-State Supramolecular Chemistry: Two and Three-dimensional Inorganic Networks*, Volume 7 of *Comprehensive Supramolecular Chemistry*, Pergamon, Elsevier Science Ltd Press, Oxford, U.K.,1996.

[2] R. Schöllhorn, *"Intercalation Compounds"* in *"Inclusion Compounds"* J. L. Atwood, J. E. D. Davies and D. D. MacNicol, eds., Academic Press, New York, 1984, Vol.1, pp. 249-349.

[3] M. S. Whittingham and A. J. Jacobson (eds.), *"Intercalation Chemistry"*, Academic Press, New York, 1982.

[4] J. Jacobson, *"Intercalation Reactions of Layered Compounds"* in *"Solid State Chemistry Compounds"*, A. K. Cheetham and P. Day, eds., Clarendon Press, Oxford, 1992, pp.182-233.

[5] W. Muller-Warmuth and R. Schöllhorn, eds, *"Progress in Intercalation Research"*, Series: Physics and Chemistry of Materials with Low-Dimensional Structures, 17, Kluwer Academic Publ., Dordrecht, 1994.

[6] R. Setton (Ed.) *"Chemical Reactions in Organic and Inorganic Constrained Systems"*, NATO ASI Series, Series C: Mathematical and Physical Sciences Vol. 165, D. Reidel Publishing Company, Dordrecht, 1986.

[7] A. J. Jacobson, *Materials Science Forum* **154-153** (1994) 1.

[8] M. Ogawa, K. Kuroda, *Chem. Rev.* **95** (1995) 399.

[9] F. Cavani, F. Trifiro, A. Vaccari, *Catal. Today* **11** (1991) 173.

[10] F. Trifirò and A. Vaccari, in *Solid-State Supramolecular Chemistry: Two and Three-dimensional Inorganic Networks*, Volume 7 of *Comprehensive supramolecular Chemistry*, G. Alberti and T Bein ,Vol Eds., Pergamon , Elsevier Science Ltd Press, Oxford, 1996, Chap. 8.

[11] A. deRoy, C. Forano, K. El Malki, and J.-P. Besse, in *"Anionic Clays: Trends in Pillaring Chemistry, Synthesis of Microporous Materials"* M. I. Occelli and H. E. Robson, Eds, Van Norstrand Reinhold, New York, 1992, Vol.2, p.108.

[12] V. Rives and M. A. Ulibarri, *Coord Chem. Rev.* **181** (1999) 61.

[13] W. Jones, S. P. Newman, *New J. Chem.* (1998) 105.

[14] R. P. Wayne, *"Principles and applications of photochemistry"*, Oxford University Press, Oxford, 1988.

[15] H. Duerr and H. Bouas-Laurent (Eds.)," *Photochromism. Molecules and Systems"*, Elsevier, Amsterdam, 1990.

[16] R. W. Munn and C. N. Ironside (Eds.), *"Principles and Applications of Nonlinear Optical Materials"*, CRC-Press, London, 1993.

[17] V. Ramamurthy, *Tetrahedron* **42** (1986) 5753.

[18] D. R. Hines, S. A. Solin, U. Costantino and M. Nocchetti, *Phys. Rev. B* (2000) 11348.

[19] S. Miyata, *Clays Clay Miner.* **31** (1983) 305

[20] F. Hellferch, *Ion Exchange,* McGraw-Hill, New York, 1962.

[21] U. Costantino, N. Coletti, M. Nocchetti, G. G. Aloisi, F. Elisei, *Langmuir* **15** (1999) 4454.

[22] W. T. Reichle, *Solid State Ionics* **22** (1986) 135.

[23] F. Rey and V. Fornes, *J. Chem. Soc. Faraday Trans.* **88** (1992) 2233.

[24] J. Rocha, M. del Arco, V. Rives and M. A. Ulibarri, *J. Mater. Chem.* **9** (1999) 2499.

[25] Y. Park, K. Kuroda and C. Kato, *J. Chem. Soc. Dalton Trans.* (1990) 3071.

[26] U. Costantino, N. Coletti, M. Nocchetti, G. G. Aloisi, F. Elisei, L. Latterini, *Langmuir* **15** (2000) 4454.

[27] H. Tagaya, A. Ogata, T. Kuwahara, S. Ogata, M. Karasu, J. Kadokawa and K Chiba, *Microporous Mater.* **7** (1996) 151.

[28] E. P. Giannelis, D. G. Nocera and T. J. Pinnavaia, *Inorg. Chem.* **26** (1987) 203.

[29] M. D. Newsham, E. P. Giannelis, T. J. Pinnavaia and D. G. Nocera, *J. Am. Chem. Soc.* **110** (1988) 3885.

[30] S. Bonnet, C. Forano, A. de Roy, J. P. Besse, P. Maillard and M. Momenteau, *Chem. Mater.* **8** (1996) 1962.

[31] Y. Park, K. Kuroda and C. Kato, *Chem. Lett.* (1989) 2057.

[32] P. K. Dutta and M. Puri, *J. Phys. Chem.* **93** (1989) 376.

[33] L. Ukrainczyk, M. Chibwe, T. J. Pinnavaia and S. A. Boyd, *J. Phys. Chem.* **98** (1994) 2668.

[34] M. Chibwe and T. J. Pinnavaia, *J. Chem. Soc., Chem. Commun.* (1993) 278

[35] V. Iliev, A. Ileva, and L. Bilyarska, *J. Mol. Catal.* **126** (1997) 99.

[36] V. Iliev, L. Prahov, L. Bilyarska, H. Fischer, G. Schulz-Ekloff, D. Wöhrle and L. Petrov, *J. Mol. Catal.* **151** (2000) 161.
[37] L. Barloy, J. P. Lallier, P. Battioni, D. Mansuy, Y. Piffard, M. Tournous, J. B. Valim and W. Jones, *New J. Chem.* **16** (1992) 71.
[38] D. S. Robins and P. K. Dutta, *Langmuir* **12** (1996) 402.
[39] P. K. Dutta and D. S. Robins, *Langmuir* **10** (1994) 1851.
[40] P. K. Dutta and D. S. Robins, *Langmuir* **10** (1994) 4681
[41] W. Y. Tseng, J. T. Lin, C. Y. Mou, S. F. Cheng, S. B. Liu, P. P. Chu and H. W. Liu, *J. Am. Chem. Soc.* **118** (1996) 4411.
[42] W. Ding, G. Gu, W. Zhong, W. Zang and Y. Du, *Chem. Phys. Lett.* **262** (1996) 259.
[43] S. Cooper and P. K. Dutta, *J. Phys. Chem.* **94** (1990) 114.
[44] T. Sato, H. Okuyama, T. Endo and M. Shimada, *React. Solids* **8** (1990) 63.
[45] J. Valim, B. M. Kariuki, J. King and W. Jones, *Mol. Cryst. Liq. Cryst. Sci. Technol., Sect. A* **211** (1992) 271.
[46] K. Takagi, T. Shichi, H. Usami and Y. Sawaki, *J. Am. Chem. Soc.* **115** (1993) 4339.
[47] T. Shichi, K. Takagi and Y. Sawaki, *Chem. Commun.* (1996) 2027.
[48] K. Takagi, E. Harata, T. Shichi, T. Kanoh and Y. Sawaki, *J. Photochem. Photobiol.* **105** (1997) 47.
[49] K. Takagi, T. Shichi and Y. Sawaki, *Chem. Lett.* **9** (1996) 781.
[50] K. Takagi, N. Saito, T. Shichi and Y. Sawaki, *Chem. Lett.* **4** (1999) 275.
[51] T. Shichi, K. Takagi and S. Yamashita, *Supramol. Science* **5** (1998) 303.
[52] H. Tagaya, S. Sato, T. Kuwahara, J. Kadokawa, M. Karasu and K Chiba, *J. Mater. Chem.* **4** (1994) 1907.
[53] S. Li, W. C. Purdy, *Chem. Rev.* **92** (1992) 1457.
[54] A. Gordon, O. S. Betler, M. Greenbaum, L. Marantz, T. Gral and M. H. Maxwell, *Trans. Am. Soc. Artif. Int. Organs* **17** (1971) 253.
[55] A. Schmassmann, A. Tarnawski, H. A. Gerber, B. Flogerzi, M. Sanner, L. Varga and F. Halter, *Gut* **35** (1994) 896.
[56] A. S. Tarnawski, R. Pai, R. Itani and F. A. Wyle, *Digestion* **60** (1999) 449.
[57] A. C. Playle, S. R. Gumming and A. F. Llewellyn, *Pharm. Acta Helv.* **49** (1974) 298.
[58] J. Pawlaczyk, Z. Kokot and A. Rafinska, *Acta Pol. Pharm.* **42** (1985) 153.
[59] M. Dreyer, D. Marwinski, N. Wolf and G. H. Dammann, *Arzneim.-Forsch.* **41** (1991) 738.
[60] M. S. Lin, P. Sun and H. Y. Yu, *J. Formos Med. Assoc.* **97** (1998) 704.
[61] A. S. Tarnawski, M. Tomikawa, M. Ohta and I. J. Sarfeh, *J. Physiol. Paris* **94** (2000) 93.
[62] Z. Kokot, *Acta Pol. Pharm.* **44** (1987) 72.
[63] Z. Kokot, *Pharmazie* **43** (1988) 249.
[64] Z. Kokot, G. Urbowicz and D. Ewertowska, *Acta Pol. Pharm.* **46** (1989) 272.
[65] J. L. Fabregas and J. Cucala, *Int. J. Pharm.* **52** (1989) 173.
[66] Z. J. Kokot, *Acta Pol. Pharm.* **48** (1991) 27.
[67] S. Linke (Bayer A.-G.) *German Patent 3,346,943* (1983); C.A. (1985), **103**, 129118u.
[68] T. Kajino,T. Okada and S. Morioka, *Jpn. Kokai Tokkyo Koho JP* 60,161,915 [85,161,915] (1984), to Sato Pharmaceutical Co., Ltd.; C.A. (1986), **104**, 39729k.
[69] W. Jettka, B. Gajdos and M. D. Benedikt, *Eur. Patent EP* 715,846 (1996), to Rhone-Poulenc Rorer Gmbh, Germany; C.A. (1996), **125**, 96121.

[70] T. Kito, N. Inatomi, S. Morimoto and T. Morimoto, *Jpn. Kokai Tokkyo Koho JP* 08,259,445 (1996), to Takeda Chemical Industries, Ltd.; C.A. (1996), **126**, 37097.
[71] M. Wanatabe, M. Nakao and K. Okuno, *Jpn. Kokai Tokkyo Koho JP* 98,163,811 (2000), to Takeda Chemical Industries, Ltd.; C.A. **132** (2000) 185410.
[72] C. Fritsch and H. G. Wetzstein (Bayer A.-G.) *German Patent* 3,612,086 (1986); C.A. **108** (1988) 173550f.
[73] R. G. W. Spickett, J. L. F. Vidal and J. C. Escoi, *Eur. Patent EP* 338,861 (1988); C.A. **113** (1990) 103388p.
[74] T. Makino, S. Marunaka and S. Imoto, *Eur. Patent EP* 638,313 (1993), to Takeda Chemical Industries, Ltd.; C.A. **122** (1995) 170248s.
[75] K. H. Bauer *German Patent* 98,19859231 (2000) to Regenold Gmbh; C.A. **133** (2000) 63981.
[76] P. Grubel and D. R. Cave, *Aliment. Pharmacol. Ther.* **12** (1998) 569.
[77] O. M. N. Al-Gohary and E. A. Hosny, *Pharm. Acta Helv.* **72** (1997) 81.
[78] O. M. N. Al-Gohary and R. S. Al-Kassas, *Pharm. Acta Helv.* **74** (2000) 351.
[79] A. Ookubo, K. Ooi and H. Hayashi, *Langmuir* **10** (1994) 407.
[80] U. costantino, M. Casciola, L. Massinelli, M. Nocchetti and R. Vivani, *Solid State Ionics* **97** (1997) 203.
[81] A. Ookubo, K. Ooi and H. Hayashi, *J. Pharm. Sciences* **81** (1992) 1139.
[82] *Jpn. Kokai Tokkyo Koho JP* 60 06,619 [85 06,619] (1983), to Kyowa Chemical Industry Co., Ltd.; C.A. **102** (1985) 226039f.
[83] Y. Matsushima, T. Kumura, I. Kawamura and T. Kawada, *Jpn. Kokai Tokkyo Koho JP* 80 38,328 (1978), to Kyowa Hakko Kogyo Co., Ltd.; C.A. **93** (1980) 155851x.
[84] T. Nakane, S. Nishiyama and T. Nanba *Jpn. Kokai Tokkyo Koho JP* 03,153,767 [91,153,767] (1989), to Shiseido Co., Ltd.; C.A. **116** (1992) 27804r.
[85] N. Koide, N. Iida, M. Nishikawa and K. Inoue, *Jpn. Kokai Tokkyo Koho JP* 98,226,290 (2000), to Lion Corp.; C.A. **132** (2000) 170868.
[86] M. Sofue and Y. Ueda, *Funtai Kogaku Kaishi*, **33** (1996) 638.
[87] N. Doi, A. Yonetani and T. Unno, *Jpn. Kokai Tokkyo Koho JP* 01,275,527 [89,275,527] (1988); C.A. **112** (1990) 145596q.
[88] J. S. Woo and H. G. Yi, *US Patent* 99,411,510 (2000), to Hanmi Pharm. Co., Ltd.; C.A. **132** (2000) 227449.
[89] M. Koide and H. Ozeki, *Jpn. Kokai Tokkyo Koho JP* 95,270,582 (1997), to Lion Corp.; C.A. **127** (1997) 23785.
[90] Y. W. Chien, in *"Novel drug delivery systems"*, M. Dekker, New York, **1992**.
[91] N. Doi, S. Nitta, M. Kusakari and N. Takahashi, *Jpn. Kokai Tokkyo Koho JP* 60,225,719 [85,225,719] (1984); C.A. **104** (1986) 155998w.
[92] K. Watanabe, Y. Sato and N. Kikuchi, *Jpn. Kokai Tokkyo Koho JP* 98,101,907 (1999), to Shiseido Co., Ltd.; C.A. **131** (1999) 303377.
[93] M. Nocchetti, *Chimica dell'intercalazione in Argille Anioniche Sintetiche e in Fosfati Lamellari*, Tesi di Dottorato in Scienze Chimiche, Università degli Studi di Perugia, Italy, **1998.**
[94] G. Fardella, L. Perioli, V. Ambrogi, G. Grandolini, U. Costantino and M. Nocchetti, *Proc. Int. Symp. Controlled. Release Bioact. Mater.*, **24**th(19979 1033.
[95] G. Fardella, L. Perioli, V. Ambrogi, G. Grandolini, *Proc. Int. Symp. Controlled. Release Bioact. Mater.* **25** (1998) 774.

[96] V. Ambrogi, G. Fardella, G. Grandolini and L. Perioli, *Int. J. of Pharmaceutics* submitted.

INDEX

#

A

B

C

D

E

F

I

J

K

L

M

N

Q

R

S

T

U

V

W

X

Y

Z